Lecture Notes in Mathematics 1678

Editors:
A. Dold, Heidelberg
F. Takens, Groningen

Springer
Berlin
Heidelberg
New York
Barcelona
Budapest
Hong Kong
London
Milan
Paris
Santa Clara
Singapore
Tokyo

Olga Krupková

The Geometry of Ordinary Variational Equations

Springer

Author

Olga Krupková
Department of Mathematics
Silesian University at Opava
Bezrucovo nám. 13
746 01 Opava, Czech Republic
e-mail: Olga.Krupkova@fpf.slu.cz

Cataloging-in-Publication Data applied for

Die Deutsche Bibliothek - CIP-Einheitsaufnahme

Krupková, Olga:
The geometry of ordinary variational equations / Olga Krupková. - Berlin ; Heidelberg ; New York ; Barcelona ; Budapest ; Hong Kong ; London ; Milan ; Paris ; Santa Clara ; Singapore ; Tokyo : Springer, 1997
(Lecture notes in mathematics ; 1678)
ISBN 3-540-63832-6

Mathematics Subject Classification (1991): Primary: 34A26, 70Hxx
Secondary: 53C15, 58F05, 58F07

ISSN 0075-8434
ISBN 3-540-63832-6 Springer-Verlag Berlin Heidelberg New York

Typesetting: Camera-ready TEX output by the author
SPIN: 10553429 46/3143-543210 - Printed on acid-free paper

To my daughters
Olga and Sonja

Preface

Ordinary differential equations which are *variational*, i.e., come from Lagrangians as their Euler-Lagrange equations, represent a class of ODE interesting both for mathematicians and physicists. From the physical point of view these equations are *equations of motion* of important mechanical systems. Mathematics deals with such equations within the range of the *calculus of variations*. On the one hand, there are various geometric structures connected with these equations and their solutions, studied by means of differential geometry and global variational analysis. On the other hand, these equations represent an interesting object for the mathematical and global analysis, namely for the theory of ordinary differential equations, since for certain classical families of variational equations there have been invented powerful *integration methods* based on the theory of canonical transformations, symmetries, Hamilton-Jacobi theory, etc.

There exist many excellent textbooks and monographs the subject of which are in fact variational ODE. A large number of them, however, deal only with very particular classes of variational equations. First of all, many restrict to the so called *regular autonomous first-order Lagrangians*, i.e., to Lagrangians depending on "positions and velocities", $L(q^\sigma, \dot{q}^\sigma)$, satisfying the classical regularity condition

$$\det\Big(\frac{\partial^2 L}{\partial \dot{q}^\sigma \partial \dot{q}^\nu}\Big) \neq 0\,;$$

this concerns, in particular, all the texts using methods of *symplectic geometry* (cf., e.g., R. Abraham and J. E. Marsden [1], C. Godbillon [1], P. Libermann and Ch.-M. Marle [1], J.-M. Souriau [1], S. Sternberg [1], A. Weinstein [1]), and works on *Finsler geometry* (M. Matsumoto [1], etc.). From the textbooks dealing also with some of the *singular first-order Lagrangians* in more detail, let us mention e.g. M. de León and P. R. Rodrigues [3], E. C. G. Sudarshan and N. Mukunda [1], and K. Sundermeyer [1]. Other books are usually general in the variational foundations of the theory but do not cover some interesting aspects, such as e.g. Hamilton equations and different integration methods for singular Lagrangians, or geometric structures connected with variational equations and their solutions (cf. C. Carathéodory [1], Th. De Donder [1], M. Giaquinta and S. Hildebrandt [1], P. A. Griffiths [1], R. Hermann [1], M. de León and P. R. Rodrigues [1], E. T. Whittaker [1], and others).

Generalizations of various structures and methods of classical mechanics to a wider class of Lagrangians are subject of an intensive research, and there exists a plenty of papers dealing with particular aspects of this problem. Much effort has been done to generalize the concept of *regularity*, and consequently of the Legendre transformation, Hamilton equations, Hamilton-Jacobi equation, etc., to *higher-order* Lagrangians on different geometric structures (tangent spaces, fibered manifolds $R \times M$ over R and Y over R, general fibered manifolds). Similarly, some of the questions of the Dirac's theory of constrained systems have been studied in higher-order situations. This seems to be important not only for the classical mechanics itself, but namely for the quantum mechanics, where many questions on quantization of Lagrangians which are not regular, or are of higher order, still remain open. Other directions of research have been oriented to study of *second-order ODE* which are explicitly solved with respect to the second derivatives (geometric structures, symmetries, existence of a Lagrangian, etc.). Bibliography included in this work can cover only a small part of existing papers on the above mentioned subjects; in fact, it is not possible to mention and discuss all contributions.

The aim of this work is to provide a *general, comprehensive and self-contained geometric theory* of ordinary differential equations (of any finite order) which are variational. We consider variational equations on *fibered manifolds over one-dimensional bases*. This underlying structure is very appropriate, since it is sufficiently general to cover all interesting physical applications, and to carry the possibility of straightforward or even obvious generalizations of many tools and constructions to partial differential equations ("field theory"). We put *no a priori restrictions* on the equations under consideration, i.e., we put no a priori restrictions on Lagrangians. This means that the presented theory is *universal*, and applies to equations explicitly solved with respect to the highest derivatives as well as to ODE which cannot be expressed in this form; from the viewpoint of Lagrangians, it covers "regular", as well as all kinds of "degenerate" Lagrangians of all finite orders, "autonomous" as well as "time-dependent" Lagrangians, variational problems such that a global Lagrangian exists, as well as (global) variational problems which do not posses this property (i.e., such that there exists only a family of local Lagrangians giving rise to a global Euler-Lagrange form), etc. We show that these equations can be represented by *distributions* (of not necessarily constant rank); we study symmetries and first integrals of these equations, their structure of solutions, introduce various integration methods for these equations, and clarify geometric structures connected with them. In this way we obtain a theory which includes Lagrangian and Hamiltonian dynamics, symmetries and first integrals, Hamilton-Jacobi theory, canonical transformations, Liouville integration theory, fields of extremals, and other aspects of the classical mechanics, generalized to *any* Lagrangian of any order. The general point of view leads to a new look at "standard" concepts (such as Lagrangean system, phase space, regularity, Hamiltonian, momenta, Legendre transformation, singular system, and many others) relating them with the *class of equivalent Lagrangians* (not with a particular Lagrangian). Consequently, this approach provides geometrically natural generalizations, which are "better adapted" to the geometry of variational systems and to the study of their dynamics.

The work is based mainly on our papers [1–12], and it is an enlarged version of [7].

I thank all my colleagues, and especially, my husband Prof. Demeter Krupka, for encouragement and support. Also, it is a pleasure to thank the students who, by their discussion, helped me to improve much of the material presented in the book. Many thanks to Dr. Michal Marvan and Ms. Petra Auerová for helping me to prepare the camera-ready version of the manuscript. I am grateful to the Czech Ministry of Education and the Czech Grant Agency for support (grants No. VS 96003 ("Global Analysis"), No. 201/93/2245, and No. 201/96/0845). Last but not least it is a pleasure to thank an anonymous referee for valuable suggestions for improvement of the manuscript, and Ms. Thanh-Ha Le Thi (Springer-Verlag) for collaboration during the preparation of the manuscript for publication.

Olga Krupková
Opava, September 1997

Table of Contents

Chapter 1.

INTRODUCTION

Problems to find an optimal solution have been already considered by the antique science. One of the best-known ones is the classical *isoperimetric problem*, i.e., the problem *to find in the plane a closed curve whose interior domain has maximal area among all interiors of closed curves of the same length.* Its formulation and solution were well known to Aristotle, Archimedes and others. Great interest in extremal problems motivated by mathematics, mechanics and philosophy has lead in the 17th, 18th and 19th centuries to rush developments of methods handling with variational functionals and their extremals. A new discipline—the classical *calculus of variations* came into being, having grown from the work of Isaac Newton, Leonhard Euler, Jacob Bernoulli, Johann Bernoulli, Jean le Rond D'Alembert, Joseph Louis Lagrange, Adrien Marie Legendre, Johann Friedrich Carl Gauss, William Rowan Hamilton, Carl Gustav Jacob Jacobi, Joseph Liouville, Amalie Emmy Noether, to mention at least some of the great founders of the theory.

Euler-Lagrange equations. The classical calculus of variations concerned in study of the so called *variational integrals*. For a first insight into the problem, consider the integral

$$I = \int_a^b L(x, y(x), y'(x))\, dx, \tag{1.1}$$

where the boudary points a, b are fixed, and the integrand L depends on x, a function $y(x)$ and its first derivative $y'(x)$. The function L is referred to as the *Lagrange function* . The problem is *to select the function $y(x)$ such that the above integral is extremized.* We shall recall a solution of this problem due to Lagrange (cf. e.g. B. Tabarrok and F. P. J. Rimrott [1]). Suppose that the required extremizing function is $y_0(x)$. Denote

$$I_0 = \int_a^b L(x, y_0(x), y_0'(x))\, dx.$$

For an admissible function $y(x)$ we have

$$I = I_0 + \Delta I.$$

We may consider admissible functions of the form

$$y(\varepsilon, x) = y_0(x) + \varepsilon u(x)$$

and

$$y'(\varepsilon, x) = y_0' + \varepsilon u'(x),$$

where ε is a small parameter which does not depend upon x, and the function $u(x)$ is an arbitrary function independent of ε vanishing at $x = a$ and $x = b$. The integrand in (1.1) can then be regarded as the function L when two of its independent variables, y_0 and y_0', are changed by amounts εu and $\varepsilon u'$, respectively. For a given x we may expand L by a Taylor series about y_0 and y_0'. Thus we write

$$L(x, y_0 + \varepsilon u, y_0' + \varepsilon u') = L(x, y_0, y_0') + \left.\frac{\partial L}{\partial y}\right|_{(y_0, y_0')} \varepsilon u + \left.\frac{\partial L}{\partial y'}\right|_{(y_0, y_0')} \varepsilon u'$$
$$+ \frac{1}{2!}\left.\frac{\partial^2 L}{\partial y^2}\right|_{(y_0, y_0')} (\varepsilon u)^2 + \frac{1}{2!}\left.\frac{\partial^2 L}{\partial y'^2}\right|_{(y_0, y_0')} (\varepsilon u')^2 + \left.\frac{\partial^2 L}{\partial y \partial y'}\right|_{(y_0, y_0')} \varepsilon^2 u u' + \ldots.$$

Now,

$$\Delta I = \varepsilon \int_a^b \left(u \left.\frac{\partial L}{\partial y}\right|_{(y_0, y_0')} + u' \left.\frac{\partial L}{\partial y'}\right|_{(y_0, y_0')} \right) dx$$
$$+ \frac{\varepsilon^2}{2} \int_a^b \left(u^2 \left.\frac{\partial^2 L}{\partial y^2}\right|_{(y_0, y_0')} + 2uu' \left.\frac{\partial^2 L}{\partial y \partial y'}\right|_{(y_0, y_0')} + u'^2 \left.\frac{\partial^2 L}{\partial y'^2}\right|_{(y_0, y_0')} \right) dx + \ldots,$$

or

$$\Delta I = \varepsilon I_1 + \frac{\varepsilon^2}{2} I_2 + \ldots,$$

where εI_1 is called the *first variation*, and $(\varepsilon^2/2) I_2$ the *second variation* of the variational integral (1.1). Evidently, sufficient conditions for I_0 to be maximum are

$$I_1 = 0, \quad I_2 < 0.$$

For I_0 to be minimum we have as sufficient conditions

$$I_1 = 0, \quad I_2 > 0.$$

Since at extremum condition we have $\partial L/\partial y|_{(y_0, y_0')} = \partial L/\partial y$, from now on we need not retain the subscript (y_0, y_0'). Consider now the integral

$$I_1 = \int_a^b \left(u \frac{\partial L}{\partial y} + u' \frac{\partial L}{\partial y'} \right) dx.$$

Integrating by parts we get

$$I_1 = \int_a^b u \left(\frac{\partial L}{\partial y} - \frac{d}{dx} \frac{\partial L}{\partial y'} \right) dx + \left(\frac{\partial L}{\partial y'} u \right)_{x=a}^{x=b}. \tag{1.2}$$

The formula (1.2) is called the *first variation formula* . Now, since $u(x)$ is equal to zero at $x = a$ and $x = b$, the second term in (1.2) vanishes. Further, recognizing that $u(x)$ is an arbitrary function we conclude that for I_1 to vanish identically we must have

$$\frac{\partial L}{\partial y} - \frac{d}{dx} \frac{\partial L}{\partial y'} = 0. \tag{1.3}$$

Equation (1.3) is called the *Euler-Lagrange equation*. This equation was originally derived also by Euler via a different scheme of reasoning.

Many variational problems involve functionals whose integrands contain also higher-order derivatives of $y(x)$. For $L(x, y, y', y^{(2)}, \ldots, y^{(r)})$, the corresponding equation for extremals, also called the Euler-Lagrange equation, takes the form

$$\frac{\partial L}{\partial y} - \frac{d}{dx}\frac{\partial L}{\partial y'} + \frac{d^2}{dx^2}\frac{\partial L}{\partial y^{(2)}} - \cdots + (-1)^r \frac{d^r}{dx^r}\frac{\partial L}{\partial y^{(r)}} = 0.$$

Higher order Lagrangians were first systematically studied by M. V. Ostrogradskii in the first half of the 19th century.

In case that L depends on one independent variable x and m dependent variables y^σ, $1 \leqslant \sigma \leqslant m$, the Euler-Lagrange equations represent a system of m second order ordinary differential equations for extremals $y(x) \equiv (y^\sigma(x))$ of the form

$$\frac{\partial L}{\partial y^\sigma} - \frac{d}{dx}\frac{\partial L}{\partial y^{\sigma\prime}} = 0. \tag{1.4}$$

The functions on the left-hand sides are called the *Euler-Lagrange expressions* of the Lagrange function L, and are denoted by $E_\sigma(L)$.

Brachystochrone. One of the best known classical problems in the calculus of variations is the *brachystochrone problem.* In fact, it was this problem which became a strong impulse for intensive theoretic studies of extremal problems.

In 1696 there appeared a treatise by Johann Bernoulli, "Problema novum, ad cujus solutionen mathematici invitantur" ("A new problem to the solutin of which mathematicians are invited") where the problem of quickest descent was posed:

Let two points A, B be given in the vertical plane. Find a line connecting them, on which a movable point M descends from A to B under the influence of gravitation in the quickest possible way.

A solution has been given by Johann Bernoulli, independent solutions have been found also by Jacob Bernoulli, Leibnitz, and an anonymous author (probably Newton).

We shall recall here the original solution provided by Johann Bernoulli, based on an analogy with the propagation of light in an optically heterogeneous medium.

In the vertical plane consider coordinates (x, y) at the point A, such that the x-axis is horizontal and the y-axis is vertical, down-oriented. Denote $A = (0, 0)$, $B = (b_1, b_2)$. Assume that the point M moves without friction, and the initial velocity is zero. This means we have the boundary conditions $y(0) = 0$, $y(b_1) = b_2$ and $v(0) = 0$. The energy equation reads

$$\tfrac{1}{2}mv^2 - mgy = \text{const},$$

where g is the gravity acceleration. At the initial point it holds $y = 0$ and $v = 0$, i.e., the above constant equals zero. Consequently, the velocity at a point $(x, y(x))$ depends only on the coordinate $y(x)$ and equals $\sqrt{2gy(x)}$. Since one needs to find the shortest time to get from A to B, one has to minimize the integral

$$T = \int \frac{ds}{v} = \int \frac{ds}{\sqrt{2gy(x)}} \tag{1.5}$$

over the arc AB, where ds is an element of the path.

The above problem is completely equivalent to the problem of finding the trajectory of the light in a two-dimensional non-homogeneous medium where the velocity at a point (x, y) equals to $\sqrt{2gy}$. The medium can be divided into parallel slices, within each of which the velocity can be considered a constant equal v_i, $i = 1, 2, \ldots$. By the Snell Law one gets

$$\frac{\sin \alpha_1}{v_1} = \frac{\sin \alpha_2}{v_2} = \ldots,$$

i.e.,

$$\frac{\sin \alpha_i}{v_i} = \text{const},$$

where α_i are angles of incidence of the light rays. In the limit

$$\frac{\sin \alpha(x)}{v(x)} = \text{const},$$

where $v(x) = \sqrt{2gy(x)}$, and $\alpha(x)$ is the angle between the tangent to the curve $y(\cdot)$ at the point $(x, y(x))$ and the axis Oy, i.e., $\sin \alpha(x) = (1 + (y'(x))^2)^{-1/2}$. Hence, the equation of the extremal curve, called the *brachystochrone*, is

$$\sqrt{1 + (y')^2}\,\sqrt{y} = C$$

which is equivalent to

$$y' = \sqrt{\frac{C - y}{y}}, \tag{1.6}$$

i.e., to

$$\frac{\sqrt{y}}{\sqrt{C - y}}\,dy = dx.$$

Integrating this equation with help of the substitution $y = C \sin^2 t/2$, hence $dx = (C \sin^2 t/2)\,dt$ we get the following *equation of the cycloide*

$$x = C_1 + \tfrac{1}{2}C(t - \sin t), \quad y = \tfrac{1}{2}C(1 - \cos t).$$

The above equations represent a family of cycloids. A unique solution of the brachystochrone problem is obtained if the boundary conditions are considered.

The brachystochrone problem can also be solved with help of the variational techniques (see e.g. M. Giaquinta and S. Hildebrand [1]). It is the problem of minimizing the action functional (1.5). We have $ds = \sqrt{dx^2 + dy^2} = \sqrt{1 + y'^2}\,dx$, hence the Larange function is

$$L = \frac{\sqrt{1 + y'^2}}{\sqrt{2gy}}.$$

From the Euler-Lagrange equations of L we obtain a first integral

$$y(1 + y'^2) = C,$$

which is equivalent to the equation (1.6).

The inverse problem of the calculus of variations. Due to intensive studies of different extremal problems during the 17th–19th centuries it turned out that *variational equations*, i.e., equations which can be expressed in form of the Euler-Lagrange equations of a Lagrange function have a fundamental meaning in theoretical physics and engineering. Consequently, an essential problem appeared:

A system of differential equations being given, one needs to find out whether they are variational or not, and in the affirmative case to construct a Lagrange function for these equations.

This problem, referred to as the *inverse problem of the calculus of variations* was first posed by H. von Helmholtz in 1887. In his paper [1] Helmholtz studied a system of m second order differential equations of the form

$$B_{\sigma\nu}(t, q^\rho, \dot{q}^\rho)\ddot{q}^\nu + A_\sigma(t, q^\rho, \dot{q}^\rho) = 0, \tag{1.7}$$

where $1 \leqslant \sigma, \nu, \rho \leqslant m$. He found necessary conditions for the existence of a Lagrange function L such that

$$B_{\sigma\nu}\ddot{q}^\nu + A_\sigma = \frac{\partial L}{\partial q^\sigma} - \frac{d}{dt}\frac{\partial L}{\partial \dot{q}^\sigma}.$$

These conditions, referred to as *Helmholtz conditions*, are of the form

$$\begin{aligned} & B_{\sigma\nu} = B_{\nu\sigma}, \quad \frac{\partial B_{\sigma\nu}}{\partial \dot{q}^\rho} = \frac{\partial B_{\sigma\rho}}{\partial \dot{q}^\nu}, \\ & \frac{\partial A_\sigma}{\partial \dot{q}^\nu} + \frac{\partial A_\nu}{\partial \dot{q}^\sigma} = 2\frac{\bar{d} B_{\sigma\nu}}{dt}, \\ & \frac{\partial A_\sigma}{\partial q^\nu} - \frac{\partial A_\nu}{\partial q^\sigma} = \frac{1}{2}\frac{\bar{d}}{dt}\left(\frac{\partial A_\sigma}{\partial \dot{q}^\nu} - \frac{\partial A_\nu}{\partial \dot{q}^\sigma}\right), \end{aligned} \tag{1.8}$$

where $1 \leqslant \sigma, \nu \leqslant m$, and the operator $\bar{d}/dt$ is defined by

$$\frac{\bar{d}}{dt} = \frac{\partial}{\partial t} + \dot{q}^\rho \frac{\partial}{\partial q^\rho}.$$

Helmholtz's result was completed in 1896 by Mayer who showed that the Helmholtz conditions (1.8) are also sufficient for (1.7) be variational (A. Mayer [1]).

In 1913 Volterra discovered that if the Helmholtz conditions are satisfied then the left-hand sides of (1.7) are Euler-Lagrange expressions of the Lagrange function

$$L = q^\sigma \int_0^1 E_\sigma(t, uq^\nu, u\dot{q}^\nu, u\ddot{q}^\nu)\, du. \tag{1.9}$$

Formula (1.9) was subsequently generalized by M. M. Vainberg [1] and E. Tonti [1]. Notice that the function L is of order two. It is called *Vainberg-Tonti Lagrangian* related to the variational expressions E_σ.

One can find the necessary and sufficient conditions of variationality very easily using methods of differential geometry. Let us recall here a proof due to O. Štěpánková [1], and L. Klapka [2].

For a system of equations (1.7) put

$$E_\sigma = A_\sigma + B_{\sigma\nu}\dot{q}^\nu,$$

and define a differential two-form E by

$$E = E_\sigma dq^\sigma \wedge dt.$$

Let us study a question under what conditions there exists a two-form F,

$$F = F_{\sigma\nu}(dq^\sigma - \dot{q}^\sigma dt) \wedge (dq^\nu - \dot{q}^\nu dt) + G_{\sigma\nu}(dq^\sigma - \dot{q}^\sigma dt) \wedge (d\dot{q}^\nu - \ddot{q}^\nu dt) \quad (1.10)$$

such that the form

$$\alpha = E + F$$

is closed. We may suppose that the functions $F_{\sigma\nu}$ are antisymmetric in σ, ν, i.e., that $F_{\sigma\nu} = -F_{\nu\sigma}$. Denote by d/dt the total derivative operator,

$$\frac{d}{dt} = \frac{\partial}{\partial t} + \dot{q}^\rho \frac{\partial}{\partial q^\rho} + \ddot{q}^\rho \frac{\partial}{\partial \dot{q}^\rho} + \dddot{q}^\rho \frac{\partial}{\partial \ddot{q}^\rho}.$$

Computing the condition $d\alpha = 0$ we get

$$\begin{aligned} &\frac{1}{2}\left(\frac{\partial E_\sigma}{\partial q^\nu} - \frac{\partial E_\nu}{\partial q^\sigma}\right) - \frac{dF_{\sigma\nu}}{dt} = 0, \\ &\frac{\partial E_\sigma}{\partial \dot{q}^\nu} - 2F_{\sigma\nu} - \frac{dG_{\sigma\nu}}{dt} = 0, \\ &\frac{\partial E_\sigma}{\partial \ddot{q}^\nu} - G_{\sigma\nu} = 0, \quad G_{\sigma\nu} = G_{\nu\sigma}, \end{aligned} \quad (1.11)$$

and since the identities (1.11) imply that

$$\frac{\partial F_{\sigma\nu}}{\partial \ddot{q}^\rho} = 0, \quad \frac{\partial G_{\sigma\nu}}{\partial \ddot{q}^\rho} = 0, \quad (1.12)$$

the remaining identities following from $d\alpha = 0$ reduce to

$$\begin{aligned} &\frac{\partial F_{\sigma\nu}}{\partial q^\rho} + \frac{\partial F_{\rho\sigma}}{\partial q^\nu} + \frac{\partial F_{\nu\rho}}{\partial q^\sigma} = 0, \\ &\frac{\partial F_{\sigma\nu}}{\partial \dot{q}^\rho} + \frac{1}{2}\left(\frac{\partial G_{\nu\rho}}{\partial q^\sigma} - \frac{\partial G_{\sigma\rho}}{\partial q^\nu}\right) = 0, \\ &\frac{\partial G_{\sigma\nu}}{\partial \dot{q}^\rho} - \frac{\partial G_{\sigma\rho}}{\partial \dot{q}^\nu} = 0 \end{aligned} \quad (1.13)$$

Now we have from (1.11)

$$G_{\sigma\nu} = \frac{1}{2}\left(\frac{\partial E_\sigma}{\partial \ddot{q}^\nu} + \frac{\partial E_\nu}{\partial \ddot{q}^\sigma}\right), \quad F_{\sigma\nu} = \frac{1}{4}\left(\frac{\partial E_\sigma}{\partial \dot{q}^\nu} - \frac{\partial E_\nu}{\partial \dot{q}^\sigma}\right), \quad (1.14)$$

and

$$\begin{aligned} &\frac{\partial E_\sigma}{\partial \ddot{q}^\nu} - \frac{\partial E_\nu}{\partial \ddot{q}^\sigma} = 0, \\ &\frac{\partial E_\sigma}{\partial \dot{q}^\nu} + \frac{\partial E_\nu}{\partial \dot{q}^\sigma} - \frac{d}{dt}\left(\frac{\partial E_\sigma}{\partial \ddot{q}^\nu} + \frac{\partial E_\nu}{\partial \ddot{q}^\sigma}\right) = 0, \\ &\frac{\partial E_\sigma}{\partial q^\nu} - \frac{\partial E_\nu}{\partial q^\sigma} - \frac{1}{2}\frac{d}{dt}\left(\frac{\partial E_\sigma}{\partial \dot{q}^\nu} - \frac{\partial E_\nu}{\partial \dot{q}^\sigma}\right) = 0. \end{aligned} \quad (1.15)$$

It is easy to see that relations (1.13) do not represent independent conditions on the functions $F_{\sigma\nu}$, $G_{\sigma\nu}$. Indeed, the third relation of (1.13) is obtained by differentiating the second relation of (1.11) by $\ddot{q}^\rho$, the second relation of (1.13) is obtained by differentiating the first relation of (1.11) by $\ddot{q}^\rho$. Finally, we show that the first relation of (1.13) is satisfied identically. Differentiating the last relation of (1.15) by $\dot{q}^\rho$, rotating the indices and summing up the resulting three identities we get

$$\frac{\partial^2 E_\sigma}{\partial q^\rho \partial \dot{q}^\nu} - \frac{\partial^2 E_\nu}{\partial q^\rho \partial \dot{q}^\sigma} + \frac{\partial^2 E_\rho}{\partial q^\nu \partial \dot{q}^\sigma} - \frac{\partial^2 E_\sigma}{\partial q^\nu \partial \dot{q}^\rho} + \frac{\partial^2 E_\nu}{\partial q^\sigma \partial \dot{q}^\rho} - \frac{\partial^2 E_\rho}{\partial q^\sigma \partial \dot{q}^\nu} = 0.$$

Applying (1.14) the first relation of (1.13) follows.

Summarizing the results, we have proved the following assertion:

If the E_σ satisfy the conditions (1.15) *then there exits a unique form F* (1.10) *such that the two-form $\alpha = E + F$ is closed. α is then expressed by*

$$\begin{aligned} \alpha = {} & E_\sigma dq^\sigma \wedge dt \\ & + \frac{1}{4}\Big(\frac{\partial E_\sigma}{\partial \dot{q}^\nu} - \frac{\partial E_\nu}{\partial \dot{q}^\sigma}\Big)(dq^\sigma - \dot{q}^\sigma dt) \wedge (dq^\nu - \dot{q}^\nu dt) \\ & + \frac{\partial E_\sigma}{\partial \ddot{q}^\nu}(dq^\sigma - \dot{q}^\sigma dt) \wedge (d\dot{q}^\nu - \ddot{q}^\nu dt). \end{aligned} \tag{1.16}$$

Conversely, if there exists a two-form F (1.10) *such that $E+F$ is closed then E_σ satisfy the conditions* (1.15).

We are ready to prove *the equivalence of the conditions* (1.15) *with variationality of E_σ.*

Suppose that E_σ satisfy the conditions (1.15). We can take the closed two-form α (1.16), and using the Poincaré Lemma we find a local one-form θ such that $\alpha = d\theta$. Denote by χ the mapping $(u, (t, q^\nu, \dot{q}^\nu, \ddot{q}^\nu)) \to (t, uq^\nu, u\dot{q}^\nu, u\ddot{q}^\nu)$ for $u \in [1, 0]$, and $(t, q^\nu, \dot{q}^\nu, \ddot{q}^\nu) \in V$, where V is a convex open set. We have

$$\begin{aligned} \theta = {} & \Big(q^\sigma \int_0^1 (E_\sigma \circ \chi)\, du\Big) dt \\ & + \Big(2q^\sigma \int_0^1 (F_{\sigma\nu} \circ \chi)\, u\, du + \dot{q}^\sigma \int_0^1 (G_{\sigma\nu} \circ \chi)\, u\, du\Big)(dq^\nu - \dot{q}^\nu dt), \end{aligned} \tag{1.17}$$

where $F_{\sigma\nu}$ and $G_{\sigma\nu}$ are given by (1.14). We put

$$L = q^\sigma \int_0^1 (E_\sigma \circ \chi)\, du. \tag{1.18}$$

Since L is a second order Lagrange function, the Euler-Lagrange expressions of L are

$$E_\sigma(L) = \frac{\partial L}{\partial q^\sigma} - \frac{d}{dt}\frac{\partial L}{\partial \dot{q}^\sigma} + \frac{d^2}{dt^2}\frac{\partial L}{\partial \ddot{q}^\sigma}.$$

Now,

$$\begin{aligned} \frac{\partial L}{\partial q^\sigma} &= \int_0^1 (E_\sigma \circ \chi)\, du + q^\nu \int_0^1 \Big(\frac{\partial E_\nu}{\partial q^\sigma} \circ \chi\Big) u\, du, \\ \frac{\partial L}{\partial \dot{q}^\sigma} &= q^\nu \int_0^1 \Big(\frac{\partial E_\nu}{\partial \dot{q}^\sigma} \circ \chi\Big) u\, du, \\ \frac{\partial L}{\partial \ddot{q}^\sigma} &= q^\nu \int_0^1 \Big(\frac{\partial E_\nu}{\partial \ddot{q}^\sigma} \circ \chi\Big) u\, du. \end{aligned}$$

Using the formula

$$\begin{aligned} E_\sigma &= \int_0^1 d\big((E_\sigma \circ \chi)\, u\big) \\ &= \int_0^1 (E_\sigma \circ \chi)\, du + q^\nu \int_0^1 \Big(\frac{\partial E_\sigma}{\partial q^\nu} \circ \chi\Big)\, u\, du \\ &\quad + \dot{q}^\nu \int_0^1 \Big(\frac{\partial E_\sigma}{\partial \dot{q}^\nu} \circ \chi\Big)\, u\, du + \ddot{q}^\nu \int_0^1 \Big(\frac{\partial E_\sigma}{\partial \ddot{q}^\nu} \circ \chi\Big)\, u\, du, \end{aligned}$$

and conditions (1.15), we get by a direct computation that $E_\sigma - E_\sigma(L) = 0$, proving that the E_σ are Euler-Lagrange expressions of the Lagrange function (1.18).

Conversely, suppose that E_σ are variational, let L be a Lagrange function for E_σ. Put $E = E_\sigma(L)\, dq^\sigma \wedge dt$ and

$$\begin{aligned} F_{\sigma\nu} &= \frac{1}{4}\Big(\frac{\partial E_\sigma(L)}{\partial \dot{q}^\nu} - \frac{\partial E_\nu(L)}{\partial \dot{q}^\sigma}\Big) = \frac{1}{2}\Big(\frac{\partial^2 L}{\partial q^\sigma \partial \dot{q}^\nu} - \frac{\partial^2 L}{\partial q^\nu \partial \dot{q}^\sigma}\Big), \\ G_{\sigma\nu} &= \frac{1}{2}\Big(\frac{\partial E_\sigma}{\partial \ddot{q}^\nu} + \frac{\partial E_\nu}{\partial \ddot{q}^\sigma}\Big) = \frac{\partial^2 L}{\partial \dot{q}^\sigma \partial \dot{q}^\nu}. \end{aligned}$$

Now it is easy to show that $\alpha = E + F$ is closed. By the above proposition, E_σ satisfy (1.15).

Note that $\alpha = d\theta_L$ where

$$\theta_L = L\, dt + \frac{\partial L}{\partial \dot{q}^\sigma}(dq^\sigma - \dot{q}^\sigma dt)$$

is the famous 1-form, usually called *Cartan form*. This form was introduced to the calculus of variations independently by E. T. Whittaker [1] and É. Cartan [1].

Although not apparent at a first sight, conditions (1.15) represent an *equivalent* form of the *Helmholtz conditions* (1.8). To check this, it is sufficient to substitute into (1.15) the relation $E_\sigma = A_\sigma + B_{\sigma\nu}\ddot{q}^\nu$. Then one gets the identities (1.15) in the form (1.8) plus one additional identity,

$$\frac{\partial B_{\sigma\rho}}{\partial q^\nu} - \frac{\partial B_{\nu\rho}}{\partial q^\sigma} = \frac{1}{2}\frac{\partial}{\partial \dot{q}^\rho}\Big(\frac{\partial A_\sigma}{\partial \dot{q}^\nu} - \frac{\partial A_\nu}{\partial \dot{q}^\sigma}\Big), \tag{1.19}$$

for all σ, ν, ρ. The latter identity, however, is not independent. To see this, let us denote

$$\phi_{\sigma\nu} = \frac{1}{2}\Big(\frac{\partial A_\sigma}{\partial \dot{q}^\nu} - \frac{\partial A_\nu}{\partial \dot{q}^\sigma}\Big), \quad \psi_{\sigma\nu} = \frac{1}{2}\Big(\frac{\partial A_\sigma}{\partial \dot{q}^\nu} + \frac{\partial A_\nu}{\partial \dot{q}^\sigma}\Big).$$

Evidently, the functions $\phi_{\sigma\nu}$ and $\psi_{\sigma\nu}$ identically obey the relation

$$\frac{\partial \phi_{\sigma\nu}}{\partial \dot{q}^\rho} = \frac{\partial \psi_{\rho\sigma}}{\partial \dot{q}^\nu} - \frac{\partial \psi_{\rho\nu}}{\partial \dot{q}^\sigma}. \tag{1.20}$$

Now, the third condition of (1.8) in terms of $\psi_{\sigma\nu}$ reads

$$\psi_{\sigma\nu} = \frac{\bar{d} B_{\sigma\nu}}{dt}$$

which implies that

$$\frac{\partial \psi_{\sigma\nu}}{\partial \dot{q}^\rho} = \frac{\partial B_{\sigma\nu}}{\partial q^\rho} + \frac{\bar{d}}{dt}\frac{\partial B_{\sigma\nu}}{\partial \dot{q}^\rho}.$$

Using the first and second identity of (1.8) we get

$$\frac{\partial \psi_{\rho\sigma}}{\partial \dot{q}^{\nu}} - \frac{\partial \psi_{\rho\nu}}{\partial \dot{q}^{\sigma}} = \frac{\partial B_{\rho\sigma}}{\partial q^{\nu}} + \frac{\bar{d}}{dt}\frac{\partial B_{\rho\sigma}}{\partial \dot{q}^{\nu}} - \frac{\partial B_{\rho\nu}}{\partial q^{\sigma}} - \frac{\bar{d}}{dt}\frac{\partial B_{\rho\nu}}{\partial \dot{q}^{\sigma}} = \frac{\partial B_{\sigma\rho}}{\partial q^{\nu}} - \frac{\partial B_{\nu\rho}}{\partial q^{\sigma}}.$$

Hence, the relation (1.19) identifies with (1.20).

The inverse problem originally posed by Helmholtz is naturally generalized to the following question, which also is referred to as *inverse problem of the calculus of variations*: *Under what conditions equations* (1.7) *are equivalent with some Euler-Lagrange equations*? In other words, one asks whether there exist Euler-Lagrange equations the solutions of which coincide with the solutions of equations (1.7). In this formulation the problem is very general and its solution is yet not known. However, one can simplify this problem restricting the class of equivalent equations, namely, to equations of the form

$$f^{\sigma}_{\rho}(A_{\sigma} + B_{\sigma\nu}\ddot{q}^{\nu}) = 0, \tag{1.21}$$

where f^{σ}_{ρ} are some functions of $(t, q^{\mu}, \dot{q}^{\mu})$. Evidently, equations (1.7) and (1.21) will have the same solutions in case that the matrix (f^{σ}_{ρ}) is everywhere regular. The inverse problem then takes the following formulation:

Given equations (1.7), *find out whether there exists an everywhere regular matrix* (f^{σ}_{ρ}) *such that the left-hand sides of* (1.21) *indentify with the Euler-Lagrange expressions of a Lagrange function.*

Such a matrix (f^{σ}_{ρ}) is then called a *variational integrating factor*, or *variational multiplier* for the equations (1.7).

It is clear that to solve this problem one can apply Helmoltz conditions (1.8) resp. (1.15). Putting $E'_{\rho} = f^{\sigma}_{\rho}(A_{\sigma} + B_{\sigma\nu}\ddot{q}^{\nu})$ and applying Helmholtz conditions to E'_{ρ} one obtains a system of first order partial differential equations for the functions f^{σ}_{ρ}. Although some concrete examples can be solved in this way, for a general analysis these equations are very complicated. This is why the problem has been extensively studied by different methods (see e.g. N. Ya. Sonin [1], G. Darboux [1], J. Douglas [1], P. Havas [1], W. Sarlet [1,2,3], M. Henneaux [1,2], I. Anderson and G. Thompson [1], M. Crampin, W. Sarlet, E. Martínez, G. B. Byrnes, and G. E. Prince [1], and many others).

However, the situation extremely simplifies if $m = 1$, i.e., if variational integrating factors for *one* ordinary (nondegenerate) second-order differential equation are studied. In this case one can consider the equation in the form

$$\ddot{q} - g = 0, \tag{1.22}$$

and the Helmholtz conditions (1.15) applied to $f(\ddot{q} - g)$ reduce to a single equation

$$\frac{\partial f}{\partial \dot{q}}g + f\frac{\partial g}{\partial \dot{q}} + \frac{\partial f}{\partial t} + \frac{\partial f}{\partial q}\dot{q} = 0$$

for $f(t, q, \dot{q}) \neq 0$, which is equivalent to

$$\frac{\partial \ln f}{\partial t} + \frac{\partial \ln f}{\partial q}\dot{q} + \frac{\partial \ln f}{\partial \dot{q}}g + \frac{\partial g}{\partial \dot{q}} = 0.$$

The general solution of such equations depends upon a single arbitrary function of any two specific solutions of the corresponding homogeneous equation. Consequently the integratig factor problem for a single equation (1.22) always has a solution, and we can say that the most general Lagrange function L for (1.22) depends upon one arbitrary function of two parameters.

Unfortunately, for a system of second-order ODE the existence of a variational integrating factor is not generally ensured. As shown by J. Douglas [1], even in the case of $m = 2$ (two equations) a solution to the variational multiplier problem may not exist.

At present, the inverse problem of the calculus of variations is understood as a more general and comprehensive problem. We shall return to this topic in Chapters 4, 6.

Trivial Lagrangians. Consider the Lagrange functions

$$\begin{aligned} L_1 &= \tfrac{1}{2}m(\dot{q}_1^2 + \dot{q}_2^2 + \dot{q}_3^2), \\ L_2 &= \tfrac{1}{2}m(\dot{q}_1^2 + \dot{q}_2^2 + \dot{q}_3^2) + mt(2q_1 + t\dot{q}_1), \\ L_3 &= -\tfrac{1}{2}m(q_1\ddot{q}_1 + q_2\ddot{q}_2 + q_3\ddot{q}_3). \end{aligned}$$

It is easy to check that L_1, L_2 and L_3 give rise to the same Euler-Lagrange expressions, namely, that

$$E_i(L_1) = E_i(L_2) = E_i(L_3) = -m\ddot{q}_i.$$

We say that Lagrange functions L, L' are *equivalent* if their Euler-Lagrange expressions are the same, i.e., if $E_\sigma(L) = E_\sigma(L')$ for all σ. We want to characterize the class of equivalent Lagrange functions. Evidently, to this end it is sufficient to describe the equivalence class of the so called *trivial Lagrangians*, i.e., Lagrangians leading to Euler-Lagrange expressions identically equal to zero.

The class of trivial Lagrangians is easily found using the above proposition on closed two-forms. Namely, suppose that $E_\sigma = 0$ for all σ. Then by (1.16) we get for the corresponding closed two-form

$$\alpha = 0.$$

Since locally $\alpha = d\theta$, the above condition means that the 1-form θ is closed, i.e., there is a (local) function F such that $\theta = dF$. Consequently, every corresponding Lagrange function L is of the form dF/dt. Conversely, if $L = dF/dt$ for some function F then we have $\theta_L = dF$, i.e., $\alpha = d\theta_L = 0$. Since α is given in terms of E_σ by (1.16), we get $E_\sigma = 0$ for all the indices σ, proving that dF/dt is a trivial Lagrangian.

Consequently, Lagrange functions L, L' are equivalent if and only if there exists a function F such that

$$L' = L + \frac{dF}{dt}.$$

Geodesics. Consider a Riemannian manifold M with the metric g. Let g in local coordinates be expressed by $g = g_{ij}\, dx^i\, dx^j$. Put

$$L = \sqrt{g_{ij}\dot{x}^i\dot{x}^j}. \tag{1.23}$$

We shall show that the geodesic curves on (M, g) are extremals of the Lagrange function (1.23). The Euler-Lagrange equations of L are

$$\tfrac{1}{2}(g_{rs}\dot{x}^r\dot{x}^s)^{-1/2}\frac{\partial g_{kl}}{\partial x^i}\dot{x}^k\dot{x}^l - \frac{d}{dt}\left((g_{rs}\dot{x}^r\dot{s}^l)^{-1/2}g_{ij}\dot{x}^j\right) = 0.$$

Choosing the parameter in such a way that $\sqrt{g_{ij}\dot{x}^i\dot{x}^j} = 1$, i.e., parametrizing by the arc length, the Euler-Lagrange equations become

$$\frac{1}{2}\frac{\partial g_{kl}}{\partial x^i} x'^k x'^l - \frac{d}{ds}(g_{ij}x'^j) = 0,$$

where we have used the notation $x'^i = dx^i/ds$. Now, introducing the Christoffel symbols

$$\Gamma_{ijk} = \frac{1}{2}\left(\frac{\partial g_{ik}}{\partial x^j} + \frac{\partial g_{ij}}{\partial x^k} - \frac{\partial g_{jk}}{\partial x^i}\right), \quad \Gamma^i_{jk} = g^{il}\Gamma_{ljk},$$

the Euler-Lagrange equations take the familiar form of the equations for geodesics,

$$g_{ij}\frac{d^2x^j}{ds^2} + \Gamma_{ijk}\frac{dx^j}{ds}\frac{dx^k}{ds} = 0,$$

or equivalently,

$$\frac{d^2x^i}{ds^2} + \Gamma^i_{jk}\frac{dx^j}{ds}\frac{dx^k}{ds} = 0.$$

Let us turn to a more general problem. Namely, let us consider equations

$$\ddot{q}^i + \Gamma^i_{jk}\dot{q}^j\dot{q}^k = 0 \tag{1.24}$$

for geodesics of a *linear torsion-free* connection Γ (this means that we suppose $\Gamma^i_{jk} = \Gamma^i_{kj}$). If g is a metric, let us ask the question under what conditions on Γ, the left-hand sides of the corresponding covariant equations,

$$g_{il}\ddot{x}^l + \Gamma_{ijk}\dot{x}^j\dot{x}^k = 0, \tag{1.25}$$

are Euler-Lagrange expressions of some Lagrange function. In other words, we want to study the existence of a time and velocity independent variational multiplier for (1.24).

Denote $E'_i = g_{il}\ddot{x}^l + \Gamma^l_{jk}\dot{x}^j\dot{x}^k$. Applying to E'_i the Helmholtz conditions (1.15) we arrive at

$$\Gamma_{ijk} + \Gamma_{jik} - \frac{\partial g_{ij}}{\partial x^k} = 0,$$

$$\frac{\partial g_{il}}{\partial x^j} - \frac{\partial g_{jl}}{\partial x^i} - \Gamma_{ijl} + \Gamma_{jil} = 0,$$

$$\frac{\partial \Gamma_{irs}}{\partial x^j} - \frac{\partial \Gamma_{jrs}}{\partial x^i} + \frac{1}{2}\left(\frac{\partial \Gamma_{jir}}{\partial x^s} + \frac{\partial \Gamma_{jis}}{\partial x^r} - \frac{\partial \Gamma_{ijr}}{\partial x^s} - \frac{\partial \Gamma_{ijs}}{\partial x^r}\right) = 0.$$

It is easy to see that the first set of the above equations is equivalent to

$$\Gamma_{ijk} = \frac{1}{2}\left(\frac{\partial g_{ik}}{\partial x^j} + \frac{\partial g_{ij}}{\partial x^k} - \frac{\partial g_{jk}}{\partial x^i}\right),$$

whereas the remaining ones turn to identities.

Summarizing the results, we have obtained that *the covariant equations* (1.25) *for geodesics of a linear torsion-free connection* Γ *are variational if and only if* Γ *is the Levi-Civita connection of the metric* g. In another formulation the result can be stated as follows: *The equations for geodesics of a linear torsion-free connection* Γ (1.24) *admit a time and velocity-independent variational integrating factor if and only if* Γ *is a metric connection*. This result is due to J. Klein [1], and D. Krupka and A. Sattarov [1]. A generalization to non-linear (semispray) connections will be discussed in Chapters 6 and 10.

Variational equations in physics. Variational techniques were first applied to physics in an exact way by Pierre de Fermat in the 17th century. Fermat proposed his famous *principle of least time* in geometrical optics stating that a ray of light, passing from a point A to a point B, describes a path for which the time of transit is a minimum with respect to all possible paths connectiong A and B. Mathematically this means to minimize the integral

$$\int_A^B \frac{ds}{v},$$

where ds is an element of the path and v is the velocity of the light. Fermat then used this principle to solve problems of reflection and refraction of light. Analogy between mechanics and geometrical optics became a motivation for mechanics to search for extremal principles which could describe motions of particles and rigid bodies. In the first half of the 19th century, W. R. Hamilton succeeded in finding a general principle, now called *Hamilton's variational principle*. This statement can be expressed as follows: *Among all kinematically possible motions of a mechanical system in the time interval* $[t_1, t_2]$, *the actual one is characterized by the stationarity of the functional*

$$\int_{t_1}^{t_2} L\,dt, \quad L = T - V, \tag{1.26}$$

where T *is kinetic energy and* $V(q^j)$ *is potential energy of the mechanical system.* The Lagrange function $L = T - V$ is called *free energy* in contrast to the *total energy* $E = T + V$ of the mechanical system. The equations of motion arising from the Hamilton's principle are then the Euler-Lagrange equations of L,

$$\frac{\partial L}{\partial q^i} - \frac{d}{dt}\frac{\partial L}{\partial \dot{q}^i} = 0,$$

which in terms of the functions T and V read

$$\frac{\partial T}{\partial q^i} - \frac{d}{dt}\frac{\partial T}{\partial \dot{q}^i} = \frac{\partial V}{\partial q^i}.$$

In the above considerations, the coordinates q^i, $1 \leqslant i \leqslant m$, called *generalized coordinates*, are *independent* coordinates, representing the configuration of a mechanical system at the time t. This is important namely in case when one considers a mechanical system subject to constraints. If (x^i), $1 \leqslant i \leqslant n$, are coordinates, referred to some fixed set of rectangular axes in R^n, and the constraints are given by the equations

$$f^j(x^i) = 0, \quad 1 \leqslant j \leqslant n - m$$

(the so called *holonomic constraints*), the mechanical system is constrained to move along the m-dimensional submanifold $M \subset R^n$ defined by the above equations. Generalized coordinates (q^i) are then local coordinates on M, and the dimension of M has the meaning of number of *degrees of freedom* of the constrained system. Thus, taking for the configuration space the manifold M, a holonomic system on R^n can be regarded as an *unconstrained system on* M.

The concept of potential energy can be extended to mechanical systems in which the acting forces depend not only on the position but also on the time, velocities and accelerations of the bodies. Consider the Lagrange's equations of motion of a holonomic system

$$\frac{\partial T}{\partial q^i} - \frac{d}{dt}\frac{\partial T}{\partial \dot{q}^i} = F_i, \tag{1.27}$$

where T is the kinetic energy of the system and F_i are the components of the sum F of external forces acting on the system. We say that the force $F(t, q^j, \dot{q}^j, \ddot{q}^j)$ is *potential* if (1.27) are variational, i.e., if there exists a Lagrange function L such that

$$\frac{\partial T}{\partial q^i} - \frac{d}{dt}\frac{\partial T}{\partial \dot{q}^i} - F_i = \frac{\partial L}{\partial q^i} - \frac{d}{dt}\frac{\partial L}{\partial \dot{q}^i}.$$

It can be seen immediately that F is potential if and only if there exists a function $V(t, q^j, \dot{q}^j)$ such that

$$F_i = \frac{\partial V}{\partial q^i} - \frac{d}{dt}\frac{\partial V}{\partial \dot{q}^i}; \tag{1.28}$$

then $L = T - V$, i.e., the mechanical system satisfies the Hamilton's variational principle.

Let us study potential forces in more detail.

If F_i depend only on (q^j) (resp. (t, q^j)), we get from (1.28) that V must be of the form

$$V = \frac{\partial \varphi}{\partial q^i}\dot{q}^i + U, \tag{1.29}$$

where $\varphi(q^j)$ and $U(q^j)$ (resp. $U(t, q^j)$) are some functions. The force F then is of the form

$$F_i = \frac{\partial U}{\partial q^i}. \tag{1.30}$$

Conversely, a potential function V (1.29) being given, the corresponding force F (1.28) does not depend on velocities and accelerations, and takes the form (1.30).

Suppose that F depends on time, positions and velocities. Then from (1.28) we get

$$\frac{\partial^2 V}{\partial \dot{q}^i \partial \dot{q}^j} = 0.$$

Denote

$$V = A_i \dot{q}^i + \phi,$$

where A_i and ϕ are functions of (t, q^k). Consequently,

$$F_i = \left(\frac{\partial A_j}{\partial q^i} - \frac{\partial A_i}{\partial q^j}\right)\dot{q}^j + \frac{\partial \phi}{\partial q^i} - \frac{\partial A_i}{\partial t}. \tag{1.31}$$

In particular, if $m = 3$ the latter formula can be written in the familiar form

$$F = v \times H + E, \tag{1.32}$$

where

$$H = \operatorname{rot} A, \quad E = -\frac{\partial A}{\partial t} + \operatorname{grad}\phi, \tag{1.33}$$

and a usual notation $r = (q^1, q^2, q^3)$, $v = \dot{r}$ is used.

It is interesting to stop at a moment to look at the Helmholtz conditions. Evidently, necessary and sufficient conditions for a force $F(t, r, v)$ be potential read

$$\frac{\partial F_i}{\partial \dot{q}^j} + \frac{\partial F_j}{\partial \dot{q}^i} = 0,$$
$$\frac{\partial F_i}{\partial q^j} - \frac{\partial F_j}{\partial q^i} + \frac{d}{dt}\frac{\partial F_j}{\partial \dot{q}^i} = 0. \tag{1.34}$$

The second set of the above identities immediately leads to

$$\frac{\partial^2 F_i}{\partial \dot{q}^j \partial \dot{q}^k} = 0.$$

Denote

$$F_i = \alpha_{ij}\dot{q}^j + \beta_i,$$

where α_{ij} and β_i are functions of (t, q^k). Substituting into (1.34) we get

$$\alpha_{ij} = -\alpha_{ji},$$
$$\frac{\partial \alpha_{ik}}{\partial q^j} + \frac{\partial \alpha_{kj}}{\partial q^i} + \frac{\partial \alpha_{ji}}{\partial q^k} = 0,$$
$$\frac{\partial \beta_i}{\partial q^j} - \frac{\partial \beta_j}{\partial q^i} = \frac{\partial \alpha_{ij}}{\partial t}.$$

Put

$$E^i = \delta^{ij}\beta_j, \quad \alpha_{ij} = -\varepsilon_{ijk}H^k,$$

where δ^{ij} is the Kronecker symbol and ε_{ijk} is the Levi-Civita symbol. Then Helmholtz conditions (1.34) for F take an equivalent form

$$F = v \times H + E,$$

with H and E satisfying

$$\operatorname{div} H = 0, \quad \operatorname{rot} E = -\frac{\partial H}{\partial t}. \tag{1.35}$$

Formula (1.31) (resp. (1.32) with (1.33) or (1.35)) provides the classification of all first-order potential forces (cf. E. Engels and W. Sarlet [1], J. Novotný [1]). Apparently, every first-order potential force comes from a "scalar potential" ϕ and a "vector potential" $A = (A_i)$, and takes a form similar to the *Lorentz force* known from electrodynamics. Notice that besides the Lorentz force, there are also other forces of this kind. As an example, consider the *Coriolis force*,

$$F_C = 2m(v \times \omega),$$

acting on a particle of mass m moving with a velocity v relative to a frame rotating with the angular velocity ω uniformly round an axis. In this case we have $E = 0$, $H = 2m\omega$, which gives us $\phi = 0$, $A = m(\omega \times r)$, i.e., $V = m(\omega \times r) \cdot v$.

Finally, suppose that F is allowed to depend on time, positions, velocities and accelerations. Applying the Helmholtz conditions (1.15), we get that F is potential if and only if

$$\frac{\partial F_i}{\partial \ddot{q}^j} = \frac{\partial F_j}{\partial \ddot{q}^i},$$
$$\frac{\partial F_i}{\partial \dot{q}^j} + \frac{\partial F_j}{\partial \dot{q}^i} = \frac{d}{dt}\left(\frac{\partial F_i}{\partial \ddot{q}^j} + \frac{\partial F_j}{\partial \ddot{q}^i}\right), \tag{1.36}$$
$$\frac{\partial F_i}{\partial q^j} - \frac{\partial F_j}{\partial q^i} = \frac{1}{2}\frac{d}{dt}\left(\frac{\partial F_i}{\partial \dot{q}^j} - \frac{\partial F_j}{\partial \dot{q}^i}\right).$$

The above conditions imply that F must be affine in the accelerations, i.e., that

$$F_i = f_i(t, q^k, \dot{q}^k) + g_{ij}(t, q^k, \dot{q}^k)\ddot{q}^j, \tag{1.37}$$

where (g_{ij}) is a symmetric matrix. An example of such a force is provided by the *Weber's electrodynamic force*

$$F_W = \frac{1}{r^2}\left(1 - \frac{\dot{r}^2 - 2r\ddot{r}}{c^2}\right),$$

where r is the distance of a particle of unit mass from the centre of force. In this case the potential function V is defined by the relation

$$V = \frac{1}{r}\left(1 + \frac{\dot{r}^2}{c^2}\right)$$

(E. T. Whittaker [1]). A general *classification* of all potential second-order forces will be provided in Chapter 6 (Theorem 6.7.2).

Besides classical mechanics and geometrical optics, variational principles have been used also in other parts of theoretical physics. The basic questions then were: *Are fundamental laws derivable from a variational principle analogous to the Hamilton's principle in classical mechanics*, and, if yes, *how to generalize the integrand* $T - V$ i.e., how, for a concrete theory, to specify the Lagrange function L. This problem has been first stated by H. Helmholtz [4]. In that paper Helmholtz showed that all known laws of reversible processes can be expressed in a form of Euler-Lagrange equations, and he generalized the Hamilton's principle to reversible thermodynamics and electrodynamics (see also Helmholtz [2,3]). He also was the first who pointed out that "it is necessary to throw out the old restrictive assumption that velocities enter in the Lagrange function L only in the kinetic energy term which moreover is assumed to be a function homogeneous of degree 2 in velocities; one should study the situation that L is an *arbitrary function* of positions and velocities" (Helmholtz [3]). Generalization of the Hamilton's variational principle to field theory stating that field equations should represent conditions for stationarity of the integral

$$\int L\, dx^1 dx^2 dx^3 dx^4,$$

where L depends on space-time variables and on the components of fields and their derivatives, lies in the foundations of modern theoretical physics. A turning point in the applicability of variational techniques to physics was the discovery that for every concrete field theory, a right Lagrange function is the one possessing *invariance property* with respect to the corresponding groups. In this way, Lagrange functions in relativistic electrodynamics must be invariant with respect to the Lorentz group (H. Poincaré [1], H. Minkowski[1]), Lagrange functions in the general relativity theory must be generally covariant (D. Hilbert [1]), etc. Besides, it turned out that Lagrange functions describing fields must possess an additional invariance, the co called *gauge invariance*. In this context, a fundamental role is played by the famous Theorem of E. Noether, stating (roughly speaking) that *every symmetry of Lagrange function gives rise to a conservation law* (E. Noether [1]). Later these results found deep applications in quantum electrodynamics and quantum field theory.

Variational equations in engineering. Newton's pioneering studies on the determination of the best shape for a solid body beeing translated through a fluid motivated interest in applications of the calculus of variations in engineering. In a variety of engineering problems the task of determining an optimal policy is encountered, and it is natural to give the mathematical statement of the problem in variational form. Thus many optimization problems, economic and design considerations, problems in hydrodynamics, elasticity, geophysics, thermodynamics can be effectively tackled by variational techniques (cf. e.g. Mikhlin [1], Rektorys [1], Schechter [1]).

In particular, the mathematical descriptions of many systems studied by engineers find their expression in the form of ordinary or partial differential equations. Typical examples are twisting of rods, bendings of rods and plates, or vibrations of strings, plates and membranes in the elasticity theory. Equations arising in this way are often of order greater than two. For example, the equation describing the bending of a nonhomogeneous rod of variable cross-section on an elastic ground is a 4th-order ordinary differential equation,

$$\frac{d^2}{dx^2}\Big(E(x)I(x)\frac{d^2u}{dx^2}\Big)+Q(x)u = f(x),$$

which is the Euler-Lagrange equation of the second-order Lagrange function

$$L = \tfrac{1}{2}E(x)I(x)\Big(\frac{d^2u}{dx^2}\Big)^2 + \tfrac{1}{2}Qu^2 - fu.$$

The existence a Lagrange function is essential, since there are methods of calculation which utilize the variational formulation of the problem directly—the integral to be extremized will serve as the basis for developing approximations to extremalizing functions. Among the methods for finding *approximate solutions* of variational problems, e.g. the *Ritz method* is frequently used. In this context, the solution of the *inverse problem of the calculus of variations* provides techniques which play an important role, and can serve as an example of a valuable contribution of "pure theory" to engineering. A similar meaning for concrete applications can have theoretical studies of *higher-order*, as well as *non-regular* variational problems.

Integration of variational equations. There is another aspect of variationality we have not mentioned so far. Namely, for certain classes of variational equations there have been developed powerful *methods of exact integration*. These methods are based on the possibility to transform the Euler-Lagrange equations to equivalent first-order *Hamilton equations*.

Let $1 \leqslant \sigma \leqslant m$, and let $L(t, q^\sigma, \dot{q}^\sigma)$ be a Lagrange function. The Euler-Lagrange equations for extremals $(q^\nu(t))$ of L read

$$\frac{\partial L}{\partial q^\sigma} - \frac{d}{dt}\frac{\partial L}{\partial \dot{q}^\sigma} = 0, \tag{1.38}$$

i.e., they represent a system of m second-order ordinary differential equations. Put

$$p_\sigma = \frac{\partial L}{\partial \dot{q}^\sigma}, \tag{1.39}$$

and

$$H = -L + \frac{\partial L}{\partial \dot{q}^\sigma}\dot{q}^\sigma = -L + p_\sigma \dot{q}^\sigma. \tag{1.40}$$

The functions p_σ are called *momenta*, and H is called *Hamilton function* or *Hamiltonian* related to the Lagrange function L.

Suppose that L satisfies the *regularity condition*

$$\det\left(\frac{\partial^2 L}{\partial \dot{q}^\sigma \partial \dot{q}^\nu}\right) = \det\left(\frac{\partial p_\sigma}{\partial \dot{q}^\nu}\right) \neq 0. \tag{1.41}$$

Then the mapping $(t, q^\sigma, \dot{q}^\sigma) \to (t, q^\sigma, p_\sigma)$ is a coordinate transformation on $R^m \times R^m$; it is called *Legendre transformation*, and the new coordinates (t, q^σ, p_σ) are called *Legendre coordinates*. Expressing L in Legendre coordinates and using (1.40) we get that

$$\frac{\partial L}{\partial q^\sigma} = -\frac{\partial H}{\partial q^\sigma}. \tag{1.42}$$

Now, we are ready to replace the Euler-Lagrange equations by equations for curves $t \to (q^\sigma(t), p_\sigma(t))$. Taking into account that

$$\frac{dq^\sigma}{dt} = \dot{q}^\sigma$$

and that, by (1.40) and (1.39)

$$\frac{\partial H}{\partial p_\sigma} = -\frac{\partial L}{\partial \dot{q}^\nu}\frac{\partial \dot{q}^\nu}{\partial p_\sigma} + \dot{q}^\sigma + p_\nu \frac{\partial \dot{q}^\nu}{\partial p_\sigma} = \dot{q}^\sigma,$$

we get

$$\frac{dq^\sigma}{dt} = \frac{\partial H}{\partial p_\sigma}. \tag{1.43}$$

Finally, substituting (1.42) into the Euler-Lagrange equations (1.38) we obtain

$$\frac{dp_\sigma}{dt} = -\frac{\partial H}{\partial q^\sigma}. \tag{1.44}$$

Equations (1.43), (1.44) are obviously *equivalent with Euler-Lagrange equations* (1.38); they are called *Hamilton equations*. We stress that while Euler-Lagrange equations are (second-order) equations for curves $t \to (q^\sigma(t)) \in R^m$, Hamilton equations are *first-order* equations for curves $t \to (q^\sigma(t), p_\sigma(t)) \in R^m \times R^m$.

It is interesting to notice that Hamilton equations (1.43), (1.44) are *variational*. Indeed, denoting $dq^\sigma/dt = q'^\sigma$, $dp_\sigma/dt = p'_\sigma$, and putting

$$E_\sigma = p'_\sigma + \frac{\partial H}{\partial q^\sigma}, \quad E_{m+\sigma} = -q'^\sigma + \frac{\partial H}{\partial p_\sigma}$$

we can see that the E's (as functions of $t, q^\sigma, p_\sigma, q'^\sigma, p'_\sigma$) satisfy the Helmholtz conditions (1.15) identically. The corresponding Vainberg-Tonti Lagrangian (1.18) is

$$\begin{aligned}\tilde{L} &= -q^\sigma \int_0^1 \left(p'_\sigma + \frac{\partial H}{\partial q^\sigma}\right) \circ \chi \, du - p_\sigma \int_0^1 \left(\frac{\partial H}{\partial p_\sigma} - q'^\sigma\right) \circ \chi \, du \\ &= \tfrac{1}{2}(p_\sigma q'^\sigma - q^\sigma p'_\sigma) - q^\sigma \int_0^1 \frac{\partial H}{\partial q^\sigma} \circ \chi \, du - p_\sigma \int_0^1 \frac{\partial H}{\partial p_\sigma} \circ \chi \, du.\end{aligned}$$

Using the identity

$$q^\sigma \int_0^1 \frac{\partial H}{\partial q^\sigma} \circ \chi \, du + p_\sigma \int_0^1 \frac{\partial H}{\partial p_\sigma} \circ \chi \, du = \int_0^1 d(H \circ \chi) = H - H(t),$$

we arrive at an equivalent Lagrangian

$$\mathcal{L} = p_\sigma q'^\sigma - H.$$

In other words this means that the solutions $t \to (q^\sigma(t), p_\sigma(t))$ of Hamilton equations are *extremals* of the Lagrangian $\mathcal{L}$.

Now, let us turn to a fundamental property of Hamilton equations. It is easy to see that Hamilton equations can be interpreted as equations for integral curves of the *vector field*

$$\frac{\partial H}{\partial p_\sigma}\frac{\partial}{\partial q^\sigma} - \frac{\partial H}{\partial q^\sigma}\frac{\partial}{\partial p_\sigma} \tag{1.45}$$

on $R^m \times R^m$, or, equivalently, as equations for integral curves of an exterior differential system generated by the *Pfaffian forms*

$$dq^\sigma - \frac{\partial H}{\partial p_\sigma} dt, \quad dp_\sigma + \frac{\partial H}{\partial q^\sigma} dt \tag{1.46}$$

on $R \times R^m \times R^m$. The above geometric interpretation makes Hamilton equations much more appropriate for exact integration than the original Euler-Lagrange equations. Applying the Frobenius Theorem we can see that the Pfaffian system (1.46) is completely integrable. This means, in particular, that in a neighborhood of every point in $R \times R^m \times R^m$ one can find an adapted chart with coordinates (t, a^σ, b_σ) such that the a's and b's are *first integrals* (i.e., functions constant along the solutions of the Hamilton equations). *In this way, the problem of integration of the original Euler-Lagrange equations* (1.38) *has been transfered to the problem of finding local systems of first integrals of* (1.46). There are different procedures how to find such systems of first integrals. One of them is the *Liouville integration method*—if one knows m appropriate first integrals (which can be found e.g. by means of the Noether's theorem) then there is an integral formula for a direct computation of the remaining m first integrals (the so called *integration by quadratures*). Another famous integration method is due to Hamilton and Jacobi. To these days, the *Hamilton-Jacobi method* remains the most powerfull method available for the exact integration of the Euler-Lagrange equations. The method is based on the fact that Hamilton equations (1.43), (1.44) can be viewed as *equations for the characteristics* of the partial differential equation,

$$\frac{\partial S}{\partial t} + H\left(t, q^\sigma, \frac{\partial S}{\partial q^\sigma}\right) = 0, \tag{1.47}$$

called the *Hamilton-Jacobi equation*. Then, knowing a *complete integral* $S(t, q^\sigma, a^\sigma)$ depending on m constant parameters $a_1, \dots, a_m$, the extremals of the original Euler-Lagrange equations are given by the implicit equations

$$b_\sigma(t, q^\nu, a^\nu) = \frac{\partial S}{\partial a^\sigma}.$$

In the expression $(q^\sigma(t, a^\nu, b_\nu))$ of the extremals, the parameters $a^1, \dots, a^m, b_1, \dots, b_m$ play thc role of *initial positions and velocities*.

We have noted that the application of the above mentioned integration methods is restricted to *first order Lagrangians which satisfy the regularity condition* (1.41). To extend the range of application of these methods, a detail study of variational equations of an arbitrary finite order—namely, their properties and geometric meaning—is necessary. This will be our primary concern throughout the present book.

Chapter 2.

BASIC GEOMETRIC TOOLS

2.1. Introduction

Modern calculus of variations is closely connected with differential geometry and global analysis, both in methodology and the range of applications. Among pioneering works due to G. Darboux, G. Ricci, T. Levi-Civita, H. Poincaré, J. L. Synge, H. Weyl, Th. de Donder, E. T. Whittaker and many others, especially É. Cartan's work motivated an impressive entrance of geometry into this field.

Foundations of the modern calculus of variations were laid in the 50's by Paul Dedecker, who was the first to use systematically modern methods of differential geometry and algebraic topology in the variational theory. He also was the first who understood the necessity to apply Ehremann's concept of *jet manifold* to be able to give the theory a correct, global and intrinsic form.

We have seen in the previous chapter that even in classical mechanics the concept of *manifold* naturally appears when the concept of configuration space is introduced. This, however, means to use from the very beginning geometric objects, such as e.g. differential forms and vector fields. The variational calculus works with Lagrangians and Euler-Lagrange equations which depend on the first, and higher derivatives of the dependent variables. To give them a correct mathematical interpreteation one has to model them via objects defined on some jet manifold. One possible formulation, which is general enough to cover also multiple integrals in the calculus of variations of an arbitrary finite order takes a *fibered manifold and its jet prolongations* for an underlying structure. This means that one considers a fibered manifold $\pi : Y \to X$ and its jet prolongations, which are manifolds of r-jets, $r \geqslant 1$, of local sections $X \to Y$. In the case of one independent variable ("higher-order mechanics") the base X is one-dimensional. Hence, when connected, the base is diffeomorphic with R or with the unit circle S^1 (the latter applies e.g. if periodic extremals are studied). Within the range of the fibered manifold scheme, solutions of variational equations are modeled via (*local*) *sections* of π. If, in particular, $Y = R \times M$ where M is a smooth manifold of dimension m, then a local section is nothing but a graph of a curve in M. Thus fibered manifolds are appropriate structures for studying the so called *time-dependent* (or *non-autonomous*) mechanics; consequently, the *time-independent* (*autonomous*) case appears as a special case in the general scheme. From a purely mathematical point of view, there is an apparent advantage of using the fibered manifold structure in mechanics: it brings the results directly into the scheme of a general global variational theory (including higher-order field theory, the variational bicomplexes or variational sequences, etc.).

The theory of variational ordinary differential equations presented in this book is based on the theory of distributions of generally non-constant rank, and on the calculus of variations on fibered manifolds. The aim of this chapter is to provide the reader with a summary of main concepts we shall need throughout the book, and to fix basic notation and background for our further exposition. However, not all of the material is needed to understand each of the following chapters, so that an individual choice from the preliminary material is possible.

If not otherwise stated, the manifolds and mappings throughout the text are smooth, and the summation convention is used. We use the symbols T for the tangent functor, J^r for the r-jet prolongation functor, id for the identity mapping, * for the pull-back, d for the exterior derivative, i for the inner product, and ∂ for the Lie derivative. Other notations are explained when first used.

2.2. Distributions

Let M be a smooth manifold. By a *distribution* on M we mean a mapping Δ assigning to every point $x \in M$ a vector subspace Δ_x of the vector space $T_x M$. The dimension of the vector space Δ_x will be called *rank* of the distribution Δ *at* x and will be denoted by $\operatorname{rank} \Delta(x)$. Evidently, $0 \leqslant \operatorname{rank} \Delta(x) \leqslant \dim M$. The function $\operatorname{rank} \Delta : M \to R$, assigning to every point $x \in M$ the number $\operatorname{rank} \Delta(x)$ will be called *rank* of the distribution Δ. If the function $\operatorname{rank} \Delta$ is constant on M, we shall say that the distribution Δ *has a constant rank.*

Two distributions Δ_1, Δ_2 on M are called *complementary* if for every $x \in M$

$$\Delta_1(x) \oplus \Delta_2(x) = T_x M.$$

Let ξ be a local vector field on M (i.e., possibly defined on an open subset of M); denote by $\operatorname{dom} \xi$ the domain of definition of ξ. We say that ξ *belongs* to a distribution Δ, and write

$$\xi \in \Delta,$$

if for every $x \in \operatorname{dom} \xi$ the vector $\xi(x)$ belongs to $\Delta(x)$.

Any system $\{\xi_\iota\}_{\iota \in I}$ (where I is an index set) of local vector fields on M, such that $\bigcup \operatorname{dom} \xi_\iota = M$, defines in an obvious way a distribution on M. Conversely, to every distribution Δ on M one can find a system $\{\xi_\iota\}$ of local vector fields such that $\xi_\iota \in \Delta$ for all ι and for each $x \in M$ the vector space $\Delta(x)$ is spanned by the system of vectors $\{\xi_\iota(x), \iota \in I\}$. In this case we say that the ξ_ι, $\iota \in I$, are *generators* of Δ, or that the distribution Δ is *spanned* by the system of vector fields $\{\xi_\iota\}$ and write

$$\Delta = \operatorname{span}\{\xi_\iota, \iota \in I\}.$$

We say that Δ is *continuous* (resp. *of class* C^r, $r \geqslant 1$, resp. *smooth*) if it can be spanned by a system of continuous (resp. of class C^r, resp. of smooth) vector fields. Note that in general a continuous distribution cannot be spanned by global vector fields.

There is another possibility how to define a distribution on M. If $\Delta : x \to \Delta_x \subset T_x M$ is a distribution on M, we can consider for every $x \in M$ the space Δ_x^* of annihilators of

Δ_x, i.e., the vector subspace of T_x^*M defined as follows: a covector η belongs to Δ_x^* iff

$$i_\xi \eta = 0 \quad \text{for all } \xi \in \Delta_x.$$

Having in mind the above duality between the vector spaces Δ_x and Δ_x^* we say, with an obvious convention, that a (local) one-form on M *belongs* to Δ if for every $x \in \operatorname{dom} \eta$, the covector $\eta(x)$ belongs to Δ_x^*, and we write

$$\eta \in \Delta.$$

Thus, to every distribution Δ on M there exist an index set K and a system $\{\eta_\kappa\}_{\kappa \in K}$ of local one-forms on M such that for each κ, η_κ belongs to Δ and at every point $x \in M$, the vector space Δ_x^* is spanned by $\{\eta_\kappa(x),\ \kappa \in K\}$. Conversely, any system of local one-forms on M, whose domains cover M, defines a distribution on M. Actually, if $\Delta^* = \operatorname{span}\{\eta_\kappa,\ \kappa \in K\}$ then $\Delta = \operatorname{span}\{\xi \mid i_\xi \eta_\kappa = 0, \forall \kappa \in K\}$. We write $\Delta \approx \Delta^*$, i.e.,

$$\Delta \approx \operatorname{span}\{\eta_\kappa,\ \kappa \in K\}.$$

The annihilator Δ^* of a distribution Δ on M is a mapping $\Delta^* : X \ni x \to \Delta^*(x) \subset T^*M$; it is called a *codistribution*, related with Δ. At each point $x \in M$,

$$\operatorname{rank} \Delta(x) + \operatorname{rank} \Delta^*(x) = \dim M,$$

i.e., $\operatorname{rank} \Delta^* = \operatorname{corank} \Delta$.

If Δ is a smooth distribution of a *constant rank* on M, $\operatorname{rank} \Delta = k$, then in a neighborhood of every point of M, Δ can be defined either by a system of k linearly independent *smooth* vector fields, or by a system of $\dim M - k$ linearly independent *smooth* one-forms. Analogous conclusions are made for distributions of locally constant rank (i.e., such that the rank is constant on each connected component of M).

The situation, however, essentially differs if the rank of Δ is not locally constant on M. Actually, the following proposition holds:

2.2.1. Proposition. *Let Δ be a continuous distribution on M, let $x_0 \in M$ be a point. There is a neighborhood U of x_0 such that for every $x \in U$,*

$$\operatorname{rank} \Delta(x) \geqslant \operatorname{rank} \Delta(x_0).$$

Equivalently one can say that the rank of a continuous distribution is a lower semicontinuous function on M. The rank of the related codistribution is therefore an upper semicontinuous function. Thus, if Δ is a continuous distribution of a non-constrant rank, the related codistribution Δ^* is not continuous (i.e., it cannot be spanned by a system of continuous 1-forms). Summarizing, we get that a smooth distribution of a non-constant rank *cannot* be defined by means of *continuous* one-forms, and conversely, a distribution of a non-constant rank defined by means of smooth one-forms *is not continuous*.

Let Δ be a distribution on M, let Q be a smooth manifold of dimension q, $f : Q \to M$ an immersion. f is called an *integral mapping* of Δ if for every $t \in Q$

$$T_t f(T_t Q) \subset \Delta(f(t)),$$

or equivalently, if

$$f^*\omega = 0 \quad \text{for all } \omega \text{ belonging to } \Delta.$$

Clearly, if $f : Q \to M$ is an integral mapping of Δ then every point t in Q has a neighborhood U such that $f(U)$ is a submanifold of M.

An integral mapping $f : Q \to M$ is said to be *of maximal dimension at a point* $t \in Q$ if $T_t f(T_t Q) = \Delta(f(t))$; it is said to be *of maximal dimension* if it is of maximal dimension at each point.

A *subset* Q of M is called a q-dimensional *integral manifold* of Δ if Q is a *connected* manifold of dimension q, and the canonical inclusion $i : Q \to M$ is an integral mapping of Δ.

If Q is an integral manifold of Δ we say that Q is of *maximal dimension at a point* $x \in Q$ if rank $\Delta(x) = \dim Q$; we say that Q is *of maximal dimension* if it is of maximal dimension at each point $x \in Q$. An integral manifold Q of Δ is called a *maximal integral manifold* if it is of maximal dimension and if any integral manifold of Δ of maximal dimension containing Q equals to Q.

Let Δ be a distribution on a manifold M. A function g, defined on an open subset U of M is called a *first integral* of Δ if the one-form dg belongs to Δ, i.e., if $i_\xi dg = 0$ for every vector field $\xi \in \Delta$. We can see that if g is a first integral of Δ on U and $\iota : Q \to U \subset M$ is an integral manifold of Δ then $d(g \circ \iota) = 0$, i.e., the function g is a constant along Q.

First integrals g_1, g_2 of Δ defined on U are called *independent* if the forms dg_1, dg_2 are linearly independent at each point of U.

Let $g_1, \dots, g_k$ be independent first integrals of Δ, defined on U. The set $\{g_1, \dots, g_k\}$ is called *a complete set of independent first integrals of* Δ *at a point* $x \in U$ (resp. *on* U) if

$$\Delta(x) \approx \text{span}\{dg_1(x), \dots, dg_k(x)\}$$

(resp. if the above condition holds at each point of U). Let $\{g_1, \dots, g_k\}$ be a complete set of independent first integrals of Δ. Then the submanifolds of U characterized by the equations $g_1 = c_1, \dots, g_k = c_k$, where $c_1, \dots, c_k$ are constants, are integral manifolds of Δ of maximal dimension, and one can find local coordinates (y^i) on M such that $y^1 = g_1, \dots, y^k = g_k$. Conversely, if Q is an integral manifold of Δ of maximal dimension, $\dim Q = q$, and if (U, ϕ), $\phi = (y^i)$ are local coordinates on M adapted to Q (such that $y^{q+1}, \dots, y^n$ are constants along Q) then $\{y^{q+1}, \dots, y^n\}$ is a complete set of independent first integrals of Δ (along $U \cap Q$).

A distribution Δ on M is called *completely integrable* if through every point $x \in M$ there passes an integral manifold of Δ of maximal dimension.

The geometric structure of solutions of *completely integrable* distributions is described by *foliations*. More precisely, maximal integral manifolds of a completely integrable distribution on M form a foliation of M, and the leaves of this foliation are *immersed* submanifolds of M. If the distribution has a constant rank then all the leaves of the corresponding foliation are manifolds of the same dimension, in general however the dimensions of the leaves may differ. The existence of a foliation means in particular that

– through every point of M there passes a unique maximal integral manifold of Δ,

– if $f : Q \to M$ is an integral mapping of *maximal dimension* then $f(Q)$ is a local diffeomorphism onto an open subset in a leaf,

– if $f : Q \to M$ is an integral mapping and rank $\Delta = \text{const}$ then f is an immersion of Q into a leaf of the foliation. One should note, however, that a completely integrable distribution of a *non-constant* rank may possess integral mappings (of dimension less than maximal) which intersect different leaves. To illustrate this by an example, consider on R^2 the distribution

$$\Delta = \operatorname{span}\left\{\frac{\partial}{\partial x}, y\frac{\partial}{\partial y}\right\}.$$

This distribution is completely integrable: the foliation $\mathcal{F}$ defined by Δ consists from three leaves, two of them 2-dimensional and one 1-dimensional, namely, the half planes $y > 0$ and $y < 0$, and the line $y = 0$. Consider the immersion $f : R \to R^2$,

$$f(t) = (t, t^3).$$

Then f is an integral mapping of Δ whose image intersects all three leaves of the foliation $\mathcal{F}$.

For distributions of a *constant* (locally constant) rank one has the following familiar necessary and sufficient conditions of complete integrability:

2.2.2. Frobenius theorem. *Let Δ be a distribution of a constant rank on M, let $k = \operatorname{rank}\Delta$, and $n = \dim M$. The following conditions are equivalent:*

(1) *Δ is completely integrable.*

(2) *The system of vector fields belonging to Δ is a Lie algebra.*

(3) *The ideal $\mathfrak{I}$ of differential forms generated by the system of one-forms belonging to Δ satisfies $d\mathfrak{I} \subset \mathfrak{I}$.*

(4) *At each point $x \in M$ there is a chart (U, φ), $\varphi = (x^i)$ on M, such that on U*

$$\Delta = \operatorname{span}\left\{\frac{\partial}{\partial x^1}, \dots, \frac{\partial}{\partial x^k}\right\} \approx \operatorname{span}\{dx^{k+1}, \dots, dx^n\}.$$

A chart (U, ϕ), $\phi = (x^i)$ on M satisfying the condition (4) of the Frobenius theorem will be called *adapted chart* to the distribution Δ. In this chart, maximal integral manifolds of Δ are given by the equations

$$x^{k+1} = c^{k+1}, \dots, x^n = c^n,$$

where $c^{k+1}, \dots, c^n$ are constants. Obviously, $\{x^{k+1}, \dots, x^n\}$ is a complete set of independent first integrals of Δ on U.

Maximal integral manifolds of a distribution of a constant rank can be found by means of *symmetries*. Let us recall the main concepts. (For more details we refer to S. V. Duzhin and V. V. Lychagin [1], and V. V. Lychagin [1]).

Let Δ be a distribution of a constant rank on M. Let ϕ be a diffeomorphism of M. ϕ is called an *invariant transformation* of Δ if for all $x \in M$,

$$T\phi(\Delta_x) \subset \Delta_{\phi(x)}.$$

Let ξ be a vector field on M, denote by $\{\phi_u^\xi\}$ its local one-parameter group of transformations. The vector field ξ is called a *symmetry* of Δ if for all u, the ϕ_u^ξ is an invariant transformation of Δ. (The reader should be aware of non-unified terminology appearing in the literature: invariant transformations are often called *finite symmetries* and symmetries are then called *infinitesimal symmetries*).

The set of all symmetries of a distribution is obviously a Lie algebra with respect to the Lie bracket of vector fields. We have the following important characterization of this Lie algebra:

2.2.3. Theorem. *Let Δ be a distribution of a constant rank. The following three conditions are equivalent:*

(1) *ξ is a symmetry of Δ,*

(2) *for every vector field ζ belonging to Δ, the Lie bracket $[\xi, \zeta]$ belongs to Δ,*

(3) *for every one-form η belonging to Δ, the Lie derivative $\partial_\xi \eta$ belongs to Δ.*

The meaning of symmetries clearly is the following: if ξ is a symmetry of Δ with the local one-parameter group $\{\phi_u^\xi\}$, and Q is an integral manifold of Δ then $\phi_u^\xi(Q)$ is also an integral manifold of Δ. In other words, the local flow of a symmetry transfers integral mappings into integral mappings, and, consequently, integral manifolds into integral manifolds.

2.2.4. Remark. Basic facts about distributions of a non-constant rank can be found e.g. in the Appendix of the book of P. Libermann and Ch.-M. Marle [1]. For a more general background we refer to the monograph by R. L. Bryant, S. S. Chern, R. B. Gardner, H. L. Goldschmidt and P. A. Griffiths [1].

In this book we shall deal with distributions defined by smooth one-forms. As mentioned above, distributions of a non-constant rank defined by smooth one-forms cannot be spanned by a system of continuous vector fields. Therefore we shall not need the known generalizations of the Frobenius theorem to differentiable distributions of a non-constant rank (see H. Sussmann [1], V. P. Viflyantsev [1,2], or the above mentioned book by Libermann and Marle).

Geometric structure of solutions of distributions of non-constant rank defined by smooth one-forms is rather complicated and has to be studied with the help of the so called *constraint algorithm* (for details we refer to Chapter 7).

2.3. Closed two-forms

Let M be a manifold of dimension n, and let α be a (smooth) closed two-form on M. Denote by $\mathcal{V}(M)$ the system of all vector fields on M.

Consider the following distribution defined by means of a system of smooth one-forms

$$\mathcal{D} \approx \text{span}\{i_\xi \alpha\}, \quad \text{where } \xi \text{ runs over } \mathcal{V}(M).$$

Taking into account the relation $i_\zeta i_\xi \alpha = -i_\xi i_\zeta \alpha$ we can see that this distribution is spanned by a system of (not necessarily continuous) vector fields as follows:

$$\mathcal{D} = \text{span}\{\zeta \in \mathcal{V}(M) \mid i_\zeta \alpha = 0\}.$$

$\mathcal{D}$ is called the *characteristic distribution* of α, and its annihilator $\mathcal{D}^* = \text{span}\{i_\xi \alpha\}$ is called the *associated system* of α. Considering the differential ideal $\mathcal{I}$ generated by the closed two-form α we can see immediately that the vector fields belonging to the distribution $\mathcal{D}$ are *Cauchy characteristic vector fields of $\mathcal{I}$*. If the distribution $\mathcal{D}$ is completely integrable then the leaves of the foliation defined by $\mathcal{D}$ are called *Cauchy characteristics*.

Recall that at each point $x \in M$ the rank of a closed two-form α is an *even* number, and that

$$\operatorname{rank} \alpha(x) = \operatorname{corank} \mathcal{D}(x) .$$

2.3.1. Darboux Theorem. *Let α be a closed two-form of a constant rank, $2p$. Then at each point $x \in M$ there is a chart (U, ϕ), $\phi = (a_1, \dots, a_p, b^1, \dots, b^p, x^{2p+1}, \dots, x^n)$ such that*

$$\alpha = \sum_{K=1}^{p} da_K \wedge db^K .$$

Any chart (a_K, b^K, x^L), $1 \leqslant K \leqslant p$, $2p+1 \leqslant L \leqslant n$, characterized by the Darboux Theorem will be called a *Darboux chart* of α. Since in a Darboux chart we get for the characteristic distribution

$$\mathcal{D} = \operatorname{span}\{\partial/\partial x^L,\ 2p+1 \leqslant L \leqslant n\} \approx \operatorname{span}\{da_K,\ db^K,\ 1 \leqslant K \leqslant p\},$$

we can see that the Darboux coordinates $a_1, \dots, a_p, b^1, \dots, b^p$ are *a complete set of independent first integrals of* $\mathcal{D}$.

As a corollary we get

2.3.2. Cartan Theorem. *Let* $\operatorname{rank} \alpha = \text{const.}$ *Then the characteristic distribution of α has a constant rank and is completely integrable.*

Let us turn to study symmetries of closed two-forms.

First, recall the concept of symmetry of a differential form. Let η be a p-form $p > 1$, on M, and let ϕ be a local diffeomorphism of M. Then ϕ is called an *invariant transformation* of η if

$$\phi^* \eta = \eta .$$

Let ξ be a vector field on M, denote by $\{\phi^\xi_u\}$ its local one-parameter group of transformations. Then ξ is called a *symmetry* of η if for all u, the ϕ^ξ_u is an invariant transformation of η. Differentiating this relation with respect to the parameter at $u = 0$, we get an equivalent condition for ξ be a symmetry of η in the form

$$\partial_\xi \eta = 0.$$

Consider now a closed two-form α on M. Then we have $\partial_\xi \alpha = d i_\xi \alpha$. This means that *if a vector field ξ on M is a symmetry of α then the one-form $i_\xi \alpha$ is closed, i.e., (locally) $i_\xi \alpha = df$, where f is a first integral of the characteristic distribution* $\mathcal{D}$. Notice that every vector field belonging to $\mathcal{D}$ is a symmetry of α; the corresponding first integrals are trivial (constant functions).

Let ρ be a one-form (possibly defined on an open subset of M) such that $\alpha = d\rho$. Since $\partial_\xi \rho = i_\xi d\rho + d i_\xi \rho$, we can see that $\partial_\xi \rho = 0$ means that $i_\xi \alpha = df$, where $f = -i_\xi \rho$; in other words, *every symmetry ξ of ρ is a symmetry of α, and $i_\xi \rho$ is a first integral of the characteristic distribution* $\mathcal{D}$.

Obviously, *a first integral corresponding to a symmetry of α is unique up to a constant function.* Conversely, to a given first integral the corresponding symmetry is not unique; all these symmetries form a class modulo the characteristic distribution $\mathcal{D}$. More precisely, we have

2.3.3. Proposition.

(1) *If ξ is a symmetry of α such that $i_\xi \alpha = df$ then for any vector field ζ belonging to $\mathcal{D}$, the $\bar{\xi} = \xi + \zeta$ is another symmetry of α satisfying $i_{\bar{\xi}}\alpha = df$.*

(2) *If ξ_1, ξ_2 are two symmetries of α such that $i_{\xi_1}\alpha = df = i_{\xi_2}\alpha$ then $\xi_1 - \xi_2$ belongs to $\mathcal{D}$.*

Let ξ_1, ξ_2 be two symmetries of a closed two-form α. Then

$$i_{[\xi_1,\xi_2]}\alpha = \partial_{\xi_1} i_{\xi_2}\alpha - i_{\xi_2}\partial_{\xi_1}\alpha = d i_{\xi_1} i_{\xi_2}\alpha.$$

Hence we get the following assertion which can be viewed as a generalization of the classical *Poisson Theorem.*

2.3.4. Proposition. *The set of all symmetries of a closed two-form α is a Lie algebra. If f_1, f_2 are first integrals of the characteristic distribution $\mathcal{D}$, and ξ_1, ξ_2 are some corresponding symmetries of α, then*

$$\{f_1, f_2\} \equiv i_{\xi_1} i_{\xi_2}\alpha = i_{\xi_1} df_2 = -i_{\xi_2} df_1$$

is a first integral of $\mathcal{D}$, corresponding to the symmetry $[\xi_1, \xi_2]$ of α.

The first integral $\{f_1, f_2\}$ will be called the *Poisson bracket* of the first integrals f_1, f_2. In particular, *for every symmetry ξ of α and every C^1-vector field ζ belonging to the characteristic distribution $\mathcal{D}$* we have $i_{[\xi,\zeta]}\alpha = d i_\xi i_\zeta \alpha = 0$, i.e., *the symmetry $[\xi, \zeta]$ of α belongs to the characteristic distribution $\mathcal{D}$*, giving rise to trivial first integrals.

Consequently, using the second condition of Theorem 2.2.3., we get the following relation between the symmetries of a closed two-form and of its characteristic distribution:

2.3.5. Corollary. *If α has a constant rank then every symmetry of α is a symmetry of the characteristic distribution $\mathcal{D}$ of α.*

Finally, we have the following interesting assertion

2.3.6. Proposition. *Let α be a closed two-form, $\mathcal{D}$ its characteristic distribution. Let ρ be defined on an open subset U of M, such that $d\rho = \alpha$ on U, and $\rho \notin \mathcal{D}^*$. Then in a neighborhood of every point $x \in U$ there is a symmetry $\bar{\xi}$ of α such that $i_{\bar{\xi}}\rho = 0$.*

Let ξ be a symmetry of α. Then for any integer $k \geqslant 0$ and vector fields $\zeta_1, \dots, \zeta_k$ belonging to the characteristic distribution $\mathcal{D}$, such that $i_{\zeta_j}\rho \neq 0$, $1 \leqslant j \leqslant k$, the vector field

$$\bar{\xi} = \xi - \frac{1}{k}\sum_{j=1}^{k} \frac{i_\xi \rho}{i_{\zeta_j}\rho}\zeta_j$$

is a required symmetry.

2.3.7. Remark. For more details on characteristic distributions of closed two-forms we refer to S. Sternberg's book [1], and on symmetries of closed two-forms e.g. to the book by Libermann and Marle [1], or to O. Krupková [6], where also applications to integration of (generally higher-order and not necessarily regular) ordinary differential equations are discussed.

A more general framework for the theory of closed two-forms is represented by the theory of exterior differential systems. To this subject, the reader may consult e.g. the book by R. L. Bryant et al. [1].

Finally in this section, let us briefly mention important concepts of *symplectic* and *presymplectic structure* (for more details see e.g. R. Abraham and J. E. Marsden [1], P. Libermann and Ch.-M. Marle [1], or A. Weinstein [1]).

2.3.8. Definition. Let ω be a closed 2-form on a manifold Q. We say that ω is a *symplectic form* if at every point $x \in Q$, $\omega(x)$ is *nondegenerate*. We then say that (Q, ω) is a *symplectic manifold*, or, that ω defines on Q a *symplectic structure*.

The condition that ω is everywhere nondegenerate in the above defintion means that the rank of ω is constant and maximal, i.e., equal to the dimension of Q. Since, as we know, the rank of a closed two-form is an *even* number, this means that a symplectic structure can exist only on manifolds of an even dimension.

An example of a symplectic manifold is the cotangent manifold T^*M of a manifold M, endowed with the so called *canonical symplectic structure*, defined in adapted local coordinates (q^σ, p_σ) (where (q^σ) are local coordinates on M), by the closed 2-form

$$\omega = dp_\sigma \wedge dq^\sigma.$$

Another example of a symplectic structure is provided e.g. by the manifold TM and the closed 2-form

$$\omega = d\left(\frac{\partial L}{\partial \dot{q}^\sigma}\right) \wedge dq^\sigma,$$

where $(q^\sigma, \dot{q}^\sigma)$ are adapted local coordinates on TM, associated with local coordinates (q^σ) on M, and L is a (global) function on TM satisfying the *regularity condition*

$$\det\left(\frac{\partial^2 L}{\partial \dot{q}^\sigma \partial \dot{q}^\nu}\right) \neq 0.$$

(In the context of the calculus of variations, L is called a *global autonomus Lagrangian* for M).

Every symplectic structure (Q, ω) gives rise to a one-to-one correspondence between vector fields and one-forms on Q, defined by

$$\xi \to i_\xi \omega.$$

2.3.9. Definition. Any closed 2-form ω on a manifold Q is called a *presymplectic form*. A manifold Q endowed with a presymplectic form is called a *presymplectic manifold*. We also say that a closed 2-form ω defines on Q a *presymplectic structure*.

The above definition of presymplectic form follows that one in P. Libermann and Ch.-M. Marle [1]. We note that often the concept of presymplectic form is understood in a more restrictive way. Namely, by a *presymplectic form* on Q one usually considers a closed two form *of a constant, nonzero rank on* Q. This, original, definition of presymplectic form is due to J.-M. Souriau [1].

If (Q, ω) is a symplectic manifold and N is a submaniflod of Q, then the form $\iota^*\omega$, where ι is the canonical injection of N into Q, is a presymplectic form on N.

We note that on a presymplectic manifold (Q, ω) the correspondence $\xi \to i_\xi \omega$ is no longer bijective: Given a one-form η on Q, there is not always a vector field ξ such that $i_\xi \omega = \eta$, and, and when it exists, it is not, in general, unique.

2.4. Jet prolongations of fibered manifolds

The theory of jets has been established by Ch. Ehresmann [1] in the late $40's$. Since that time, jet manifolds have been used as a suitable background for the calculus of variations, theory of differential equations, and many problems of mathematical physics. Basic concepts of the jet theory are explained in the book by D. J. Saunders [3], where also some applications to the calculus of variations can be found.

In what follows, we shall utilize the concept of *jets of sections of a fibered manifold.*

A *fibered manifold* π with *base* X, *total space* Y and *projection* π is a surjective submersion $\pi : Y \to X$. This means that to each point $y \in Y$ there exists a chart (V, ψ) where V is a neighborhood of this point, $\psi = (z^1, \dots, z^n, y^1, \dots, y^m)$, where $n = \dim X$ and $m = \dim Y - n$, and a unique chart (U, φ), $\varphi = (x^1, \dots, x^n)$ at the point $x = \pi(y)$ on X such that $U = \pi(V)$ and $z^i = x^i \circ \pi$, $1 \leqslant i \leqslant n$. Evidently, we have $\varphi \circ \pi = pr_1 \circ \psi$ where pr_1 denotes the first canonical projection $R^n \times R^m \to R^n$. The chart (V, ψ) on Y is called a *fiber chart* for π, and the chart (U, φ) on X is called *associated* with (V, ψ). To simplify notations we shall usually write (U, φ), $\varphi = (x^i)$, and (V, ψ), $\psi = (x^i, y^\sigma)$ or $\psi = (\varphi\pi, \mu)$.

A (smooth) mapping $\gamma : W \to Y$, where W is an open subset of X is called a (*local*) *section* of the fibered manifold $\pi : Y \to X$ if $\pi \circ \gamma = \mathrm{id}_W$.

Two sections γ_1, γ_2 defined on an open set $W \subset X$ are called *s-equivalent* at a point $x \in W$ if

$$\gamma_1(x) = \gamma_2(x),$$

and if there is a fiber chart around $\gamma_1(x) = \gamma_2(x)$ such that the derivatives of the components $\gamma_1^\sigma = y^\sigma \gamma_1 \varphi^{-1}$ and $\gamma_2^\sigma = y^\sigma \gamma_2 \varphi^{-1}$ of the sections γ_1 and γ_2 at the point x coincide up to the order s, i.e., if

$$D_{i_1} \dots D_{i_k} \gamma_1^\sigma (\varphi(x)) = D_{i_1} \dots D_{i_k} \gamma_2^\sigma (\varphi(x)), \quad 1 \leqslant k \leqslant s,$$

for every $\sigma = 1, \dots, m$ and $i_1, \dots, i_k = 1, 2, \dots, n$, where D_l is the l-th partial derivative operator. One can check directly that the latter condition does not depend on the choice of fibered coordinates. The equivalence class containing a section γ is called the *s-jet* of γ at x and is denoted by $J^s_x \gamma$. Roughly speaking, $J^s_x \gamma$ is the s-th order Taylor polynomial of γ around x.

Denote by $J^s_x Y$ the set of all s-jets at x and put

$$J^s Y = \bigcup_{x \in X} J^s_x Y.$$

The set $J^s Y$ has the structure of a smooth manifold, it is called the *manifold of s-jets of local sections of* π, or briefly, the *s-jet prolongation* of π. If there is a danger of confusion, the s-jet prolongation of π is denoted by $J^s \pi$. The manifold $J^s Y$ can be viewed as the total space of a fibered manifold $\pi_s : J^s Y \to X$ with the base X and projection $\pi_s(J^s_x \gamma) = x$, or as the total space of the fibered manifold $\pi_{s,k} : J^s Y \to J^k Y$

with the base J^kY and projection $\pi_{s,k}(J^s_x\gamma) = J^k_x\gamma$, where k is fixed, $0 \leqslant k < s$ (within this notation, $J^0Y = Y$ and $J^0_x\gamma = \gamma(x)$). Given a fixed fibered manifold π, one often makes use of these canonically associated fibered manifolds.

If γ is a section of π then the mapping $x \to J^s_x\gamma$ is a section of π_s; it is called the *s-jet prolongation* of the section γ and denoted by $J^s\gamma$.

A section δ of π_s is called *holonomic* if there exists a section γ of π such that $\delta = J^s\gamma$.

To each fiber chart (V, ψ), $\psi = (x^i, y^\sigma)$ on π there exists the so called *associated fiber chart* on J^sY, denoted by (V_s, ψ_s), $\psi_s = (x^i, y^\sigma, y^\sigma_{j_1}, \ldots, y^\sigma_{j_1\cdots j_s})$, where the coordinates $y^\sigma_{j_1\cdots j_k}$, $1 \leqslant k \leqslant s$, $1 \leqslant j_1 \leqslant j_2 \leqslant \cdots \leqslant j_k \leqslant n$, are defined by the formula

$$y^\sigma_{j_1\cdots j_k}(J^s_x\gamma) = D_{j_1}\cdots D_{j_k}(y^\sigma\gamma\varphi^{-1})(\varphi(x)).$$

A jet prolongation of a fibered manifold π_s, $s > 0$, is often called an *anholonomic prolongation* of π. Note that if $J^{r+s}\pi$, $r, s > 0$, is the $(r+s)$-jet prolongation of a fibered manifold $\pi : Y \to X$ and $J^r\pi_s$ is the r-jet prolongation of the fibered manifold π_s then there is the *canonical embedding* $\iota : J^{r+s}\pi \to J^r\pi_s$; it is defined by $\iota(J^{r+s}_x\gamma) = J^r_xJ^s\gamma$.

Let $\pi : Y \to X$, $\bar\pi : \bar Y \to \bar X$ be two fibered manifolds, $V \subset Y$ an open set. Recall that $\phi : V \to \bar Y$ is called a *homomorphism of fibered manifolds* if there exists a mapping $\phi_0 : \pi(V) \to \bar X$ such that

$$\bar\pi \circ \phi = \phi_0 \circ \pi.$$

Notice that if ϕ_0 exists, it is unique.

A homomorphism ϕ is called an *isomorphism* if the mappings ϕ, ϕ_0 are local diffeomorphisms. Isomorphisms of a fibered manifold transfer *sections into sections*; more precisely, if (ϕ, ϕ_0) is an isomorphism of $\pi : Y \to X$, and $\gamma : U \to Y$ is a section, then $\bar\gamma = \phi \circ \gamma \circ \phi_0^{-1} : \phi_0(U) \to Y$ is a section.

Throughout the book, we shall use most frequently fibered manifolds over bases of dimension 1. In this case fibered coordinates on Y will be denoted by

$$(t, q^\sigma), \quad 1 \leqslant \sigma \leqslant m,$$

and the associated coordinates on J^sY by $(t, q^\sigma, q^\sigma_1, \cdots, q^\sigma_s)$, or shortly by

$$(t, q^\sigma_j), \quad 1 \leqslant \sigma \leqslant m,\ 0 \leqslant j \leqslant s,$$

where $q^\sigma_0 = q^\sigma$. In the case of jet prolongations of lower-order we shall also use the notation

$$(t, q^\sigma, \dot q^\sigma, \ddot q^\sigma)$$

etc. It is worth to write down the transformation rule for higher fibered coordinates. If (V, ψ), $\psi = (t, q^\sigma)$ and $(\bar V, \bar\psi)$, $\bar\psi = (\bar t, \bar q^\sigma)$ are two fiber charts on Y such that $V \cap \bar V \neq \emptyset$ then $\bar t = \bar t(t)$, $\bar q^\sigma = \bar q^\sigma(t, q^\nu)$, and

$$\bar q^\sigma_1 = \frac{dt}{d\bar t}\Big(\frac{\partial \bar q^\sigma}{\partial t} + \frac{\partial \bar q^\sigma}{\partial q^\nu}q^\nu_1\Big),$$

$$\cdots$$

$$\bar q^\sigma_s = \frac{dt}{d\bar t}\Big(\frac{\partial \bar q^\sigma_{s-1}}{\partial t} + \sum_{j=0}^{s-1}\frac{\partial \bar q^\sigma_{s-1}}{\partial q^\nu_j}q^\nu_{j+1}\Big).$$

We shall also consider fibered manifolds $\pi_{r,k}$, where $r > k > 0$, with the base J^kY; $\pi_{r,k}$ is obviously a fibered manifold with the $(mk+1)$-dimensional base.

In case we shall work with the first jet prolongation of the fibered manifold π_{s-1}, the associated fibered coordinates on $J^1(J^{s-1}Y)$ will be denoted by

$$(t, q_j^\sigma, q_{j,1}^\sigma), \quad 1 \leqslant \sigma \leqslant m,\ 0 \leqslant j \leqslant s-1 .$$

2.4.1. Remark. To describe time-independent higher-order mechanics, many authors use another jet structure, namely T^sM, the *manifold of s-velocities* (a concept due to Ch. Ehresmann [1]). Consider a manifold M (it has the meaning of a configuration space), and let U be an open neighborhood of the origin $0 \in R$. Two mappings $f, g : U \to M$ are called *s-equivalent* if $f(x) = g(x)$, and if the derivatives of f and g at 0 coincide up to the order s. The point 0 is called the *source*, and $y = f(x)$ the *target* of the s-jet of the mapping f. T^sM is then defined to be the manifold of all s-jets of mappings $U \to M$, with source at the origin 0 of R and target in M, i.e.,

$$T^sM = J_0^s(R, M).$$

In particular, T^1M is the usual tangent manifold TM of M.

The underlying structure for time-dependent mechanics then can be modeled by means of a fibered manifold where one takes R for the base manifold, $R \times M$ for the total space, and the first canonical projection of $R \times M$ onto R for the projection π. In this case, however, $J^s(R \times M)$, $s \geqslant 1$, identifies with $R \times T^sM$. This means that in the general fibered manifold structure there are incorporated both time-independent mechanics on T^sM and time-dependent mechanics on $R \times T^sM$.

On fibered manifolds and their jet prolongations there arise many specific geometric objects, such as vector fields, differential forms, distributions, etc., which are "adapted" to the fibered and prolongation structures. In the sequel of this chapter some of them will be recalled.

2.5. Projectable vector fields

A vector field ξ on J^sY is called *π_s-projectable* if there exists a vector field ξ_0 on X such that

$$T\pi_s.\xi = \xi_0 \circ \pi.$$

The vector field ξ_0 is then called the projection of ξ. A π_s-projectable vector field ξ is called π_s-vertical if

$$T\pi_s.\xi = 0.$$

Hence, in fibered coordinates, π_s-projectable vector fields have their components at $\partial/\partial x^i$ dependent on x^i only, and π_s-vertical vector fields have these components equal to zero. The bundle of π_s-vertical vectors is obviously a subbundle of the tangent bundle TJ^sY; it will be denoted by $V_{\pi_s}J^sY$.

In an analogous way one defines the concept of a *$\pi_{s,k}$-projectable* and *$\pi_{s,k}$-vertical* vector field on J^sY for $0 \leqslant k < s$.

Let ξ be a π-projectable vector field on Y, ξ_0 its π-projection. Denote $\{\phi_u\}$ (resp. $\{\phi_{0u}\}$) the local one-parameter group of ξ (resp. ξ_0) where u is the parameter. For each u, the mapping ϕ_u is an isomorphism of the fibered manifold π. If γ is a (local) section of π then obviously $\phi_u \gamma \phi_{0u}^{-1}$ is a (local) section of π. Putting

$$J^s\phi_u(J^s_x\gamma) = J^s_{\phi_{0u}(x)}(\phi_u\gamma\phi_{0u}^{-1})$$

we get a mapping $J^s\phi_u : \pi_{s,0}^{-1}(\operatorname{dom}\phi_u) \to J^sY$ called the *s-jet prolongation* of ϕ_u. For each u, the mapping $J^s\phi_u$ is an isomorphism of the fibered manifolds π_s, and $\{J^s\phi_u\}$ is a local one-parameter group of transformations of J^sY. Denote by $J^s\phi$ the corresponding local flow. We set for each $J^s_x\gamma \in \operatorname{dom} J^s\phi$

$$J^s\xi\,(J^s_x\gamma) = \left(\frac{d}{du}\,J^s\phi_u(J^s_x\gamma)\right)_{u=0}.$$

$J^s\xi$ is a vector field on J^sY, called the *s-jet-prolongation* of a projectable vector field ξ. The vector field $J^s\xi$ is π_s-projectable (resp. $\pi_{s,k}$-projectable, $0 \leqslant k < s$) and its π_s-projection (resp. $\pi_{s,k}$-projection) is the vector field ξ_0 (resp. $J^k\xi$).

In case of $\dim X = 1$, the fiber-chart expression of a jet prolongation of a π-projectable vector field is the following: if

$$\xi = \xi^0(t)\,\frac{\partial}{\partial t} + \xi^\sigma(t, q^\nu)\,\frac{\partial}{\partial q^\sigma} \tag{2.1}$$

then

$$J^s\xi = \xi^0(t)\,\frac{\partial}{\partial t} + \xi^\sigma(t, q^\nu)\,\frac{\partial}{\partial q^\sigma} + \sum_{i=1}^{s}\xi_i^\sigma\,\frac{\partial}{\partial q_i^\sigma}, \tag{2.2}$$

where the functions ξ_i^σ are defined by the recurrent formula

$$\xi_i^\sigma = \frac{d\xi_{i-1}^\sigma}{dt} - q_i^\sigma\,\frac{d\xi^0}{dt}, \quad 1 \leqslant i \leqslant s. \tag{2.3}$$

A distribution on Y is called *π-vertical* if it is spanned by π-vertical vector fields only; if there is no danger of confusion we simply call such a distribution *vertical*. The vertical distribution spanned by *all* vertical vector fields is called the *maximal vertical distribution*. A distribution on Y is called *horizontal* if it is complementary to the maximal vertical distribution. Horizontal distributions have a constant rank, equal to the dimension of the base manifold X.

In an obvious way one defines the concept of a *π_s-vertical* (*the maximal π_s-vertical*, resp. *π_s-horizontal distribution*), and of a *$\pi_{s,k}$-vertical* (*the maximal $\pi_{s,k}$-vertical*, resp. *$\pi_{s,k}$-horizontal distribution*), $0 \leqslant k < s$, on J^sY.

2.6. Calculus of horizontal and contact forms on fibered manifolds

Differential forms on fibered manifolds inherit a rich structure from the fibered and prolongation structures of the underlying manifolds. In the calculus of variations namely the properties of *horizontality* and different kinds of *contactness* are fundamental. Horizontal and contact forms have been studied and used by many authors (cf. e.g. I. M. Anderson and T. Duchamp [1], H. Goldschmidt and S. Sternberg [1], M. Gotay [2], B. Kupershmidt [1], W. M. Tulczyjew [1], to mention at least a few of them); systematically the calculus of horizontal and contact forms has been developed in the work of D. Krupka since the early 70's (see e.g. [1,2,5], and namely [8,11] for a relatively complete exposition).

Let $\pi : Y \to X$ be a fibered manifold, $\dim X = n$, let $s, q \geqslant 1$.

Denote by $\Lambda^q(J^sY)$ the module of q-forms on J^sY over the ring of functions. A form $\eta \in \Lambda^q(J^sY)$ is called *π_s-projectable* (resp. *$\pi_{s,k}$-projectable*, $0 \leqslant k < s$, if there exists a form η_0 on X (resp. on J^kY) such that $\eta = \pi_s^*\eta_0$ (resp. $\eta = \pi_{s,k}^*\eta_0$). If η_0 exists it is unique and it is called the *π_s-projection* (resp. *$\pi_{s,k}$-projection*) of η. If there is no danger of confusion *we shall often identify projectable forms with their projections.*

A form $\eta \in \Lambda^q(J^sY)$ is called *π_s-horizontal* if

$$i_\xi \eta = 0 \quad \text{for every } \pi_s\text{-vertical vector field } \xi \text{ on } J^sY.$$

A form $\eta \in \Lambda^q(J^sY)$ is called *$\pi_{s,k}$-horizontal*, $0 \leqslant k < s$, if

$$i_\xi \eta = 0 \quad \text{for every } \pi_{s,k}\text{-vertical vector field } \xi \text{ on } J^sY.$$

The module of π_s-horizontal (resp. of $\pi_{s,k}$-horizontal) q-forms on J^sY is a submodule of $\Lambda^q(J^sY)$ and will be denoted by $\Lambda^q_X(J^sY)$ (resp. $\Lambda^q_{J^kY}(J^sY)$). It should be noticed that for $\dim X = 1$ we get from the definition that a form $\eta \neq 0$ is π_s-horizontal if and only if in fibered coordinates it is represented by $\eta = f\,dt$, where $f = f(t, q^\sigma, \cdots, q^\sigma_s)$; similarly, a $\pi_{s,k}$-horizontal q-form η is expressed by means of $dt, dq^\sigma, \cdots, dq^\sigma_k$ only (with the components dependent on $t, q^\sigma, \cdots q^\sigma_s$).

Let $\eta \in \Lambda^q(J^sY)$, $q \geqslant 1$, be a form. For every point $y = J^{s+1}_x\gamma \in J^{s+1}Y$, and every system of vector fields $\xi_1, \cdots, \xi_q \in T_yJ^{s+1}Y$ we set

$$h\eta(J^{s+1}_x\gamma)(\xi_1, \cdots, \xi_q) = \eta(J^s_x\gamma)(T_xJ^s\gamma \cdot T\pi_{s+1} \cdot \xi_1, \cdots, T_xJ^s\gamma \cdot T\pi_{s+1} \cdot \xi_q).$$

Evidently, $h\eta$ is a π_{s+1}-horizontal q-form on $J^{s+1}Y$. If f is a function on J^sY we set

$$hf(J^{s+1}_x\gamma) = f(J^s_x\gamma).$$

The mapping h is called *horizontalization with respect to π* or simply *π-horizontalization*.

The following properties of the mapping h will be frequently used throughout the text:

2.6.1. Proposition. *Let $\rho, \eta \in \Lambda^q(J^sY)$, $\omega \in \Lambda^p(J^sY)$.*

(1) $h(\rho + \eta) = h\rho + h\eta, \quad h(\rho \wedge \omega) = h\rho \wedge h\omega.$

(2) *If $q > \dim X$ then $h\rho = 0$.*

(3) *If ρ is $\pi_{s,s-1}$-horizontal then $h\rho$ is $\pi_{s+1,s}$-projectable.*

(4) *$h\rho$ is a unique horizontal form such that for every section γ of π the condition*

$$J^s\gamma^*\rho = J^{s+1}\gamma^*h\rho$$

is satisfied.

(5) *Let ϕ be an isomorphism of π. Then*

$$h\,J^s\phi^*\rho = J^{s+1}\phi^*h\rho.$$

For $\dim X = 1$ we get the following useful formulas for computing horizontalization of forms on J^sY:

$$h\,dt = dt, \quad h\,dq^\sigma_j = q^\sigma_{j+1}\,dt, \quad 0 \leqslant j \leqslant s, \quad hf = f \circ \pi_{s+1,s},$$

$$h\,df = \frac{df}{dt}\,dt, \quad \frac{df}{dt} = \frac{\partial f}{\partial t} + \sum_{j=0}^{s} \frac{\partial f}{\partial q^\sigma_j} q^\sigma_{j+1}.$$

Note that in the case $\dim X = 1$ we easily get that *if* $h\rho = 0$ *for a* 1*-form* ρ *on* J^rY *then* $\rho = 0$.

Now, let us return to the general case of $\pi : Y \to X$, $\dim X = n$. Let $q \geqslant 0$. A form $\eta \in \Lambda^q(J^sY)$ is called π*-contact*, or *contact* if $h\eta = 0$. Obviously, η is contact iff

$$J^s\gamma^* \eta = 0$$

for every section γ of π. Obviously, a function f is contact iff $f = 0$.

For $\eta \in \Lambda^q(J^sY)$ we set

$$p\eta = \pi^*_{s+1,s}\eta - h\eta.$$

It is easy to see that $p\eta$ is a contact q-form on $J^{s+1}Y$. The mapping p is called *contactization with respect to* π.

The mapping p has the following elementary properties:

2.6.2. Proposition. *Let* $\rho, \eta \in \Lambda^q(J^sY)$, $\omega \in \Lambda^p(J^sY)$.

(1) $p(\rho + \eta) = p\rho + p\eta$, $\quad p(\rho \wedge \omega) = p\rho \wedge p\omega + p\rho \wedge h\omega + h\rho \wedge p\omega$. *In particular,* $p(f \cdot \rho) = f \cdot p\rho$ *for a function* f.

(2) *If* $q > \dim X$ *then* $p\rho = \pi^*_{s+1,s}\rho$.

(3) *If* ρ *is* $\pi_{s,s-1}$*-horizontal then* $p\rho$ *is* $\pi_{s+1,s}$*-projectable.*

(4) *For every section* γ *of* π *the condition* $J^{s+1}\gamma^*\, p\rho = 0$ *is satisfied.*

(5) $p\rho$ *is* $\pi_{s+1,s}$*-projectable if and only if* $h\rho$ *is* $\pi_{s+1,s}$*-projectable.*

(6) ρ *is horizontal* (*resp. contact*) *if and only if* $p\rho = 0$ (*resp.* $h\rho = 0$).

(7) *Let* ϕ *be an isomorphism of* π. *Then*

$$p\, J^s\phi^*\, \rho = J^{s+1}\phi^*\, p\rho.$$

In particular, if ρ *is contact then* $J^{s+1}\phi^*\rho$ *is contact.*

Notice that the definition of the mapping p gives us

$$p\, dx^i = 0,$$

for $i = 1, 2, \dots, n$, and

$$p\, dy^\sigma_{j_1\cdots j_k} = dy^\sigma_{j_1\cdots j_k} - y^\sigma_{j_1\cdots j_k i}\, dx^i,$$

for $1 \leqslant \sigma \leqslant m$, $0 \leqslant k \leqslant r$, $j_1, \dots, j_k = 1, 2, \dots n$. Denote

$$\omega^\sigma_{j_1\cdots j_k} = dy^\sigma_{j_1\cdots j_k} - y^\sigma_{j_1\cdots j_k i}\, dx^i\,. \tag{2.4}$$

These contact 1-forms on $J^{r+1}Y$ define a distribution, called the *Cartan distribution*, and denoted by $\mathcal{C}_{\pi_{r+1}}$. If $\dim X = 1$ we shall write

$$\mathcal{C}_{\pi_{r+1}} \approx \operatorname{span}\{\omega^\sigma_j = dq^\sigma_j - q^\sigma_{j+1}\, dt,\ 1 \leqslant \sigma \leqslant m,\ 0 \leqslant j \leqslant r\}\,. \tag{2.5}$$

Equivalently, for $s \geqslant r$ we get for the Cartan distribution $\mathcal{C}_{\pi_{r+1}}$ on $J^{s+1}Y$ that

$$\mathcal{C}_{\pi_{r+1}} = \operatorname{span}\left\{\frac{\partial}{\partial t} + \sum_{j=0}^{r} q^\sigma_{j+1}\frac{\partial}{\partial q^\sigma_j}\,,\ \frac{\partial}{\partial q^\sigma_l}\,,\ 1 \leqslant \sigma \leqslant m\,,\ r+1 \leqslant l \leqslant s+1\right\}.$$

Cartan distributions play an important role in the theory of higher order differential equations on manifolds, since they "recognize" *holonomic sections*, i.e., sections which are prolongations of sections of π. More precisely, *a section δ of π_s is an integral section of the Cartan distribution $\mathcal{C}_{\pi_s}$ if and only if*

$$\delta = J^s\gamma$$

for some section γ of π.

Non-vertical vector fields belonging to Cartan distributions are called *semisprays*. More precisely, a *semispray of order s and type r, $s \geqslant r$,* is defined as a vector field ζ on J^sY belonging to the Cartan distribution $\mathcal{C}_{\pi_r}$. For example, if $\dim X = 1$, we get for a semispray ζ of order s and type r the following fiber chart expression:

$$\zeta = \zeta_0\Big(\frac{\partial}{\partial t} + \sum_{j=0}^{r-1} q^\sigma_{j+1}\frac{\partial}{\partial q^\sigma_j} + \sum_{j=r}^{s} \zeta^\sigma_j \frac{\partial}{\partial q^\sigma_j}\Big),$$

where $\zeta_0 \neq 0$ and ζ^σ_j, $1 \leqslant \sigma \leqslant m$, $r \leqslant j \leqslant s$, are functions on an open set of J^sY.

It is worthwhile to note that the contact 1-forms (2.4) together with the forms dx^i and $dy^\sigma_{j_1\dots j_s}$ form a *basis of linear forms* on $V_s \subset J^sY$. In our notation for $\dim X = 1$, this basis is denoted by

$$dt, \omega^\sigma, \omega^\sigma_1, \dots, \omega^\sigma_{s-1}, dq^\sigma_s. \tag{2.6}$$

Further note that for all the σ and $j \geqslant 0$,

$$d\omega^\sigma_j = -dq^\sigma_{j+1} \wedge dt = -\omega^\sigma_{j+1} \wedge dt. \tag{2.7}$$

We have seen that every form $\eta \in \Lambda^q(J^sY)$ can be uniquely decomposed into a sum of its horizontal and contact part, namely

$$\pi^*_{s+1,s}\,\eta = h\eta + p\eta.$$

However, if $q > 1$ then $h\eta = 0$ and we would like to have a further decomposition of the contact part $p\eta$. Let us now turn to do this.

To simplify notations, we shall consider the module $\Lambda^q_{J^{s-1}Y}(J^sY)$ of $\pi_{s,s-1}$-horizontal q-forms. Note that the fiber chart expressions of the elements of $\Lambda^q_{J^{s-1}Y}(J^sY)$ contain only dx^i's and the contact one-forms omega's, i.e., they do not contain "the highest dy's". This means, in particular, that the mappings h and p restricted to $\Lambda^q_{J^{s-1}Y}(J^sY)$ *save the order* (they map forms on J^sY to forms on J^sY). An important property of this module is its relatively simple algebraic structure. To see this, let us recall the following definition. Let $q \geqslant 1$, and let $\eta \in \Lambda^q_{J^{s-1}Y}(J^sY)$ be a *contact* form. We say that η is *one-contact* if for each π_s-vertical vector field ξ on J^sY the $(q-1)$-form $i_\xi\eta$ is π_s-horizontal; we say that η is *k-contact*, $2 \leqslant k \leqslant q$, if $i_\xi\rho$ is $(k-1)$-contact. In this context, π_s-horizontal forms are also called 0-*contact*. For each k, $0 \leqslant k \leqslant q$, we denote by $\Lambda^{q-k,k}(J^sY)$ the submodule of $\Lambda^q_{J^{s-1}Y}(J^sY)$, consisting from all k-contact q-forms on J^sY. We can say that a form is i-contact if and only if each term in its (fibered) coordinate expression contains exactly i of the 1-contact linear forms "omega".

2.6.3. Proposition (Decomposition theorem). *The module $\Lambda^q_{J^{s-1}Y}(J^sY)$ is the direct sum of the submodules of i-contact forms, $0 \leqslant i \leqslant q$, i.e.,*

$$\Lambda^q_{J^{s-1}Y}(J^sY) = \Lambda^{q,0}(J^sY) \oplus \Lambda^{q-1,1}(J^sY) \oplus \cdots \oplus \Lambda^{0,q}(J^sY).$$

In other words, every $\eta \in \Lambda^q_{J^{s-1}Y}(J^sY)$ is uniquely decomposable in the form

$$\eta = \eta_0 + \eta_1 + \cdots + \eta_q\,,$$

where η_i, $0 \leqslant i \leqslant q$, is a i-contact form on J^sY.

If η_i is the i-contact part of η, we write $\eta_i = p_i\eta$. In this way, for *every* q-form η on J^sY we obtain the unique decomposition

$$\pi^*_{s+1,s}\,\eta = h\eta + p_1\eta + \cdots + p_q\eta$$

into a sum of a horizontal form and i-contact q-forms, $1 \leqslant i \leqslant q$.

We say that a form $\eta \in \Lambda^q_{J^{s-1}Y}(J^sY)$ has the *order of contactness* k if, in the notation of Proposition 2.6.3, $\eta_i = 0$ for all $i > k$. This means that in a fiber chart expression of η each term contains the wedge product of *at most* k of the contact linear forms "omega".

We shall need the *Poincaré Lemma* for closed two-forms on fibered manifolds. To fix the notations, let us recall the Lemma and its proof here.

2.6.4. Poincaré Lemma. *Let $\tau : W \to U$ be a fibered manifold such that $W = U \times V$, where $U \subset R$ is an open interval and $V \subset R^m$ is an open ball with the center at the origin. Let α be a closed two-form on J^sW. Then there exists a one-form ρ on J^sW such that $d\rho = \alpha$.*

Proof. Without loss of generality, we can suppose that $\alpha \in \Lambda^2_{J^{s-1}W}(J^sW)$. (Indeed, for any p-form η on J^rY we have $\pi^*_{r+1,r}\,\eta \in \Lambda^p_{J^rY}(J^{r+1}Y)$, and the exterior derivative operator d commutes with the projection operator $\pi_{r+1,r}$).

Let (t, q^σ) denote the canonical coordinates on W. Define a mapping $\chi_s : [0,1] \times J^sW \to J^sW$ by

$$\chi_s(u, (t, q^\sigma, \ldots, q^\sigma_s)) = (t, uq^\sigma, \ldots, uq^\sigma_s)\,. \tag{2.8}$$

By the Decomposition Theorem we have

$$\alpha = \sum_{j=0}^{s-1} E^j_\sigma\, \omega^\sigma_j \wedge dt + \sum_{j,k=0}^{s-1} F^{jk}_{\sigma\nu}\, \omega^\sigma_j \wedge \omega^\nu_k\,, \quad F^{jk}_{\sigma\nu} = -F^{kj}_{\nu\sigma}\,.$$

Now, since

$$\chi^*_s\omega^\sigma_j = q^\sigma_j\, du + u\,\omega^\sigma_j\,, \quad 0 \leqslant j \leqslant s-1\,,$$

we obtain

$$\chi^*_s\alpha = du \wedge \rho + \alpha'$$

where

$$\rho = a_0\, dt + a^j_\sigma\, \omega^\sigma_j \tag{2.9}$$

with a_0, a^j_σ defined by

$$a_0 = \sum_{j=0}^{s-1} q^\sigma_j (E^j_\sigma \circ \chi_s)\,, \quad a^j_\sigma = 2\sum_{j,k=0}^{s-1} (q^\nu_k F^{kj}_{\nu\sigma}) \circ \chi_s\,, \tag{2.10}$$

and α' is a two-form not containing du. Put

$$A\alpha = \left(\int_0^1 a_0\, du\right) dt + \left(\int_0^1 a_\sigma^j\, du\right) \omega_j^\sigma . \tag{2.11}$$

$A\alpha$ is a 1-form on $J^s W$. By a straightforward computation we get

$$dA\alpha + Ad\alpha = \alpha ,$$

and since, by assumption, $d\alpha = 0$, we are done. □

Note that the operator A defined by (2.11) is "adapted" to the decomposition of forms into i-contact parts: namely, A maps the module of 1-contact (resp. 2-contact) two-forms on $J^s W$ into the module of horizontal (resp. 1-contact) one-forms on $J^s W$.

2.6.5. Remark (Lagrangians and dynamical forms). In this book we shall frequently meet the module $\Lambda_X^1(J^s Y)$, $s \geqslant 1$, of π_s-horizontal one-forms on $J^s Y$, the elements of which are called *Lagrangians of order* s, and the module $\Lambda_Y^{1,1}(J^s Y)$, $s \geqslant 1$, of one-contact two-forms on $J^s Y$, horizontal with respect to the projection $\pi_{s,0}$, the elements of which will be called *dynamical forms of order* s. Note that fiber-chart expressions of a Lagrangian λ of order s, and of a dynamical form E of order s are

$$\lambda = L\, dt, \qquad E = E_\sigma\, dq^\sigma \wedge dt, \tag{2.12}$$

where L and the E_σ's are functions of $(t, q^\nu, \ldots, q_s^\nu)$.

2.7. Jet fields, connections, semispray connections and generalized connections on fibered manifolds

The concept of a (nonlinear) connection on a fibered manifold, introduced in 1950 by Ch. Ehresmann [2], and further developed mainly by L. Mangiarotti and M. Modugno [1], and M. Modugno [1], has numerous applications in the calculus of variations, the theory of differential equations, and mathematical physics. In the present theory of variational equations, connections and related structures, such as jet fields, semispray connections or generalized connections, play a fundamental role. Let us turn now to recall basic concepts.

Throughout this section, $\pi : Y \to X$ is a fibered manifold, $\dim X = n \geqslant 1$.

Let $k \geqslant 0$, $s > k$. A (local) section $\varphi : J^k Y \to J^s Y$ of the fibered manifold $\pi_{s,k}$, defined on an open subset $U \subset J^k Y$, is called a *jet field* (of *order* s and *type* k).

Jet fields are identified with certain horizontal distributions. Namely, jet fields of order s and type k are in one-to-one correspondence with horizontal subdistributions of the Cartan distribution $\mathcal{C}_{\pi_k}$ on open subsets of $J^{s-1}Y$ (i.e., such distributions are spanned by semisprays of order $s-1$ and type k). Let $\varphi : J^k Y \to J^s Y$ be a jet field, defined on an open set $U \subset J^k Y$. A section γ of π, defined on an open set $I \subset X$ is called an *integral section* of the jet field φ if $\gamma(I) \subset U$, and

$$\varphi \circ J^k \gamma = J^s \gamma .$$

Notice that integral sections of a jet field coincide with holonomic integral sections of the corresponding distribution. Writing this condition in fibered coordinates, we get a system of differential equations (ordinary if $\dim X = 1$, and partial if $\dim X > 1$) of order $k+1$ for the components of γ.

Let us write down fiber chart expressions concerning jet fields for the case $\dim X = 1$. Let $\varphi : J^k Y \to J^s Y$ be a jet field, (t, q^σ) local fiber coordinates on Y. Then φ is represented in the form

$$(t, q^\sigma, \ldots, q_k^\sigma) \to (t, q^\sigma, \ldots, q_k^\sigma, \varphi_{k+1}^\sigma, \ldots, \varphi_s^\sigma),$$

where the *components* $\varphi_{k+1}^\sigma, \ldots, \varphi_s^\sigma$ are defined by

$$\varphi_{k+1}^\sigma = q_{k+1}^\sigma \circ \varphi, \quad \ldots, \quad \varphi_s^\sigma = q_s^\sigma \circ \varphi.$$

For the corresponding distribution Δ_φ we get

$$\begin{aligned}\Delta_\varphi &= \operatorname{span}\left\{\frac{\partial}{\partial t} + \sum_{i=0}^{k-1} q_{i+1}^\sigma \frac{\partial}{\partial q_i^\sigma} + \sum_{i=k}^{s-1} \varphi_{i+1}^\sigma \frac{\partial}{\partial q_i^\sigma}\right\} \\ &\approx \operatorname{span}\{\omega_i^\sigma, dq_j^\sigma - \varphi_{j+1}^\sigma dt, 0 \leqslant i \leqslant k-1, k \leqslant j \leqslant s-1, 1 \leqslant \sigma \leqslant m\}.\end{aligned}$$

The equations for integral sections of φ are of the form

$$\begin{aligned}\frac{d^{k+1}\gamma^\sigma}{dt^{k+1}} &= \varphi_{k+1}^\sigma\left(t, \gamma^\nu, \frac{d\gamma^\nu}{dt}, \ldots, \frac{d^k\gamma^\nu}{dt^k}\right) \\ \frac{d^{k+2}\gamma^\sigma}{dt^{k+2}} &= \varphi_{k+2}^\sigma\left(t, \gamma^\nu, \frac{d\gamma^\nu}{dt}, \ldots, \frac{d^k\gamma^\nu}{dt^k}\right) \\ &\ldots \\ \frac{d^s\gamma^\sigma}{dt^s} &= \varphi_s^\sigma\left(t, \gamma^\nu, \frac{d\gamma^\nu}{dt}, \ldots, \frac{d^k\gamma^\nu}{dt^k}\right),\end{aligned} \tag{2.13}$$

where $\gamma^\sigma = q^\sigma \circ \gamma$, $1 \leqslant \sigma \leqslant m$, are the components of γ. Notice that the first m equations of (2.13) are *ODE of order $k+1$ explicitly solved with respect to the highest derivatives*, and the remaining ones are *compatibility conditions*.

Jet fields which are local sections of $\pi_{1,0}$ are called *connections* on π. Integral sections of a connection are also called *paths*. Equations for integral sections of a connection take the form of *nm first-order differential equations explicitly solved with respect to the derivatives* (they are ODE if $n = 1$ and PDE for $n > 1$).

There is an obvious *one-to-one correspondence between connections and horizontal distributions on π*, therefore these two kinds of objects are usually *identified*.

To every connection Γ on π there exists a unique vector-valued 1-form $h_\Gamma : Y \to T^*X \otimes TY$ projectable onto the canonical form $\mathrm{id} : X \to T^*X \otimes TX$, i.e., such that $\mathrm{id}(\xi) = \xi$ for all vector fields ξ on X; h_Γ is defined by $h_\Gamma(\xi) = \zeta \in \Gamma$ such that $T_\pi.\zeta = \xi$, and is called the *horizontal form* of the connection Γ. The vector field $h_\Gamma(\xi)$ on J^1Y is then called the *Γ-prolongation* of the vector field ξ.

The *curvature* of Γ is defined as a vector-valued two-form $R_\Gamma : Y \to V_\pi Y \otimes \Lambda^2 T^*X$, defined by $R_\Gamma = \frac{1}{2}[h_\Gamma, h_\Gamma]$, where $[\cdot, \cdot]$ is the Frölicher-Nijenhuis bracket (for the definition of this bracket see e.g. Kolář, Michor, and Slovák [1]). A connection Γ is called *flat* if $R_\Gamma = 0$. *A connection is flat if and only if it is completely integrable.* Clearly, connections on fibered manifolds over one-dimensional bases are locally spanned by one nowhere-zero vector field, i.e., they are completely integrable.

It is worthwhile to recall the structure of the fibered manifold $\pi_{1,0} : J^1Y \to Y$. This manifold is an affine bundle modeled on the associated vector bundle $V_\pi Y \otimes \pi^*(T^*X) \to Y$ where T^*X is the cotangent bundle of X and $\pi^*(T^*X)$ is by definition the fibered product $Y \times_X T^*X$. (Local) sections of this vector bundle (which are π-vertical vector valued 1-forms) are called *soldering forms* along π. Soldering forms have the meaning of "differences of connections". More precisely, if Γ is a connection and σ is a soldering form then $\bar{\Gamma}$ defined by $h_{\bar{\Gamma}} = h_\Gamma + \sigma$ is another connection on π.

Notice that a (global) connection on a fibered manifold $\pi : Y \to X$ in fact is a way of choosing an *origin* of each affine fiber of $J^1Y \to Y$.

Let us recall some fiber-chart expressions concerning connections for the case of $\dim X = 1$. Consider a connection $\Gamma : Y \to J^1Y$, let (t, q^σ) be a fiber chart on Y. Denote by Γ^σ the components of Γ in this chart. Then the equations of Γ are $q_1^\sigma \circ \Gamma = \Gamma^\sigma(t, q^\nu)$. The horizontal form is

$$h_\Gamma = \left(\frac{\partial}{\partial t} + \Gamma^\sigma \frac{\partial}{\partial q^\sigma}\right) \otimes dt.$$

For the corresponding distribution we have

$$\Delta_\Gamma = \operatorname{span}\left\{\frac{\partial}{\partial t} + \Gamma^\sigma \frac{\partial}{\partial q^\sigma}\right\} \approx \operatorname{span}\{dq^\sigma - \Gamma^\sigma dt, 1 \leqslant \sigma \leqslant m\},$$

and the equations for integral sections of Γ are of the form

$$\frac{d\gamma^\sigma}{dt} = \Gamma^\sigma(t, \gamma^\nu),$$

where (γ^σ) are the components of a section γ of π.

Let $s \geqslant 2$. Jet fields, which are local sections of $\pi_{s,s-1}$ are called *semispray connections (of order s)* on π. It is easy to see that they are a particular case of connections on π_{s-1}. Semispray connections of order s represent a generalization from Euclidean spaces to manifolds of the concept of a *system of differential equations of order s explicitly solved with respect to the highest derivatives*, ordinary if $n = 1$, and partial if $n > 1$, respectively.

As it is clear from the theory of jet fields, semispray connections of order s are identified with the n-dimensional horizontal subdistributions of the Cartan distribution $\mathcal{C}_{\pi_{s-1}}$ on open subsets of $J^{s-1}Y$; such distributions are called *semispray distributions*.

If $\dim X = 1$, we get for a semispray connection $\Gamma : J^{s-1}Y \to J^sY$ the following fiber chart expression: The components of Γ in a fiber chart (t, q^σ) are defined by $\Gamma^\sigma(t, q^\nu, \ldots, q^\nu_{s-1}) = q^\sigma_s \circ \Gamma$, and we get

$$\Delta_\Gamma = \operatorname{span}\left\{\frac{\partial}{\partial t} + \sum_{i=0}^{s-2} q^\sigma_{i+1} \frac{\partial}{\partial q^\sigma_i} + \Gamma^\sigma \frac{\partial}{\partial q^\sigma_{s-1}}\right\}$$
$$\approx \operatorname{span}\{\omega^\sigma_i, dq^\sigma_{s-1} - \Gamma^\sigma dt, 0 \leqslant i \leqslant s-2, 1 \leqslant \sigma \leqslant m\}.$$

The corresponding equations for integral sections (paths) of Γ are then of the form

$$\frac{d^s\gamma^\sigma}{dt^s} = \Gamma^\sigma\left(t, \gamma^\nu, \frac{d\gamma^\nu}{dt}, \ldots, \frac{d^{s-1}\gamma^\nu}{dt^{s-1}}\right).$$

We have seen that connections/semispray connections represent differential equations explicitly solved with respect to the highest derivatives. In this book, however, we shall work with a more general class of differential equations. They will be represented by more general geometric objects, which also correspond to certain distributions on fibered manifolds.

Let $\pi : Y \to X$ be a fibered manifold ($\dim X \geqslant 1$), Δ a distribution on Y, $y \in Y$ a point. We say that Δ is *weakly horizontal at* y if $\Delta(y)$ is complementary to a subspace of the vertical vector space $V_\pi Y$ at y. We say that Δ is *weakly horizontal* if it is weakly horizontal at each point $y \in Y$. In other words, a weakly horizontal distribution is a distribution which is complementary to a subbundle of the vertical bundle. Notice that a weakly horizontal distribution can be of a non-constant rank. A weakly horizontal distribution of constant rank will be called a *generalized connection*. Clearly, every connection on π trivially is a generalized connection.

A generalized connection will be called *flat* if it is completely integrable.

Let us turn to generalized connections in more detail. Let Δ be a generalized connection on π. Denote $\Delta^V = \Delta \cap V_\pi Y$. The Δ^V is a vertical distribution (of constant rank $l = \operatorname{rank} \Delta - \dim X$). Consider the quotient distribution Δ/Δ^V; obviously, at each point $y \in Y$, the vector space $\Delta/\Delta^V(y)$ is isomorphic to $T_{\pi(y)}X$ via the canonical isomorphism $T\pi(y)$. Hence, the quotient distribution has the meaning of a *class of connections*. We can see that a generalized connection can be equivalently characterized by a vertical subbundle of $V_\pi Y$ and a connection on π.

For generalized connections one can introduce generalizations of similar concepts as for connections; for example, the *generalized horizontal form* is defined as a *class* $\{h_\Delta\}$ of vector-valued one-forms $Y \to T^*X \otimes TY$ projectable onto $\mathrm{id} : X \to T^*X \otimes TX$, defined by $\{h_\Delta\}(\xi) = \{\zeta\} \in \Delta/\Delta^V$ such that $T_\pi.\zeta = \xi$, $\zeta \in \{\zeta\}$, i.e., $\operatorname{Im}\{h_\Delta\} = \Delta/\Delta^V$. The class $\{h_\Delta\}(\xi)$ of vector fields on Y is then called the Δ-*prolongation* of the vector field $\xi \in X$.

If $\dim X = 1$ then there is a one-to-one correspondence between generalized connections Δ such that $\operatorname{rank} \Delta^V = l$ and the so called *generalized jet fields* of rank l; a generalized jet field of rank l is defined as a mapping assigning to every $y \in Y$ an l-dimensional affine subspace of $\pi_{1,0}^{-1}(y)$.

2.7.1. Remark. For a detailed exposition of the theory of jet fields and connections the reader can consult the book by D. J. Saunders [3]. Semispray connections on fibered manifolds and their related distributions have been studied extensively namely by A. Vondra [1,2] (cf. also e.g. L. C. de Andres, M. de León and P. R. Rodrigues [1]), generalized connections and generalized jet fields were introduced in O. Krupková [4]. For the concepts of semispray and spray on tangent bundles we refer to the book by C. Godbillon [1].

Chapter 3.

LAGRANGEAN DYNAMICS ON FIBERED MANIFOLDS

3.1. Introduction

The beginnings of the modern calculus of variations go back to the 50's when modern methods of differential geometry were applied, and the meaning of the Ehresmann's theory of jets for building a rigorous variational theory has been recognized. Specifically, fibered manifolds and their jet prolongations, which systematically started to be used in the early 70's, represented an appropriate and sufficiently general background for study of global and higher order parametrized variational problems. Alternatively, one can also develop the calculus of variations on manifolds of contact elements ("jets of submanifolds"), suitable for study of non-parametric variational problems.

Foundations of a systematic global variational theory are connected namely with the work of P. Dedecker, H. Goldschmidt and S. Sternberg, D. Krupka, P. L. Garcia, V. Aldaya and J. de Azcárraga, B. Kupershmidt, W. M. Tulczyjew, and A. M. Vinogradov. Since the 80's, when foundations of the first-order theory were almost established, there has been a permanent interest namely in the higher-order calculus of variations. To these days, however, the global variational theory, namely higher-order field theory, is not closed.

The aim of this chapter is to introduce key-concepts of Lagrange theory on fibered manifolds, such as the *first variation*, *Lepagean n-form* (where n is the dimension of the base manifold X), *Euler-Lagrange mapping* and *Euler-Lagrange form*, and to derive the *first variation formula* and the *Euler-Lagrange equations* in an intrinsic (coordinate independent) form. Our exposition will be, of course, adapted to the case of one-dimensional base manifolds. The exposition is taken from D. Krupka [8], where a systematic lay-out of the geometric theory of Lagrangean structures is presented; there, and in the papers [1–7] of D. Krupka, also more details, and a general setting for the theory including higher-order field theory can be found.

3.2. Lepagean one-forms and the first variation

By a *Lagrangian of order* r on a fibered manifold $\pi : Y \to X$, $\dim X = 1$, we shall mean a horizontal one-form λ on J^rY. This means that in fibered coordinates (t, q^σ) a Lagrangian of order r is expressed as follows:

$$\lambda = L(t, q^\nu, q_1^\nu, \dots, q_r^\nu)\, dt.$$

The function L is called *Lagrange function*. On the overlap of two fiber charts the corresponding Lagrange functions satisfy the transformation rule

$$\bar{L} = L\,\frac{dt}{d\bar{t}}.$$

Let $r \geqslant 1$, and consider a Lagrangian λ of order r. Let Ω be a piece of the base manifold X, i.e., a compact submanifold of X with boundary $\partial\Omega$ (without loss of generality we can assume $\Omega = [a, b] \subset R$, $a < b$). Denote by $\mathcal{S}_\Omega(\pi)$ the set of sections γ of π such that $\operatorname{dom}\gamma$ is a neighborhood of Ω. The function

$$\mathcal{S}_\Omega(\pi) \ni \gamma \to \lambda_\Omega(\gamma) = \int_\Omega J^r\gamma^*\lambda \in R \tag{3.1}$$

is called the *variational function*, or the *action function* of the Lagrangian λ over Ω. Similarly one can introduce the action function for *any* one-form on J^rY. Since for any *contact* one-form η,

$$\int_\Omega J^r\gamma^*\lambda = \int_\Omega J^r\gamma^*(\lambda + \eta), \tag{3.2}$$

we can see that the action function does not change if one takes in place of the Lagrangian a general one-form belonging to $\Lambda^1_{J^{r-1}Y}(J^rY)$.

Let ξ be a π-projectable vector field on Y, ξ_0 its π-projection. Let $\{\phi_u\}$ (resp. $\{\phi_{0u}\}$) be the local one-parameter group of ξ (resp. ξ_0). Let $\gamma \in \mathcal{S}_\Omega(\pi)$ be a section. There exists an $\varepsilon > 0$ such that for each $u \in (-\varepsilon, \varepsilon)$ the section $\gamma_u = \phi_u\gamma\phi_{0u}^{-1}$ is defined in a neighborhood of $\phi_{0u}(\Omega)$. The one-parameter family $\{\gamma_u\}$ of sections of π is called the *deformation of the section γ induced by ξ*. Consider the real valued function

$$(-\varepsilon, \varepsilon) \ni u \to \lambda_{\phi_{0u}(\Omega)}(\phi_u\gamma\phi_{0u}^{-1}) = \int_{\phi_{0u}(\Omega)} J^r(\phi_u\gamma\phi_{0u}^{-1})^*\lambda \in R,$$

where $\varepsilon > 0$ is a suitable number. This function is differentiable; differentiating it with respect to u at $u = 0$ we obtain

$$\begin{aligned}
\left(\frac{d}{du}\lambda_{\phi_{0u}(\Omega)}(\phi_u\gamma\phi_{0u}^{-1})\right)_{u=0} &= \left(\frac{d}{du}\int_{\phi_{0u}(\Omega)} (J^r\phi_u \circ J^r\gamma \circ \phi_{0u}^{-1})^*\lambda\right)_{u=0} \\
&= \left(\frac{d}{du}\int_\Omega J^r\gamma^*(J^r\phi_u^*\lambda)\right)_{u=0} \\
&= \int_\Omega J^r\gamma^*\left(\frac{d}{du}J^r\phi_u^*\lambda\right)_{u=0} \\
&= \int_\Omega J^r\gamma^*\,\partial_{J^r\xi}\lambda.
\end{aligned} \tag{3.3}$$

The function

$$\mathcal{S}_\Omega(\pi) \ni \gamma \to \left(\partial_{J^r\xi}\lambda\right)_\Omega(\gamma) = \int_\Omega J^r\gamma^*\,\partial_{J^r\xi}\lambda \in R \tag{3.4}$$

is the action function of the Lagrangian $\partial_{J^r\xi}\lambda$ over Ω; it is called the *first variation of the action function λ_Ω, induced by ξ*.

Let us stop for a moment to look at the first variation (3.4). Our aim now will be to get the first variation formula, i.e., to decompose the integral on the right-hand side of (3.4) into two terms: one—having the meaning of the Euler-Lagrange equations—depending on the vector field ξ (not on its prolongation), and another one, a boundary term. The procedure, however, must be intrinsic. Notice that to this purpose, it is not sufficient to utilize the formula

$$\partial_{J^r\xi}\lambda = i_{J^r\xi}d\lambda + di_{J^r\xi}\lambda,$$

since the term $i_{J^r\xi}d\lambda$ depends on the prolongation $J^r\xi$ of ξ (indeed, expressing $i_{J^r\xi}d\lambda$ we can see that it contains derivatives of the components of the vector field ξ up to the order r). We can, however, make use of the relation (3.2). Then the first variation of the action function takes the form

$$\gamma \to \int_\Omega J^r\gamma^* \partial_{J^r\xi}(\lambda + \eta)$$

and we can pose the following problem:

Find a contact one-form η such that

$$J^r\gamma^* \partial_{J^r\xi}(\lambda + \eta) = J^r\gamma^* i_{J^r\xi}d(\lambda + \eta) + J^r\gamma^* di_{J^r\xi}(\lambda + \eta) \tag{3.5}$$

would become the first variation formula.

3.2.1. Example. Let us make explicit computations to solve the above problem in a simple situation, namely in the first-order case ($r = 1$). Keeping the notations of the previous chapter, put

$$\lambda = L\,dt, \quad \eta = \eta_\sigma\,\omega^\sigma,$$

and

$$\xi = \xi^0\frac{\partial}{\partial t} + \xi^\sigma\frac{\partial}{\partial q^\sigma}.$$

Then

$$\begin{aligned} d(\lambda+\eta) &= dL \wedge dt + d\eta_\sigma \wedge \omega^\sigma - \eta_\sigma\,\omega_1^\sigma \wedge dt \\ &= \left(\left(\frac{\partial L}{\partial q^\sigma} - \frac{d\eta_\sigma}{dt}\right)\omega^\sigma + \left(\frac{\partial L}{\partial q_1^\sigma} - \eta_\sigma\right)\omega_1^\sigma\right) \wedge dt + \left(\frac{\partial \eta_\sigma}{\partial q^\nu}\omega^\nu + \frac{\partial \eta_\sigma}{\partial q_1^\nu}\omega_1^\nu\right)\wedge\omega^\sigma, \end{aligned}$$

and we have

$$\begin{aligned} i_{J^1\xi}d(\lambda+\eta) = &\left(\left(\frac{\partial L}{\partial q^\sigma} - \frac{d\eta_\sigma}{dt}\right)(\xi^\sigma - q_1^\sigma\xi^0) + \left(\frac{\partial L}{\partial q_1^\sigma} - \eta_\sigma\right)(\xi_1^\sigma - q_2^\sigma\xi^0)\right)dt \\ &- \left(\left(\frac{\partial L}{\partial q^\sigma} - \frac{d\eta_\sigma}{dt}\right)\omega^\sigma + \left(\frac{\partial L}{\partial q_1^\sigma} - \eta_\sigma\right)\omega_1^\sigma\right)\xi^0 + \left(\frac{\partial \eta_\sigma}{\partial q^\nu}(\xi^\nu - q_1^\nu\xi^0)\right. \\ &\left. + \frac{\partial \eta_\sigma}{\partial q_1^\nu}(\xi_1^\nu - q_2^\nu\xi^0)\right)\omega^\sigma - \left(\frac{\partial \eta_\sigma}{\partial q^\nu}\omega^\nu + \frac{\partial \eta_\sigma}{\partial q_1^\nu}\omega_1^\nu\right)(\xi^\sigma - q_1^\sigma\xi^0), \end{aligned}$$

where

$$\xi_1^\sigma = \frac{d\xi^\sigma}{dt} - q_1^\sigma\frac{d\xi^0}{dt}.$$

Since contact forms are annihilated by $J^1\gamma$, we obtain

$$\begin{aligned} J^1\gamma^* i_{J^1\xi}\, d(\lambda+\eta) &= \Big(\Big(\frac{\partial L}{\partial q^\sigma} - \frac{d\eta_\sigma}{dt}\Big)(\xi^\sigma - q_1^\sigma \xi^0) + \Big(\frac{\partial L}{\partial q_1^\sigma} - \eta_\sigma\Big)(\xi_1^\sigma - q_2^\sigma \xi^0)\Big)\, dt \\ &= \Big[-\xi^0\Big(\Big(\frac{\partial L}{\partial q^\sigma} - \frac{d\eta_\sigma}{dt}\Big)q_1^\sigma + \Big(\frac{\partial L}{\partial q_1^\sigma} - \eta_\sigma\Big)q_2^\sigma\Big) \\ &\quad + \xi^\sigma\Big(\frac{\partial L}{\partial q^\sigma} - \frac{d\eta_\sigma}{dt}\Big) + \xi_1^\sigma\Big(\frac{\partial L}{\partial q_1^\sigma} - \eta_\sigma\Big)\Big]\, dt. \end{aligned} \tag{3.6}$$

Since we require that in the above formula the ξ_1^σ would not appear, we are lead to put

$$\eta_\sigma = \frac{\partial L}{\partial q_1^\sigma}, \tag{3.7}$$

i.e.,

$$\eta = \frac{\partial L}{\partial q_1^\sigma}\,\omega^\sigma.$$

Then (3.6) becomes

$$J^1\gamma^* i_{J^1\xi}\, d(\lambda+\eta) = \Big(\frac{\partial L}{\partial q^\sigma} - \frac{d}{dt}\frac{\partial L}{\partial q_1^\sigma}\Big)(\xi^\sigma - \xi^0 q_1^\sigma)\, dt. \tag{3.8}$$

Denote

$$\theta_\lambda = L\, dt + \frac{\partial L}{\partial q_1^\sigma}\,\omega^\sigma. \tag{3.9}$$

The form θ_λ (3.9) represents the unique solution to the above problem for the case $r = 1$. Thus, the first variation formula takes the form

$$J^1\gamma^*\, \partial_{J^1\xi}\theta_\lambda = J^1\gamma^*\, i_{J^1\xi} d\theta_\lambda + J^1\gamma^*\, d i_{J^1\xi}\theta_\lambda$$

for every local section γ; this is equivalent to

$$h\, \partial_{J^1\xi}\theta_\lambda = h\, i_{J^1\xi} d\theta_\lambda + h\, d i_{J^1\xi}\theta_\lambda.$$

The integral form of the above formula then reads

$$\int_\Omega J^1\gamma^*\, \partial_{J^1\xi}\theta_\lambda = \int_\Omega J^1\gamma^*\, i_{J^1\xi} d\theta_\lambda + \int_\Omega J^1\gamma^*\, d i_{J^1\xi}\theta_\lambda.$$

Motivated by the above example, we shall be able to find out an intrinsic characterization a form $\rho = \lambda + \eta$ such that (3.5) would turn to become the first variation formula. We deduce that ρ must satisfy the following condition:

For every $\pi_{s,0}$-projectable vector field ξ on J^sY, $h\, i_{J^s\xi} d\rho$ depends only on the $\pi_{s,0}$-projection of ξ.

In fact, we shall show now that there are more (equivalent) conditions characterizing such forms. Before doing that, recall that a *dynamical form* E on J^rY is defined to be a one-contact element of $\Lambda^2_Y(J^rY)$. In fibered coordinates,

$$E = E_\sigma(t, q^\nu, q_1^\nu, \ldots, q_r^\nu)\, dq^\sigma \wedge dt.$$

3.2.2. Proposition. *Let $s \geqslant 0$ and let ρ be a one-form on J^sY. The following conditions are equivalent:*

(1) *$d\rho$ is decomposable in the form*

$$\pi^*_{s+1,s}\, d\rho = E + F, \tag{3.10}$$

where E is a dynamical form of order $s+1$ and F is a two-contact two-form on $J^{s+1}Y$.

(2) *The one-contact part $p_1\, d\rho$ of $d\rho$ is a dynamical form.*

(3) *For every $\pi_{s,0}$-projectable vector field ξ on J^sY the form $h\, i_\xi d\rho$ depends on the $\pi_{s,0}$-projection of ξ only.*

(4) *For every $\pi_{s,0}$-vertical vector field ξ on J^sY, $h\, i_\xi d\rho = 0$.*

(5) *In each fiber chart on Y,*

$$\pi^*_{s+1,s}\, \rho = L\, dt + \sum_{i=0}^{s} f^{i+1}_\sigma\, \omega^\sigma_i, \tag{3.11}$$

where

$$f^{i+1}_\sigma = \sum_{k=0}^{s-i} (-1)^k \frac{d^k}{dt^k} \frac{\partial L}{\partial q^\sigma_{i+1+k}}, \quad 0 \leqslant i \leqslant s. \tag{3.12}$$

Let us sketch the proof. The equivalence of (1)–(4) is obvious, it follows directly from the definitions of the operators h and p_1. It remains to show the equivalence of (5) with one of the conditions (1)–(4).

Denote

$$\pi^*_{s+1,s}\, \rho = L\, dt + \sum_{i=0}^{s} \rho^i_\sigma\, \omega^\sigma_i .$$

Then computing $\pi^*_{s+1,s} d\rho$ and omitting the two-contact part we get

$$\begin{aligned}
p_1\, d\rho &= \left(\sum_{i=0}^{s+1} \frac{\partial L}{\partial q^\sigma_i} \omega^\sigma_i - \sum_{i=0}^{s} \frac{d\rho^i_\sigma}{dt} \omega^\sigma_i - \sum_{i=0}^{s} \rho^i_\sigma\, \omega^\sigma_{i+1} \right) \wedge dt \\
&= \left(\frac{\partial L}{\partial q^\sigma} - \frac{d\rho_\sigma}{dt} \right) \omega^\sigma \wedge dt + \left(\sum_{i=1}^{s} \frac{\partial L}{\partial q^\sigma_i} - \frac{d\rho^i_\sigma}{dt} - \rho^{i-1}_\sigma \right) \omega^\sigma_i \wedge dt \\
&+ \left(\frac{\partial L}{\partial q^\sigma_{s+1}} - \rho^s_\sigma \right) \omega^\sigma_{s+1} \wedge dt .
\end{aligned}$$

Thus, the assumption (2) leads to the following formulas defining the contact part of ρ:

$$\begin{aligned}
\rho^s_\sigma &= \frac{\partial L}{\partial q^\sigma_{s+1}} \\
\rho^{s-1}_\sigma &= \frac{\partial L}{\partial q^\sigma_s} - \frac{d\rho^s_\sigma}{dt} = \frac{\partial L}{\partial q^\sigma_s} - \frac{d}{dt} \frac{\partial L}{\partial q^\sigma_{s+1}}, \\
&\dots \\
\rho_\sigma &= \frac{\partial L}{\partial q^\sigma_1} - \frac{d\rho^1_\sigma}{dt} = \frac{\partial L}{\partial q^\sigma_1} + \sum_{k=1}^{s} (-1)^k \frac{d^k}{dt^k} \frac{\partial L}{\partial q^\sigma_{k+1}} .
\end{aligned}$$

Hence, (3.11) and (3.12) follow.

Conversely, let a form ρ be given in a fiber chart by (3.11) and (3.12). Computing $d\rho$ and omitting the two-contact part we get

$$p_1\, d\rho = E = \left(\frac{\partial L}{\partial q^\sigma} + \sum_{k=1}^{s+1}(-1)^k \frac{d^k}{dt^k}\frac{\partial L}{\partial q^\sigma_k}\right)\omega^\sigma \wedge dt. \tag{3.13}$$

Hence, $p_1\, d\rho$ is a dynamical form, and we are done.

A form ρ satisfying any of the equivalent conditions of the above proposition is called a *Lepagean one-form.*

Note that for $s = 0$ all the conditions of Proposition 3.2.2 are satisfied identically, i.e., *every one-form on Y is a Lepagean one-form.*

If ρ is a Lepagean one-form then the dynamical form E defined by (3.10), or equivalently, by

$$i_{J^{s+1}\xi}\, E = h\, i_{J^s\xi}\, d\rho \tag{3.14}$$

for every π-vertical vector field ξ on Y, is called the *Euler-Lagrange form.* In a fiber chart where ρ is expressed by (3.11) and (3.12) we get E expressed by (3.13), i.e.,

$$E = E_\sigma(L)\,\omega^\sigma \wedge dt,$$

where

$$E_\sigma(L) = \sum_{k=0}^{s+1}(-1)^k \frac{d^k}{dt^k}\frac{\partial L}{\partial q^\sigma_k}, \quad 1 \leqslant \sigma \leqslant m. \tag{3.15}$$

The above functions $E_\sigma(L)$, $1 \leqslant \sigma \leqslant m$, are called the *Euler-Lagrange expressions* of L.

Let λ be a *Lagrangian of order r*, $r \geqslant 1$. A Lepagean one-form ρ such that $h\rho = \lambda$ is called a *Lepagean equivalent* of the Lagrangian λ.

Taking into account Proposition 3.2.2 it is seen that every Lepagean 1-form ρ is uniquely determined by its horizontal part $h\rho$. Consequently, to every Lagrangian there exists a *unique* Lepagean equivalent; it will be denoted by θ_λ. In fibered coordinates, the expression of θ_λ is given by the formulas (3.11) and (3.12). Looking at these formulas, we can easily see that for a Lagrangian of order r its Lepagean equivalent is in general of order $2r-1$ (and horizontal with respect to the projection π_{r-1}). Thus, in the case of a first-order Lagrangian, the θ_λ is also of order one. In general, however, the Lepagean equivalent of a Lagrangian λ is of order greater than λ.

We have the mapping

$$\mathfrak{Lep}_1 : \Lambda^1_X(J^rY) \ni \lambda \to \theta_\lambda \in \Lambda^1(J^{2r-1}Y),$$

which we shall call the *Lepage mapping of the first kind.*

The Euler-Lagrange form defined by the Lepagean equivalent θ_λ of λ depends only on the horizontal part of θ_λ, i.e., on the Lagrangian λ. Therefore it is referred to as the *Euler-Lagrange form of the Lagrangian* λ, and is denoted by E_λ. If λ is of order r then E_λ is generally of order $2r$. The mapping

$$\mathcal{E} : \lambda \to E_\lambda, \tag{3.16}$$

assigning to every Lagrangian its Euler-Lagrange form is called the *Euler-Lagrange mapping*.

By definition of the Lepagean equivalent of a Lagrangian,

$$\int_\Omega J^r\gamma^*\lambda = \int_\Omega J^{2r-1}\gamma^*\theta_\lambda,$$

which means that the action functions of λ and of its Lepagean equivalent θ_λ over Ω coincide.

Now, we are prepared to formulate the basic assertion characterizing the meaning of the concept of Lepagean one-form. To this purpose we use Proposition 3.2.2, the relation $\lambda = h\theta_\lambda$, and we utilize that

$$\partial_{J^r\xi}\, h\rho = h\,\partial_{J^{r+1}\xi}\rho$$

for every one-form ρ on J^rY.

3.2.3. Proposition. *Let λ be a Lagrangian of order r, let θ_λ be its Lepagean equivalent.*

(1) *For every π-projectable vector field ξ on Y*

$$\partial_{J^r\xi}\lambda = h\, i_{J^{2r-1}\xi}\, d\theta_\lambda + h\, d\, i_{J^{2r-1}\xi}\theta_\lambda. \tag{3.17}$$

(2) *For every π-projectable vector field ξ on Y and every section γ of π*

$$J^r\gamma^*\,\partial_{J^r\xi}\lambda = J^{2r-1}\gamma^*\, i_{J^{2r-1}\xi}\, d\theta_\lambda + d\, J^{2r-1}\gamma^*\, i_{J^{2r-1}\xi}\theta_\lambda. \tag{3.18}$$

(3) *For every π-projectable vector field ξ on Y, every closed interval $[a,b] \in R$ and every section γ of π*

$$\begin{aligned}\int_a^b J^r\gamma^*\,\partial_{J^r\xi}\lambda = &\int_a^b J^{2r-1}\gamma^*\, i_{J^{2r-1}\xi}\, d\theta_\lambda \\ &+ J^{2r-1}\gamma^*\, i_{J^{2r-1}\xi}\theta_\lambda\,(b) - J^{2r-1}\gamma^*\, i_{J^{2r-1}\xi}\theta_\lambda\,(a).\end{aligned} \tag{3.19}$$

Since for every π-projectable vector field ξ on Y and every section γ of π we have

$$J^{2r-1}\gamma^*\, i_{J^{2r-1}\xi}\, d\theta_\lambda = J^{2r}\gamma^*\, i_{J^{2r}\xi}\, \pi^*_{2r,2r-1} d\theta_\lambda = J^{2r}\gamma^*\, i_{J^{2r}\xi}\, E_\lambda,$$

we can express (3.18) and (3.19) equivalently by means of the Euler-Lagrange form E_λ of the Lagrangian λ as follows:

$$J^r\gamma^*\,\partial_{J^r\xi}\lambda = J^{2r}\gamma^*\, i_{J^{2r}\xi}\, E_\lambda + d\, J^{2r-1}\gamma^*\, i_{J^{2r-1}\xi}\theta_\lambda,$$

$$\begin{aligned}\int_a^b J^r\gamma^*\,\partial_{J^r\xi}\lambda = &\int_a^b J^{2r}\gamma^*\, i_{J^{2r}\xi}\, E_\lambda \\ &+ J^{2r-1}\gamma^*\, i_{J^{2r-1}\xi}\theta_\lambda\,(b) - J^{2r-1}\gamma^*\, i_{J^{2r-1}\xi}\theta_\lambda\,(a).\end{aligned}$$

If we consider π-*vertical* vector fields only, we get (3.17) in the form

$$\partial_{J^r\xi}\lambda = i_{J^{2r}\xi}E_\lambda + h\, d\, i_{J^{2r-1}\xi}\theta_\lambda.$$

Thus, with help of the Lepagean equivalent of a Lagrangian, any of the formulas (3.17)–(3.19) represents a decomposition of the action function of the (by means of ξ) transformed Lagrangian λ into a sum of two terms—a term containing the vector field ξ but *not* its higher order prolongations, and a *boundary term*.

Accordingly, (3.17) or (3.18) is referred to as the *infinitesimal first variation formula*, (3.19) is called the *integral first variation formula*.

3.2.4. Remarks. The concept of Lepagean one-form in mechanics represents an intrinsic definition of the classical *Cartan form* (called also *Poincaré-Cartan form*),

$$\theta_\lambda = L\,dt + \frac{\partial L}{\partial \dot{q}^\sigma}\,(dq^\sigma - \dot{q}^\sigma\,dt).$$

Such a form was considered by É. Cartan [1], and earlier in analytical mechanics by E. T. Whittaker [1], as a more or less heuristic object; it was used in Hamilton theory and in the theory of contact transformations. Higher-order generalization of this form appear in the work of Th. De Donder [1], Gelfand and Dikii [1], Sternberg [2], Dedecker [7], and others.

Lepagean n-forms (for arbitrary $n = \dim X$) have been introduced in the early 70's by D. Krupka [1,2], inspired by the work of Th. Lepage [1], and P. Dedecker [3]. In Krupka's definition the Lepage's idea that the Euler-Lagrange equations should arise from the exterior derivative of differential forms, found a rigorous expression. Later the concept of Lepagean n-form was re-discovered by B. Kupershmidt [1]; an equivalent definition is due to L. Mangiarotti and M. Modugno [2]. Global existence of Lepagean equivalents was proved by D. Krupka [6], another proof is due to M. Marvan [1].

We have seen that in mechanics the definition of Lepagean equivalent of a Lagrangian leads to a *unique* Lepagean onc form θ_λ such that $h\theta_\lambda = \lambda$. Contrary to mechanics, in field theory ($\dim X > 1$) a Lagrangian has many Lepagean equivalents with different properties. Indeed, considering the condition (3) of Proposition 3.2.2. as definition of Lepagean n-form in the general case, we shall see immediately that the definition ensures that Lagrangian enters in the horizontal and the one-contact part of ρ, the remaining parts (from two-contact to n-contact) letting undetermined (hence arbitrary).

In particular, in the first-order case one gets for Lepagean equivalents of a Lagrangian λ the expression

$$\rho = L\,\omega_0 + \frac{\partial L}{\partial y^\sigma_j}\,\omega^\sigma \wedge \omega_j + \nu \tag{3.20}$$

where

$$\omega_0 = dx^1 \wedge \cdots \wedge dx^n, \quad \omega_j = i_{\partial/\partial x^j}\,\omega_0,$$

and ν is an arbitrary n-form of order of contactness $\geqslant 2$. The form $\rho - \nu$ is uniquely determined by the Lagrangian. In analogy with mechanics, it is denoted by θ_λ and called the *Cartan form*. Thus, explicitly,

$$\theta_\lambda = L\,\omega_0 + \frac{\partial L}{\partial y^\sigma_j}\,\omega^\sigma \wedge \omega_j \tag{3.21}$$

(see J. Sniatycki [1], P. L. Garcia and A. Pérez-Rendón [1], T. Nôno and F. Mimura [1], H. Goldschmidt and S. Sternberg [1], P. L. Garcia [1], D. Krupka [1], and others). While studying field theories many authors restrict themselves to consider the form (3.21). However, the Cartan form in field theory does not possess some fundamental properties shared by the Cartan form in mechanics: Namely, in mechanics, $d\theta_\lambda = 0$ if and only if the Euler-Lagrange equations of λ vanish identically, which is not true for (3.21). To save the latter property, one must pick up another n-form from the class (3.20) of Lepagean equivalents of λ, namely

$$\rho_\lambda = L\,\omega_0 + \sum_{k=1}^{n} \sum_{j_1<\cdots<j_k} \frac{\partial^k L}{\partial y_{j_1}^{\sigma_1} \cdots \partial y_{j_k}^{\sigma_k}} \omega^{\sigma_1} \wedge \omega^{\sigma_k} \wedge \omega_{j_1 \cdots j_k}, \tag{3.22}$$

where $\omega_{ij} = i_{\partial/\partial x^j}\,\omega_i$, etc. The Lepagean equivalent (3.22) of λ was discovered by D. Krupka ([3], see also [8], where a geometric construction for obtaining (3.22) was given), and by D. Betounes [1]. It is referred to as the *fundamental Lepagean equivalent* of λ. Another disinguished Lepagean n-form considerd by Carathéodory [1] is

$$\rho_C = \frac{1}{L^{n-1}} \left(L\,dx^1 + \frac{\partial L}{\partial y_1^{\sigma_1}} \omega^{\sigma_1} \right) \wedge \cdots \wedge \left(L\,dx^n + \frac{\partial L}{\partial y_n^{\sigma_n}} \omega^{\sigma_n} \right),$$

and it has the property of being invariant under contact transformations of the Lagrangian. For a discussion on different distinguished Lepagean equivalents of a Lagrangian, the reader may consult D. Krupka [8], H. Rund [1], and M. J. Gotay [2]; for a discussion from the point of view of physics we refer to H. A. Kastrup [1].

If $n > 1$ and $r > 1$, the situation becomes even more complicated. First of all, a straightforward extension of the formula (3.21) to higher-order field theory (considered e.g. by Th. de Donder [1], V. Aldaya and J. de Azcárraga [2], D. Krupka [5], W. F. Shadwick [1]) does not result to a well-defined form. A correct generalization of the classical Cartan form to higher-order field theory is provided by the theory of Lepagean n-forms. It turns out that the desired form is obtained if Lepagean equivalents of a Lagrangian of order of contactness 2 are considered. However, such a *generalized Poincré-Cartan form* is *not uniquely determined by the Lagrangian.* These forms have been studied by many authors; to this topic the reader may consult e.g. D. Krupka [5,8], D. J. Saunders [1], P. J. Olver [1], P. L. Garcia and J. Muñoz [1], M. Ferraris [1], M. Ferraris and M. Francaviglia [1], M. Horák and I. Kolář [1], I. Kolář [1,2], M. J. Gotay [2], Dedecker [4], etc. A generalization of fundamental Lepagean equivalents (3.22) to higher-order field theories is due to D. Krupka [11]. Again, the higher order generalization is not a simple straightforward extension of the first order case, and it possesses some unexpected properties.

Another fundamental object in this section, the Euler-Lagrange form, representing a "globalization" of the concept of Euler-Lagrange equations to fibered manifolds, has been introduced by D. Krupka in [1], and independently by I. Anderson and T. Duchamp [1]. Alternatively, the Euler-Lagrange equations have been interpreted in terms of a vector-valued form by H. Goldschmidt and S. Sternberg [1], or P. L. Garcia [1]. It should be stressed that although for $\dim X > 1$ Lepagean equivalent of a Lagrangian is not unique, the Euler-Lagrange form *is* unique. The reason is that the Euler-Lagrange form of a Lepagean n-form ρ depends on $h\rho$ (i.e., on the Lagrangian) only.

3.3. Extremals of a Lagrangian

Let $\varepsilon > 0$, and $W \subset X$ be an open set. By a *one-parameter family of sections* of the fibered manifold π we shall mean a mapping $(-\varepsilon, \varepsilon) \times W \ni (u, x) \to \gamma(u, x) \in Y$ such that for each $u \in (-\varepsilon, \varepsilon)$ the mapping $\gamma_u : W \to Y$ defined by $\gamma_u(x) = \gamma(u, x)$ is a section of π over W. A one-parameter family of sections will be denoted by $\{\gamma_u\}$. If

$\gamma : W \to Y$ is a section over a piece $\Omega \subset W$, then by a *deformation of* γ we shall mean a one-parameter family $\{\gamma_u\}$ such that $\gamma_0 = \gamma$. Putting

$$\zeta(x) = \left(\frac{d}{du}\gamma_u(x)\right)_{u=0}$$

we obtain a π-vertical vector field *along* γ, called the *variation of the section* γ. This vector field can be extended into a π-vertical vector field ξ, defined in a neighborhood of $\gamma(W)$, such that the deformation induced by ξ coincides with $\{\gamma_u\}$. Conversely, if ξ is a π-vertical vector field defined in a neighborhood of γ then $\{\phi_u\gamma\}$, where $\{\phi_u\}$ is the local one-parameter group of ξ, is a deformation of γ. In other words, to describe *all deformations* of sections from $\mathcal{S}_\Omega(\pi)$ it is sufficient to consider π*-vertical* vector fields only.

For a deformation $\{\gamma_u\}$ of γ defined on $(-\varepsilon, \varepsilon) \times W$ we put $\operatorname{supp}\{\gamma_u\} = \operatorname{cl}\{x \in W \mid \zeta(x) \neq 0\}$, where cl is the topological closure, and $\zeta(x)$ is the variation of γ. We call this set the *support* of the deformation $\{\gamma_u\}$.

Let λ be a Lagrangian of order r on π, let Ω be a piece of X. Consider the action function of λ over Ω (3.1). A section $\gamma \in \mathcal{S}_\Omega(\pi)$ is called a *critical section*, or an *extremal of* λ *on* Ω if

$$\left(\frac{d}{du}\lambda_\Omega(\gamma_u)\right)_{u=0} = 0 \tag{3.23}$$

for every "fixed-endpoints" deformation $\{\gamma_u\}$ of γ, i.e. such that $\operatorname{supp}\{\gamma_u\} \subset \Omega$. A section γ of the fibered manifold π is called a *critical section*, or an *extremal of the Lagrangian* λ if the restriction of γ to any piece Ω of X such that $\Omega \subset \operatorname{dom}\gamma$, is an extremal of λ on Ω. By (3.3), the condition (3.23) can be written in the form

$$\left(\frac{d}{du}\int_\Omega J^r\gamma_u^*\lambda\right)_{u=0} = \left(\frac{d}{du}\int_\Omega J^r(\phi_u\gamma)^*\lambda\right)_{u=0} = \int_\Omega J^r\gamma^*\partial_{J^r\xi}\lambda = 0,$$

where ξ is a π-vertical vector field defined by the deformation $\{\gamma_u\}$, and $\{\phi_u\}$ is its local one-parameter group. Using the above considerations we obtain

3.3.1. Proposition. *A section* γ *of* π *is an extremal of a Lagrangian* λ *of order* r *on a piece* $\Omega \subset X$ *if and only if*

$$\int_\Omega J^r\gamma^*\partial_{J^r\xi}\lambda = 0 \tag{3.24}$$

for every π*-vertical vector field* ξ *defined in a neighborhood of* $\gamma(\Omega)$ *such that* $\operatorname{supp}\xi \subset \pi^{-1}(\Omega)$.

3.3.2. Theorem. *Let* λ *be a Lagrangian of order* r *on* π, E_λ *and* θ_λ *its Euler-Lagrange form and Lepagean equivalent, respectively. Let* γ *be a section of* π. *The following conditions are equivalent:*

(1) γ *is an extremal of* λ.

(2) *For every vector field* ζ *on* $J^{2r-1}Y$

$$J^{2r-1}\gamma^*\, i_\zeta\, d\theta_\lambda = 0. \tag{3.25}$$

(3) *For every* π*-projectable vector field* ξ *on* Y

$$J^{2r-1}\gamma^*\, i_{J^{2r-1}\xi}\, d\theta_\lambda = 0. \tag{3.26}$$

(4) *For every* π*-vertical vector field* ξ *on* Y

$$J^{2r-1}\gamma^* \, i_{J^{2r-1}\xi} \, d\theta_\lambda = 0 \,. \tag{3.27}$$

(5) *The Euler-Lagrange form* E_λ *vanishes along* $J^{2r}\gamma$, *i.e.,*

$$E_\lambda \circ J^{2r}\gamma = 0 \,. \tag{3.28}$$

(6) *In every fiber chart,* γ *satisfies the system of ODE of order* $2r$

$$\left(\sum_{k=0}^{r} (-1)^k \frac{d^k}{dt^k} \frac{\partial L}{\partial q_k^\sigma} \right) \circ J^{2r}\gamma = 0, \quad 1 \leqslant \sigma \leqslant m \,. \tag{3.29}$$

Proof. Using the integral first variation formula we get immediately that (1) and (4) are equivalent. The equivalence of (5) and (6) is obvious. It remains to show that (2), (3), (4) and (5) are equivalent.

(3) follows from (2), and (4) follows from (3) trivially.

Suppose (4). The form $\pi^*_{2r,2r-1} \, d\theta_\lambda$ is decomposed into the one-contact part (the Euler-Lagrange form E_λ), and the two-contact part F_λ. Since the contraction of F_λ by a vertical vector field is a one-contact form, it vanishes along $J^{2r}\gamma$. In this way we get

$$\begin{aligned} 0 &= J^{2r-1}\gamma^* \, i_{J^{2r-1}\xi} \, d\theta_\lambda = J^{2r}\gamma^* \, i_{J^{2r}\xi} \, \pi^*_{2r,2r-1} \, d\theta_\lambda \\ &= J^{2r}\gamma^* \, i_{J^{2r}\xi} \, E_\lambda + J^{2r}\gamma^* \, i_{J^{2r}\xi} \, F_\lambda = J^{2r}\gamma^* \, i_{J^{2r}\xi} \, E_\lambda \,. \end{aligned}$$

Denoting the components of E_λ by $E_\sigma(L)$, we get that $J^{2r}\gamma^*(E_\sigma(L)\xi^\sigma \, dt) = 0$ for every π-vertical vector field ξ on Y, i.e., that $E_\sigma(L) \circ J^{2r}\gamma = 0$, proving (5).

Suppose (5). Then (by similar arguments as above) for every π-projectable vector field ξ on Y,

$$J^{2r-1}\gamma^* \, i_{J^{2r-1}\xi} \, d\theta_\lambda = J^{2r}\gamma^* \, i_{J^{2r}\xi} \, E_\lambda = J^{2r}\gamma^*(\xi^\sigma \, dt - \xi^0 \, dq^\sigma)(E_\sigma \circ J^{2r}\gamma) = 0 \,,$$

proving (3).

Finally, (2) follows from (3). Indeed, since $J^{2r-1}\gamma^* \, i_{J^{2r-1}\xi} \, d\theta_\lambda$ depends only on the projection ξ of $J^{2r-1}\xi$, the components of ζ at $\partial/\partial q_j^\sigma$, $j \geqslant 1$ do not enter into the expression $J^{2r-1}\gamma^* \, i_\zeta \, d\theta_\lambda$. □

Any of the equivalent equations (3.25)–(3.29) are called the *Euler-Lagrange equations* of the Lagrangian λ. Hence, a section of π is an extremal (critical section) of a Lagrangian λ if and only if it satisfies the Euler-Lagrange equations of λ.

4. VARIATIONAL EQUATIONS

4.1. Introduction

Let us turn to study the family of (systems) of ordinary differential equations which are *variational*, i.e., come from Lagrangians as their Euler-Lagrange equations. On fibered manifolds such equations can be represented by *locally variational* or *globally variational forms* (D. Krupka [4,5,8], I. Anderson and T. Duchamp [1], I. Anderson [1,2]. To be more concrete, we shall be interested in the following class of ODE

$$E_\sigma\left(t, \gamma^\nu, \frac{d\gamma^\nu}{dt}, \ldots, \frac{d^s\gamma^\nu}{dt^s}\right) = 0, \quad 1 \leqslant \sigma \leqslant m$$

for sections $\gamma = (\gamma^\nu)$ of a fibered manifold $\pi : Y \to X$, $\dim X = 1$, $\dim Y = m + 1$, such that E_σ identify with the *Euler-Lagrange expressions* of a Lagrange function L (cf. (3.15)).

In the class of all ordinary differential equations, variational equations represent a large family which includes equations of any finite order $s \geqslant 1$, time-independent or time-dependent, solvable or non-solvable with respect to the highest derivatives. A detailed look at various examples of variational equations leads to a conclusion that this class of ODE is so heterogeneous regarding the form of the equations and properties of their solutions, that within the range of the classical theory of differential equations, it could hardly be characterized by means of some simple general properties.

In this chapter, our first aim is to provide an effective *intrinsic characterization* of the class of equations representable by locally variational forms. We shall show that these equations are in *one-to-one* correspondence with certain closed two-forms, called *Lepagean two-forms* (O. Krupková [2]). This property is fundamental, since it enables one to *transfer all the problems concerning variational ODE and their solutions to the study of a family of closed two-forms*. Consequently, we shall have at hands a powerful tool to investigate dynamical properties, as well as various methods for exact integration of such equations—in these topics we shall be interested in detail in the next chapters.

To be comparable with other approaches to classical and higher-order mechanics, it should be emphasized that the concept of *Lepagean two-form on a fibered manifold* is an *extension* and *generalization* of the well-known concept of *symplectic structure*, carrying however no *a priori* restrictions on Lagrangians, and representing a *universal geometric structure adapted to variational problems*, covering all (time-independent/time-dependent, local/global, regular/singular, first order/higher order, etc.) Lagrangians.

As a result, the theory based on Lepagean two-forms is applicable to *any* system of ordinary variational equations of any order. It brings new results significant for applications, namely for *mathematical physics*. One of the most important is a new understanding of a *Lagrangean system* as a Lepagean two-form, or equivalently, as the *class of all equivalent Lagrangians* (not as a particular Lagrangian), a new and geometrically clear concept of *regularity* which will be discussed in detail in Chapter 6, a new look at the concepts of Hamiltonian and momenta, which are related to variational equations rather than to individual Lagrangians, generalization of Legendre transformation (Chapters 6, 7), new understanding and generalization of the Hamilton theory (Chapters 5–7), and of integration methods for variational equations (Chapter 9).

Last but not least it should be stressed that the Lepagean two-forms approach to higher-order mechanics covers also *odd-order differential equations*, and enables one to treat in a completely new and consistent way the related Lagrangean systems.

In this chapter we introduce Lepagean two-forms and discuss their properties. As a consequence of the definition of Lepagean two-form one obtains the necessary and sufficient conditions for a form to coincide with an Euler-Lagrange form, therefore we also discuss the *local inverse problem of the calculus of variations.*

Another important result in this chapter is Theorem on the canonical form of a Lepagean two-form. It is a key-theorem for Hamilton theory of variational equations, and for local order-reducibility of Lagrangians.

Finally in this chapter, we study Lagrangean systems subject to holonomic constraints.

4.2. Locally variational forms

Consider a fibered manifold $\pi : Y \to X$ with $\dim X = 1$ and $\dim Y = m + 1$. We have seen in the previous chapter that to transfer the concept of the Euler-Lagrange equations to a fibered manifold one is lead to consider them as *components of a differential two-form*, namely, the Euler-Lagrange form, which belongs to the family of dynamical forms. It is worthwhile to recall that fiber-chart representation of a dynamical form E of order s, $s \geqslant 1$, is

$$E = E_\sigma(t, q^\nu, \dots, q^\nu_s)\, \omega^\sigma \wedge dt. \tag{4.1}$$

Let E be a dynamical form of order s. We say that a (local) section γ of the fibered manifold π is an *integral section* of E if

$$E \circ J^s\gamma = 0. \tag{4.2}$$

The above equation locally represents a system of m ordinary differential equations of order s for the components (γ^ν) of a section γ,

$$E_\sigma\left(t, \gamma^\nu, \frac{d\gamma^\nu}{dt}, \dots, \frac{d^s\gamma^\nu}{dt^s}\right) = 0. \tag{4.3}$$

4.2.1. Definition. Let E be a dynamical form of order s, where $s \geqslant 1$. E is called *variational* or *globally variational* if there exists an integer r and a Lagrangian λ defined on J^rY such that $E = E_\lambda$ (up to the projection $\pi_{2r,s}$ or $\pi_{s,2r}$). E is called *locally variational* if there exists a covering of J^sY by open sets such that the restriction of E to any of the elements of this covering is a variational form.

If a dynamical form is locally (or even globally) variational then the equations for its integral sections (4.2), resp. (4.3) are precisely the *Euler-Lagrange equations*. Therefore, integral sections of a locally variational form E will be called *extremals of* E. The definition of locally variational form now gives a natural relation between the extremals of E and the extremals of the corresponding Lagrangians: every extremal of any Lagrangian of E is an extremal of E, and, conversely, if γ is an extremal of E then every point of $\operatorname{dom}\gamma$ has a neighborhood on which γ is an extremal of a (local) Lagrangian of E.

Let us stop for a moment at the concepts of local and global variationality, since they are basic to the understanding of the material to follow. It is important to distinguish carefully between a globally and a locally variational form (the latter concept being more general).

By Definition 4.2.1, a *globally variational form* E of order s is a dynamical form defined on J^sY which comes from a global Lagrangian as its Euler-Lagrange form. There is no a priori requirement regarding the order of this Lagrangian; also one can see that the definition admits the existence of other, possibly local Lagrangians for E. On the other hand, a *locally variational form* E of order s is a global (i.e., defined on J^sY) dynamical form such that one can find a family of Lagrangians of possibly different orders, defined on open sets, whose Euler-Lagrange forms can be "glued together" to produce the form E; however, it may happen that there exists no family of Lagrangians which could be "glued together" to a global Lagrangian producing E. It should be emphasized that local existence of Lagrangians (local variationality) does *not* imply existence of a global Lagrangian (global variationality); the obstructions relate to the topology of the total space Y (Takens [1], Tulczyjew [3], Vinogradov [1], I. Anderson and T. Duchamp [1], I. Anderson [1,2], D. Krupka [9,10]). Locally variational systems are considered by mathematicians only rarely (in standard mathematical formulations of Lagrangean and Hamiltonian dynamics and field theory the existence of a global Lagrangian is an a priori assumption). However, the extension to locally variational problems is important for physics, since many physical problems are nontrivially of this kind—in particular, this concerns Lagrangean systems described by the familiar Newton equations. For illustration, let us mention two easy examples.

4.2.2. Example (Newtonian mechanics). Consider a particle of mass m moving in a force field F, where F is a vector field on an open subset $U \subset R^3$. The corresponding underlying structure is the fibered manifold $\pi : R \times U \to R$. Denote by (t, x^i) the (global) canonical chart on $R \times U$, and put $\omega^i = dx^i - \dot{x}^i dt$, $\dot{\omega}^i = d\dot{x}^i - \ddot{x}^i dt$. The Newton's equations of motion are ODE for local sections γ of π,

$$(m\ddot{x}^i - F^i) \circ J^2\gamma = 0,$$

where F^i are the components of F. Putting

$$E = \delta_{ij}(m\ddot{x}^i - F^i)\, dx^j \wedge dt,$$

where δ_{ij} are the components of the Euclidean metric on U, we get the corresponding dynamical form on $J^2(R \times U) = R \times T^2U$. By the Helmholtz conditions (1.15), E is locally variational iff the force F is *potential*, i.e., of the form $F = \operatorname{grad} V$ (cf. Chapter 1). Note that although F is defined on U, *the function V need not be defined on U* (we only get from the Poincaré Lemma that U can be covered by open sets in such a way that on each of these sets there exists a potential V for F).

4.2.3. Example (Particle in an electromagnetic field). Consider a particle of mass m moving in an electromagnetic field defined by the vector fields

$$E = 0, \quad B = \frac{I}{2\pi((x^1)^2 + ((x^2)^2)}(x^2, -x^1, 0),$$

where I is a current and (x^i) are the canonical coordinates on R^3; this electromagnetic field corresponds to the magnetic field created by I traveling along the x^3 axis (Grigore [1]). Now, the fibered manifold is $\pi : R \times U \to R$, where U is the open set $R^3 - \{(0, 0, x^3)\}$. The motion is described by a dynamical form

$$E = (\delta_{ij} m \ddot{x}^j - \epsilon_{ijk} v^j B^k)\, dx^i \wedge dt$$

which is locally variational, satisfying the Helmholtz conditions (1.15) (cf. also (1.32). E arises from *local Lagrangians* of the form

$$L = \tfrac{1}{2} m v^2 - v.A + \Phi,$$

where the potentials A and Φ are defined on an open subset of U by $B = \operatorname{rot} A$ and $\operatorname{grad} \Phi = \partial A / \partial t$.

4.3. Lepagean two-forms, Lagrangean systems

By the above, "variational equations on manifolds" can be correctly understood in terms of locally variational forms on jet prolongations of fibered manifolds. A crucial role in our study of geometry of variational equations and their solutions will be played by Lepagean two-forms, which are *global closed* counterparts of (global) locally variational forms. Consequently, one can use Lepagean two-forms to get all information about variational equations and the geometric structure of extremals. It should be noticed that the theory of Lepagean two-forms covers also an important class of global variational problems representable only by local Lagrangians (i.e., such that no global Lagrangian can be found).

The idea is to introduce a mapping $\mathfrak{Lep}_2$ such that the following diagram would be commutative:

$$\begin{array}{ccc} \lambda & \xrightarrow{\mathfrak{Lep}_1} & \theta_\lambda \\ {\scriptstyle \mathcal{E}}\big\downarrow & & \big\downarrow{\scriptstyle d} \\ E_\lambda & \xrightarrow[\mathfrak{Lep}_2]{} & d\theta_\lambda \end{array} \tag{4.4}$$

Let $\pi : Y \to X$ be a fibered manifold, $\dim X = 1$. Let $s \geqslant 1$.

4.3.1. Definition. A *closed* two-form α on $J^{s-1}Y$ is called *Lepagean two-form* if $p_1\alpha$ is a dynamical form.

In other words, α is a Lepagean two-form iff (1) $d\alpha = 0$, and (2) $\pi^*_{s,s-1}\alpha = E + F$ where E is a dynamical form and F is a two-contact two-form.

The condition (2) above is essential, it does not suffice for $p_1\alpha$ to be merely 1-contact. The reason is that we want to define closed counterparts to locally variational forms, belonging to the class of dynamical forms.

Note that, in particular, *any closed two-form on Y is a Lepagean two-form.* On the other hand, for $s \geqslant 2$ there are closed two-forms which are not Lepagean.

Taking into account the definition of Lepagean one-form we can see immediately that (in a neighborhood of every point of $J^{s-1}Y$) $\alpha = d\rho$ where ρ is a *Lepagean one-form* (defined in this neighborhood). This means that *for every Lepagean two-form α the form $E = p_1\alpha$ is locally variational.* Using the Poincaré Lemma and setting

$$\rho = A(\pi^*_{s,s-1}\alpha), \tag{4.5}$$

where A is the operator defined by (2.11), we get a Lepagean one-form ρ defined on an open subset of J^sY. Hence,

$$\lambda = h\rho = AE = \left(q^\sigma \int_0^1 (E_\sigma \circ \chi_s)\, du\right) dt \tag{4.6}$$

is a Lagrangian (of order s) for $E = p_1\alpha$; this Lagrangian is called the *Vainberg-Tonti Lagrangian.* Notice that the formula (4.6) ensures the existence of local Lagrangians of the same order as is that of the corresponding Euler-Lagrange equations. Since, as apparent from elementary examples, Lagrangians of possibly different orders can give rise to the same Euler-Lagrage expressions, one could ask a question about the existence of a Lagrangian of a lowest possible order (the so called *order-reduction problem*). This problem is solved affirmatively, and we shall consider it in Section 4.5.

Now, we shall find fiber-chart expressions for Lepagean two-forms.

4.3.2. Proposition. *Let α be a two-form on $J^{s-1}Y$, $s \geqslant 1$. α is a Lepagean two-form if and only if in each fiber chart*

$$\pi^*_{s,s-1}\alpha = E_\sigma\, \omega^\sigma \wedge dt + \sum_{j,k=0}^{s-1} F^{jk}_{\sigma\nu}\, \omega^\sigma_j \wedge \omega^\nu_k\,, \tag{4.7}$$

where

$$F^{jk}_{\sigma\nu} = -F^{kj}_{\nu\sigma}\,, \tag{4.8}$$

the functions E_σ, $1 \leqslant \sigma \leqslant m$, satisfy for all $0 \leqslant l \leqslant s$ and $1 \leqslant \sigma,\ \nu \leqslant m$ the identities

$$\frac{\partial E_\sigma}{\partial q^\nu_l} - (-1)^l \frac{\partial E_\nu}{\partial q^\sigma_l} - \sum_{k=l+1}^{s} (-1)^k \binom{k}{l} \frac{d^{k-l}}{dt^{k-l}} \frac{\partial E_\nu}{\partial q^\sigma_k} = 0, \tag{4.9}$$

and the $F^{jk}_{\sigma\nu}$'s are given by means of the components of $E = p_1\alpha$ by the formulas

$$\begin{aligned} F^{jk}_{\sigma\nu} &= \frac{1}{2} \sum_{l=0}^{s-j-k-1} (-1)^{j+l} \binom{j+l}{l} \frac{d^l}{dt^l} \frac{\partial E_\sigma}{\partial q^\nu_{j+k+l+1}}\,, \quad 0 \leqslant j+k \leqslant s-1\,, \\ F^{jk}_{\sigma\nu} &= 0\,, \quad s \leqslant j+k \leqslant 2s-2\,. \end{aligned} \tag{4.10}$$

Though the proof is straightforward, the computations are rather complicated; therefore we recall here the basic steps.

Proof. Let α be a Lepagean two-form on $J^{s-1}Y$. This means that in fibered coordinates α is expressed in the form (4.7), where we can suppose (4.8). Now, the condition $d\alpha = 0$

is equivalent with the following identities:

$$\frac{\partial E_\sigma}{\partial q^\nu} - \frac{\partial E_\nu}{\partial q^\sigma} + 2\frac{d}{dt}F_{\nu\sigma} = 0\,, \tag{4.11}$$

$$\frac{\partial E_\sigma}{\partial q_k^\nu} - 2\frac{d}{dt}F_{\sigma\nu}^{0k} - F_{\sigma\nu}^{0,k-1} = 0\,, \quad 1 \leqslant k \leqslant s-1\,, \tag{4.12}$$

$$\frac{\partial E_\sigma}{\partial q_s^\nu} - 2F_{\sigma\nu}^{0,s-1} = 0\,, \tag{4.13}$$

$$\frac{d}{dt}F_{\sigma\nu}^{jk} + F_{\sigma\nu}^{j-1,k} + F_{\sigma\nu}^{j,k-1} = 0\,, \quad 1 \leqslant j,k \leqslant s-1\,, \tag{4.14}$$

$$F_{\sigma\nu}^{s-1,k} = 0\,, \quad 1 \leqslant k \leqslant s-1\,, \tag{4.15}$$

$$\frac{\partial F_{\sigma\nu}^{jk}}{\partial q_l^\rho} + \frac{\partial F_{\rho\sigma}^{lj}}{\partial q_k^\nu} + \frac{\partial F_{\nu\rho}^{kl}}{\partial q_j^\sigma} = 0\,, \quad 0 \leqslant j,k,l \leqslant s-1\,. \tag{4.16}$$

From (4.13) and (4.12) we can get the $F_{\sigma\nu}^{0k}$, $0 \leqslant k \leqslant s-1$, expressed by means of the E_ρ's in the following form

$$F_{\sigma\nu}^{0k} = \frac{1}{2}\sum_{l=0}^{s-k-1}(-1)^l \frac{d^l}{dt^l}\frac{\partial E_\sigma}{\partial q_{k+l+1}^\nu}\,, \quad 0 \leqslant k \leqslant s-1\,. \tag{4.17}$$

From (4.15) and (4.14) we get the second set of (4.10) and the following relations:

$$F_{\sigma\nu}^{jk} = \sum_{l=0}^{s-j-k-1}(-1)^{k+l}\binom{k+l-1}{l}\frac{d^l}{dt^l}F_{\sigma\nu}^{j+k+l,0}\,, \quad 2 \leqslant j+k \leqslant s-1\,, \tag{4.18}$$

and

$$F_{\sigma\nu}^{0k} = \sum_{l=0}^{s-k-1}(-1)^{k+l}\binom{k+l-1}{l}\frac{d^l}{dt^l}F_{\sigma\nu}^{k+l,0}\,, \quad 1 \leqslant k \leqslant s-1\,. \tag{4.19}$$

Now, (4.17) and (4.18) prove the first set of (4.10). Next, substituting (4.17) into (4.19) and into (4.11) we get after straightforward calculations the identities (4.9).

Finally, we shall show that the relations (4.16) are fulfilled identically. Put

$$G_{\sigma\nu\rho}^{jkl} = \frac{\partial F_{\sigma\nu}^{jk}}{\partial q_l^\rho} + \frac{\partial F_{\rho\sigma}^{lj}}{\partial q_k^\nu} + \frac{\partial F_{\nu\rho}^{kl}}{\partial q_j^\sigma}\,, \quad 0 \leqslant j,k,l \leqslant s\,, \quad 1 \leqslant \sigma,\nu,\rho \leqslant m\,. \tag{4.20}$$

Differentiating the relations (4.11), (4.12), and (4.14) with respect to q_{s+1}^ρ we obtain

$$G_{\sigma\nu\rho}^{jks} = \frac{\partial F_{\sigma\nu}^{jk}}{\partial q_s^\rho} = 0\,, \quad 0 \leqslant j,k \leqslant s\,, \quad 1 \leqslant \sigma,\nu,\rho \leqslant m\,. \tag{4.21}$$

Using (4.14) we get for $1 \leqslant j,k,l \leqslant s$, $1 \leqslant \sigma,\nu,\rho \leqslant m$, the relation

$$\frac{d}{dt}G_{\sigma\nu\rho}^{jkl} + G_{\sigma\nu\rho}^{j-1,k,l} + G_{\sigma\nu\rho}^{j,k-1,l} + G_{\sigma\nu\rho}^{j,k,l-1} = 0\,. \tag{4.22}$$

Now, proceeding by induction starting from (4.21) the desired relations

$$G_{\sigma\nu\rho}^{jkl} = 0\,, \quad 0 \leqslant j,k,l \leqslant s\,, \quad 1 \leqslant \sigma,\nu,\rho \leqslant m \tag{4.23}$$

are obtained.

The converse statement of the proposition is proved in an obvious way. □

The reader surely has noted that the identities (4.9) should refer to the Helmholtz conditions (1.15). Indeed, as we shall prove a little bit later, (4.9) are a higher-order generalization of (1.15).

4.3.3. Definition. Let E be a locally variational form. A Lepagean two-form α such that $p_1\alpha = E$ will be called a *Lepagean equivalent* of E. The mapping $\mathfrak{Lep}_2$ assigning to a locally variational form its Lepagean equivalent is called *Lepage mapping of the second kind*.

Now, we shall clarify the relation between locally variational forms and Lepagean two-forms. Almost by definition, the one-contact part of a Lepagean two-form is a locally variational form. On the other hand, every locally variational form has (in a neighborhood of every point) a Lepagean equivalent: if E is a locally variational form (of order s) and λ is a Vainberg-Tonti Lagrangian of E defined on an open set then $d\theta_\lambda$ is clearly a (local) Lepagean two-form (of order $2s-1$) such that $p_1 d\theta_\lambda = E$. But what one can say about the global existence of Lepagean equivalents, and about their uniqueness?

The following key theorem says that in fact *Lepagean two-forms are in one-to-one correspondence with locally variational forms.* Moreover, according to this theorem, *the theory of Lepagean two-forms leads to assigning a natural proper order to every variational problem.* This is crucial for understanding the geometry of solutions of variational equations, since (as we shall see in the next chapter) *the order of Lepagean two-form defines the order of the jet-space, where the dynamics takes place.*

4.3.4. Theorem. *Let E be a locally variational form of order s. Then there exists a unique Lepagean two-form α such that $p_1\alpha = E$; this Lepagean two-form is projectable onto $J^{s-1}Y$.*

Proof. If E is a locally variational form on J^sY, we shall show that (1) there is a unique two-contact two-form F such that $\alpha = E + F$ is a Lepagean two-form, and (2) that α is projectable onto $J^{s-1}Y$.

By definition of locally variational form, there exists an open covering $\{W_\iota\}$ of J^sY such that (i) for each index ι, $W_\iota \subset V_s$, where (V, ψ) is a fiber chart on Y, (ii) the restriction $E|_{W_\iota}$ of E to W_ι is variational. Let ι, κ be arbitrary such that $W_\iota \cap W_\kappa \neq \emptyset$. Denote by α_ι (resp. α_κ) a Lepagean equivalent of $E|_{W_\iota}$ (resp. $E|_{W_\kappa}$) (we can take e.g. $\alpha_\iota = d\theta_{\lambda_\iota}$ where λ_ι is a Lagrangian for $E|_{W_\iota}$, and similarly for κ). Now, on the intersection of the corresponding domains, we have $\alpha_\iota = E + F_\iota$ and $\alpha_\kappa = E + F_\kappa$, where F_ι, F_κ are 2-contact two-forms. Hence, $\alpha_\iota - \alpha_\kappa = F_\iota - F_\kappa$. The form $\alpha_\iota - \alpha_\kappa$ is a Lepagean two-form such that $p_1(\alpha_\iota - \alpha_\kappa) = 0$, i.e., by Proposition 4.3.2, $\alpha_\iota - \alpha_\kappa = 0$. Thence, on the intersection of the domains it holds $\alpha_\iota = \alpha_\kappa$, proving global existence of a Lepagean equivalent of E. Uniqueness is a direct consequence of Proposition 4.3.2.

It remains to show that the Lepagean equivalent of E is projectable onto $J^{s-1}Y$. Let $E = E_\sigma\, \omega^\sigma \wedge dt$ be a fiber-chart expression of E. Then the Lepagean equivalent α is given by the formulas (4.7), (4.8), (4.10). Since the functions $F^{ik}_{\sigma\nu}$ are of order $2s-1-i-k$, α is of order $2s-1$, but it is obviously horizontal with respect to the projection onto $J^{s-1}Y$.

Let us express α in the basis $(dt, dq^\sigma, \dots, dq^\sigma_{s-1})$. Now (4.7) takes the form

$$\alpha = \Big(E_\sigma - \sum_{k=0}^{s-1} 2F^{0k}_{\sigma\nu} q^\nu_{k+1}\Big)\, dq^\sigma \wedge dt - \sum_{i=1}^{s-1}\sum_{k=0}^{s-1-i} 2F^{ik}_{\sigma\nu} q^\nu_{k+1}\, dq^\sigma_i \wedge dt + \sum_{j,k=0}^{s-1} F^{jk}_{\sigma\nu}\, dq^\sigma_j \wedge dq^\nu_k \tag{4.24}$$

and we can see that it is enough to show that the functions $F^{ik}_{\sigma\nu}$, $E_\sigma - 2F^{0,s-1}_{\sigma\nu} q^\nu_s$, where $1 \leqslant \sigma, \nu \leqslant m$, $0 \leqslant i, k \leqslant s-1$, are of order $s-1$. Differentiating the relations (4.11) and (4.12) consecutively with respect to $q^\rho_{2s}, q^\rho_{2s-1}, \dots, q^\rho_{s+1}$, and taking into account that the E_σ's are of order s, we obtain that $F^{0k}_{\sigma\nu}$'s are of order $s-1$. Similar conclusions are made for the remaining $F^{ik}_{\sigma\nu}$'s with the help of (4.14). To finish the proof we have to show that the functions $E_\sigma - 2F^{0,s-1}_{\sigma\nu} q^\nu_s$, $1 \leqslant \sigma \leqslant m$, do not depend upon the q^ρ_s's. To this end let us turn to the relations (4.9). Differentiating the relation for $l = s-1$ with respect to q^ρ_{s+1} we get

$$\frac{\partial^2 E_\sigma}{\partial q^\nu_s \partial q^\rho_s} = 0\,, \quad 1 \leqslant \sigma, \nu, \rho \leqslant m\,, \tag{4.25}$$

which means that the functions E_σ, $1 \leqslant \sigma \leqslant m$, are affine in the highest derivatives, i.e., they are of the form

$$E_\sigma = A_\sigma + B_{\sigma\nu} q^\nu_s\,, \tag{4.26}$$

where the functions A_σ, $B_{\sigma\nu}$ are of order $s-1$. Now, since

$$B_{\sigma\nu} = \frac{\partial E_\sigma}{\partial q^\nu_s} = 2F^{0,s-1}_{\sigma\nu}\,, \tag{4.27}$$

we are done. □

4.3.5. Corollary. *The mapping $\mathfrak{Lep}_2$ of the set of locally variational forms to the set of Lepagean two-forms is bijective and inverse to the mapping p_1.*

In what follows, the Lepagean equivalent of a locally variational form E will be denoted by α_E.

4.3.6. Remark. Let E be a locally variational form on J^sY. Then by (4.25), E can be represented in fibered coordinates in the form

$$E = E_\sigma\, \omega^\sigma \wedge dt\,, \qquad E_\sigma = A_\sigma + B_{\sigma\nu} q^\nu_s\,,$$

where the A_σ's and $B_{\sigma\nu}$'s do not depend upon q^ρ_s, $1 \leqslant \rho \leqslant m$, and, by (4.9), the matrix $\mathrm{B} = (B_{\sigma\nu})$ is *symmetric* if s is even and *antisymmetric* if s is odd.

Consequently, for the Lepagean equivalent α_E of E the following equivalent fiber-chart expressions can be used:

$$\pi^*_{s,s-1}\, \alpha_E = E_\sigma\, \omega^\sigma \wedge dt + \sum_{j,k=0}^{s-1} F^{jk}_{\sigma\nu}\, \omega^\sigma_j \wedge \omega^\nu_k\,,$$

or, in the basis $(dt, \omega^\sigma_j, 0 \leqslant j \leqslant s-2, dq^\sigma_{s-1})$ on $J^{s-1}Y$ (relevant for $s \geqslant 2$),

$$\alpha_E = A_\sigma \omega^\sigma \wedge dt + \sum_{j,k=0}^{s-2} F^{jk}_{\sigma\nu} \omega^\sigma_j \wedge \omega^\nu_k + B_{\sigma\nu} \omega^\sigma \wedge dq^\nu_{s-1}$$

or, in the basis $(dt, dq^\sigma_j, 0 \leqslant j \leqslant s-1)$ on $J^{s-1}Y$ $(s \geqslant 1)$,

$$\alpha_E = \Big(E_\sigma - \sum_{k=0}^{s-1} 2F^{0k}_{\sigma\nu} q^\nu_{k+1}\Big) dq^\sigma \wedge dt$$
$$- \sum_{j=1}^{s-1} \sum_{k=0}^{s-1-j} 2F^{jk}_{\sigma\nu} q^\nu_{k+1}\, dq^\sigma_j \wedge dt + \sum_{j,k=0}^{s-1} F^{jk}_{\sigma\nu}\, dq^\sigma_j \wedge dq^\nu_k,$$

where the $F^{jk}_{\sigma\nu}$'s are given by (4.10).

In particular, *second order variational equations* have the Lepagean equivalent α_E defined on J^1Y and it holds

$$\alpha_E = A_\sigma \omega^\sigma \wedge dt + \frac{1}{4}\left(\frac{\partial A_\sigma}{\partial q^\nu_1} - \frac{\partial A_\nu}{\partial q^\sigma_1}\right) \omega^\sigma \wedge \omega^\nu + B_{\sigma\nu} \omega^\sigma \wedge dq^\nu_1.$$

First order variational equations have their Lepagean equivalent defined on Y, and it is of the form

$$\alpha_E = A_\sigma\, dq^\sigma \wedge dt + \tfrac{1}{2} B_{\sigma\nu}\, dq^\sigma \wedge dq^\nu = dq^\sigma \wedge (A_\sigma\, dt + \tfrac{1}{2} B_{\sigma\nu}\, dq^\nu).$$

Lepagean two-forms have been introduced and studied by O. Krupková in the 80's ([2], see also [4,5], where some other properties of Lepagean two-forms can be found as well). Recently, the idea to represent the class of variational equations equivalently by appropriate closed two-forms appeared in the work of D. R. Grigore [1,2], who introduced Lepagean two-forms (he calls them Souriau-Lagrange forms) for the case of a different underlying structure—one-dimensional contact elements (jets of one-dimensional submanifolds). Closed two-forms in higher-order mechanics have been studied also by L. Klapka [2]. Formulas (4.10) have been first obtained independently by L. Klapka [2], O. Štěpánková [1], and I. Anderson [1].

From the point of view of *symplectic geometry*, Lepagean two-forms represent a universal generalization of the concepts of the symplectic, presymplectic, cosymplectic and precosymplectic form adapted to *general* higher order variational problems of one independent variable. In the last chapter we shall discuss relations of the theory of Lepagean two-forms to the symplectic geometry in more detail.

Theorem 4.3.4 has the following important consequences which will be used throughout the text:

4.3.7. Corollary. *Let r, s be integers. Let E be a locally variational form of order s, α_E its Lepagean equivalent. Let λ be a Lagrangian for E, defined on an open subset of J^rY. Then the form $d\theta_\lambda$ is projectable onto an open subset $W \subset J^{s-1}Y$ and $\pi^*_{2r-1,s-1} d\theta_\lambda = \alpha_E|_W$.*

The above corollary says in fact that locally the Lepagean equivalent of E can be obtained if one takes the exterior derivative of the Cartan form of any Lagrangian for E.

Next, we obtain a well-known property of variational equations which is known as the *Theorem on local triviality of a Lagrangian.*

4.3.8. Corollary. *Consider the Euler-Lagrange mapping* $\mathcal{E}$, *denote by* $\ker \mathcal{E}$ *its kernel. A Lagrangian* λ, *defined on an open subset* W *of* J^rY *belongs to* $\ker \mathcal{E}$ *if and only if there exists a closed one-form* ρ *on* $\pi_{r,r-1}W$ *such that* $\lambda = h\rho$.

Proof. Let $\lambda \in \ker \mathcal{E}$, i.e., let $E_\lambda = 0$. Then $\alpha_{E_\lambda} = 0 = d\theta_\lambda$. Since $\lambda = h\theta_\lambda$, the form $h\theta_\lambda$ is projectable onto W, and, in a neighborhood of every point in W, $h\theta_\lambda = hdf$ where f is a function. Since hdf is of order r, we conclude that f does not depend upon q_r^ρ, $1 \leqslant \rho \leqslant m$. Now, $\theta_\lambda = df$, i.e., θ_λ is projectable onto $\pi_{r,r-1}W$.

Conversely, if ρ is a closed one-form on $\pi_{r,r-1}W$, such that $\lambda = h\rho$, we get $\theta_\lambda = \theta_{h\rho} = \rho$, i.e., $d\theta_\lambda = d\rho = 0$. Thence, $E_\lambda = 0$. □

Since the kernel of the Euler-Lagrange mapping $\mathcal{E}$ is not trivial, we have the following equivalence relation on (local) Lagrangians: two Lagrangians λ_1 of order r and λ_2 of order $k \geqslant r$ (such that, up to projection, $\operatorname{dom}\lambda_1 \cap \operatorname{dom}\lambda_2 \neq \emptyset$) are called *equivalent* if $\lambda_1 - \lambda_2 \in \ker \mathcal{E}$; in other words, two local Lagrangians (of possibly different orders) are equivalent if (up to a projection) their Euler-Lagrange forms coincide on the common domain of definition. This means that Lagrangians of order r and k, $k \geqslant r$, are equivalent iff they differ locally by a total derivative hdf where f is a function depending on $t, q^\sigma, \ldots, q^\sigma_{k-1}$. We shall call f a *gauge function of order* $k - 1$.

4.3.9. Corollary. *A dynamical form of order* s *is locally variational if and only if its fiber-chart components* E_σ, $1 \leqslant \sigma \leqslant m$, *satisfy the identities* (4.9).

Proof. Let E be a dynamical form on J^sY. If E is locally variational then there exists the Lepagean equivalent α_E. By Proposition 4.3.2 the form $p_1\alpha_E = E$ satisfies (4.9).

Conversely, if E satisfies (4.9), then we can construct a two-contact two-form F putting

$$F = \sum_{j,k=0}^{s-1} F^{jk}_{\sigma\nu}\, \omega^\sigma_j \wedge \omega^\nu_k, \tag{4.28}$$

where the functions $F^{jk}_{\sigma\nu} = -F^{kj}_{\nu\sigma}$ are defined by (4.10). Now, the form $\alpha = E + F$ is a Lepagean two-form, the one-contact part of which is E. Hence, E is locally variational. □

Corollary 4.3.9 brings necessary and sufficient conditions for a system of higher-order ordinary differential equations to come from a Lagrangian as a system of its Euler-Lagrange equations. This problem, known as the *local inverse problem of the calculus of variations*, was first studied by H. Helmholtz [1] for second order ordinary differential equations. Solution for higher-order ODE is due to A. L. Vanderbauwhede [1] by means of Vainberg's potential operator method. The explicit solution for the general situation (higher order PDE) has been first obtained independently by D. Krupka [4] (see also [5] for more details), who used the properties of Lepagean n-forms (n = dimension of the base manifold), and by I. Anderson and T. Duchamp [1], who applied the techniques the variational bicomplex (cf. P. Dedecker and W. M. Tulczyjew [1], F. Takens [1], W. M. Tulczyjew [1–3], and A. M. Vinogradov [1]). Necessary and sufficient conditions of variationality will be referred to as *Anderson-Duchamp-Krupka conditions*. Let us show that in case of dim $X = 1$ and $s = 2$ (classical mechanics) they turn to be the familiar *Helmholtz conditions* ((1.8) resp. equivalent (1.15)), discussed in Chapter 1. Writing down (4.9) explicitly for $s = 2$ we get them in the form

$$\frac{\partial E_\sigma}{\partial \ddot{q}^\nu} - \frac{\partial E_\nu}{\partial \ddot{q}^\sigma} = 0, \quad \frac{\partial E_\sigma}{\partial \dot{q}^\nu} + \frac{\partial E_\nu}{\partial \dot{q}^\sigma} - 2\frac{d}{dt}\frac{\partial E_\nu}{\partial \ddot{q}^\sigma} = 0,$$
$$\frac{\partial E_\sigma}{\partial q^\nu} - \frac{\partial E_\nu}{\partial q^\sigma} + \frac{d}{dt}\frac{\partial E_\nu}{\partial \dot{q}^\sigma} - \frac{d^2}{dt^2}\frac{\partial E_\nu}{\partial \ddot{q}^\sigma} = 0. \tag{4.29}$$

We obtained these conditions without making any additonal assumptions on the functions $E_\sigma(t, q^\nu, \dot{q}^\nu, \ddot{q}^\nu)$ while Helmholtz supposed from the very beginning that E_σ's are affine in the accelerations, i.e., of the form (1.7). However, the arguments in proof of Theorem 4.3.4 show that, in fact, this is no true restriction, since (4.29) *imply* (1.7). Now, the equivalence of (4.29) with (1.15), and hence (1.8), is apparent.

We remind the reader that generally, if a dynamical form E of order s is locally variational then its components E_σ are *affine in the highest derivatives* (see (4.26) and the surrounding arguments). It can be also easily seen (e.g., looking at the Euler-Lagrange expressions) that a locally variational form coming from a Lagrangian of order r (which generally is of order $2r$) has its components *polynomial* in the derivatives q^ν_{r+j}, $1 \leqslant j \leqslant r$.

It is worthwhile to note that by Proposition 4.3.2, *for first order ODE* the conditions (4.9) can be geometrically interpreted as *closedness conditions* for (general) two-forms on Y. Namely, for $s = 1$, (4.9) read

$$\frac{\partial E_\sigma}{\partial \dot{q}^\nu} + \frac{\partial E_\nu}{\partial \dot{q}^\sigma} = 0, \quad \frac{\partial E_\sigma}{\partial q^\nu} - \frac{\partial E_\nu}{\partial q^\sigma} + \frac{d}{dt}\frac{\partial E_\nu}{\partial \dot{q}^\sigma} = 0, \tag{4.30}$$

which is equivalent to

$$B_{\sigma\nu} = -B_{\nu\sigma}, \quad \frac{\partial B_{\sigma\rho}}{\partial q^\nu} + \frac{\partial B_{\nu\sigma}}{\partial q^\rho} + \frac{\partial B_{\rho\nu}}{\partial q^\sigma} = 0, \quad \frac{\partial A_\sigma}{\partial q^\nu} - \frac{\partial A_\nu}{\partial q^\sigma} = \frac{\partial B_{\sigma\nu}}{\partial t}.$$

The latter are conditions for the two-form

$$\alpha = A_\sigma \, dq^\sigma \wedge dt + \tfrac{1}{2} B_{\sigma\nu} \, dq^\sigma \wedge dq^\nu$$

be closed.

For higher-order ODE, (4.9) represent the closedness conditions for two-forms, the one-contact part of which is a dynamical form.

Theorem 4.3.4 also leads to a result on global variationality of a dynamical form E.

4.3.10. Corollary. *A dynamical fom E on J^sY is globally variational if and only if its Lepagean equivalent α_E is exact.*

Indeed, if α_E is exact, $\alpha_E = d\rho$, then ρ is a Lepagean 1-form and $\lambda = h\rho$ is a global Lagrangian (of order s) for E. Conversely, if E is globally variational and λ is a global Lagrangian for E then the form $d\theta_\lambda$ identifies (poosibly up to a projection) with the Lepagean equivalent α_E od E.

Finally, by the following Corollary, Lepagean two-forms can be used to study solutions of variational equations.

4.3.11. Corollary. *Let E be a locally variational form on J^sY, α_E its Lepagean equivalent. A (local) section γ of $\pi : Y \to X$ is an extremal of E if and only if*

$$J^{s-1}\gamma^* i_{J^{s-1}\xi}\, \alpha_E = 0 \tag{4.31}$$

for every π-vertical vector field ξ on Y.

We have seen that instead of the class of locally variational forms one can *equivalently* consider the class of Lepagean two-forms. This leads us naturally to the following definitions which reflect the relation to physics.

4.3.12. Definition. By a *Lagrangean system of order r* ($r \geqslant 0$) for a fibered manifold $\pi : Y \to X$ we shall understand a Lepagean two-form α on the r-jet prolongation of π. The manifold Y will then be called the *configuration space*, and J^rY will be referred to as the *phase space* for the Lagrangean system α.

The above concept of Lagrangean system, introduced by O. Krupková [9], essentially differs from the usual one. Recall that usually, a Lagrangean system of order r is identified with a Lagrangian of order r. This means, in particular, that two equivalent Lagrangians (i.e., differing by total derivative of a function) in the usual understanding give rise to two different Lagrangean systems (possibly of different orders). Since these Lagrangians may possess essentially different properties (for example, one is time-independent and "regular" (in the usual sense of regularity), and the other time-dependent and "singular") the conventional approach results to study two essentially different Lagrangean systems, which, however, as we know by Corollary 4.3.8, give rise to the *same* Euler-Lagrange form, hence must possess *the same dynamical properties*. On the contrary, by Definition 4.3.12, a Lagrangean system of order r is a Lepagean two-form of order r, and can be equivalently represented e.g. by an *Euler-Lagrange form* of order $r + 1$, or by the *equivalence class of Lagrangians* differing by total derivative, and containing, as we shall see in Section 4.5, Lagrangians of any finite order k where $k \geqslant c$ (resp. $k \geqslant c + 1$) if r is odd $= 2c + 1$ (resp. if r is even $= 2c$); equivalently, a Lagrangean system can be characterized by the equivalence class of the so-called *minimal-order Lagrangians*, the existence of which will be proved in Section 4.5. We remind the reader that among the Lagrangians characterizing a (global) Lagrangean system there need not exist a global Lagrangian. This approach to Lagrangean systems is very close to the recent work of D. R. Grigore [1,2], and D. R. Grigore and O. T. Popp [1], where also *field theory* is considered.

The understanding of Lagrangean system as the *class of equivalent Lagrangians* (i.e., differing by total derivative) seems to have a significant meaning in *quantum mechanics*: it has been proved recently that the total derivative terms (even if they lead to Lagrangians of different orders) do not affect the quantum properties of the system (see C. A. P. Galvão and N. A. Lemos [1], Y. Kaminaga [1]).

Consequently, the above concept of Lagrangean system results in a natural definition of the *order* of a Lagrangean system, which is better adapted than the conventional one to the geometry of variational systems, to the study of their dynamics, and, as it seems from the recent research, also to quantization (cf. C. A. P. Galvão and N. A. Lemos [1], Y. Kaminaga [1], R. Jackiw [1]).

Notice that in the light of the terminology introduced above, Examples 4.2.2 and 4.2.3 refer to the following first-order Lagrangean systems:

$$\begin{aligned}\alpha &= \left(m\delta_{ij}\ddot{x}^j - \frac{\partial V}{\partial x^j}\right)\omega^i \wedge dt + m\delta_{ij}\,\omega^i \wedge \dot{\omega}^j \\ &= -\frac{\partial V}{\partial x^i} dx^i \wedge dt + m\delta_{ij}\dot{x}^j\, d\dot{x}^i \wedge dt + m\delta_{ij}\, dx^i \wedge d\dot{x}^j,\end{aligned}$$

for the Newtonian mechanics, and

$$\begin{aligned}\alpha &= (m\delta_{ij}\ddot{x}^j - \epsilon_{ijk}v^j B^k)\, dx^i \wedge dt + \tfrac{1}{2}\epsilon_{ijk}B^k\,\omega^i \wedge \omega^j + m\delta_{ij}\,\omega^i \wedge \dot{\omega}^j \\ &= m\delta_{ij}v^i\, d\dot{x}^i \wedge dt + \tfrac{1}{2}\epsilon_{ijk}B^k\, dx^i \wedge dx^j + m\delta_{ij}\, dx^i \wedge d\dot{x}^j\end{aligned}$$

for the particle in electromagnetic field, respectively.

4.4. A few words on global properties of the Euler-Lagrange mapping

Studying properties of the Euler-Lagrange mapping, one is encountered with the problem to describe its *kernel*, i.e., to find all trivial Lagrangians, and *image*, i.e., to find out necessary and sufficient conditions of variationality. The latter problem is closely connected with the order-reduction problem, i.e., to find the lowest possible order for Lagrangians of a given Euler-Lagrange form. In the previous section we have seen that the main results concerning *local* properties of the Euler-Lagrange mapping in higher-order mechanics can be easily formulated in terms of Lepagean two-forms and can be proved by elementary methods—cf. Corollary 4.3.8 for the local triviality of Lagrangians, Corollary 4.3.9 for the local inverse problem, and in the next section Theorem 4.5.5 for the local order-reduction of Lagrangians (for a more detailed and selfcontained exposition of this subject the reader can consult O. Krupková [5]). To all of the above mentioned local problems one can consider their *global* counterparts. In this way, the problem of global triviality of Lagrangians means to find all global Lagrangians belonging to the kernel of the Euler-Lagrange mapping (i.e., giving rise to the zero Euler-Lagrange form). The global inverse problem of the calculus of variations then means to characterize obstructions for existence of global Lagrangians, and the global form of the order-reduction problem means to find the lowest possible order for global Lagrangians, resp. to characterize obstructions for existence of global Lagrangians of the minimal possible order.

Both global and local problems associated with the Euler-Lagrange mapping can be studied with help of the *variational bicomplex*, or with help of the *variational sequence*.

The idea to find a suitable "variational complex", analogous to the well-known de Rham complex, which would contain the Euler-Lagrange mapping as one of its morphisms, belongs to P. Dedecker. Explicit constructions of variational bicomplex appear in the papers by P. Dedecker and W. M. Tulczyjew [1], L. Haine [1], Yu. I. Manin [1], F. Takens [1], W. M. Tulczyjew [2,3], B. Kupershmidt [1], I. Anderson and T. Duchamp [1], A. M. Vinogradov [1,2], I. Anderson [1,2], and others. Further studies in the theory of variational bicomplex have shown its importance not only for the calculus of variations, but also for differential geometry, theory on differential equations, and for mathematical physics. For a survey of the theory of variational bicomplex and its applications in the calculus of variations and the theory of differential equations we refer to I. Anderson [2], A. M. Vinogradov [3] and T. Tsujishita [1].

The variational bicomplex is defined to be a certain double complex of differential forms on the *infinite jet prolongation* of a fibered manifold. One of the differentials in the variational bicomplex identifies with the Euler-Lagrange operator, another one leads to an intrinsic formulation of the Helmholtz conditions. The reason for the choice of infinite jets in the theory of variational bicomplex consists in a relative simplicity of these spaces in comparison with finite-order jet spaces. In particular, on infinite jets, mappings and differential oprators which increase the order of differential forms, can be used freely. Also, the infinite jet structure enables one to use bases of one-forms consisting of semibasic and contact elements only; such bases then generate canonical decompositions of the spaces of differential forms into direct sums of certain horizontal and contact subspaces. It was probably due to the non-existence of analogous constructions on finite-order jet spaces that one did not succeed to introduce a "finite-order variational complex". At the same time, infinite jet constructions are not suitable for study of some important problems, such as e.g. the order-reduction of Lagrangians. The solution of the latter problem is important not only for the calculus of variations, but also for the theory of differential equations (boundary value problems) and for physics (kinetic energy in mechanics). In 1990, D. Krupka [9] introduced the *variational sequence* on *finite-order* jet spaces, admitting subtle order considerations. The variational sequence differs from the constructions of the variatonal bicomplex, being a *quotient sequence* of the de Rham sequence of sheaves. The Euler-Lagrange mapping then arises as a quotient mapping of the exterior derivative operator. Investigations concerning the variational sequence can be found in the papers by D. Krupka [9–13], J. F. Pommaret [1], J. Štefánek [1–3], R. Vitolo [1], and J. Musilová [1].

The global inverse problem of the calculus of variations was first solved in the papers by F. Takens [1], and A. M. Vinogradov [2], proving that the obstructions to the existence of global Lagrangians lie in $H^{n+1}(Y)$, the $(n+1)$-st de Rham cohomology group of the total space Y, where n is the dimension of the base. If a form is globally variational there arises the question about the *order* of the global Lagrangian. It has been shown in I. Anderson and T. Duchamp [1], I. Anderson [2], and D. Krupka [9], that if E on J^sY is globally variational then there exists a global Lagrangian of order s for E. Papers on the global inverse problem contain an implicit solution of the local inverse problem, based on the local exactness of the corresponding sequence. The explicit solution, however, was obtained by other methods, independently by I. Anderson and T. Duchamp [1], and D. Krupka [4]. It seems that in the general case to get explicit local results is more difficult than to obtain their global counterparts. This concerns also both the local triviality and the local order-reducibility problems mentioned above. For the former we refer to D. Krupka and J. Musilová [2], and D. R. Grigore [3], a contribution to the latter can be found in D. Krupka [10].

It should be pointed out that a more general setting of the inverse problem of the calculus of variations is possible, namely, studying under what conditions given equations are *equivalent* with some Euler-Lagrange equations (equivalence in the sense that they have the same set of solutions, or even more generally, in the sense that their solutions are in 1-1 correspondence). A discussion of this problem is postponed to Chapter 6.

4.5. Canonical form of Lepagean two-form and minimal-order Lagrangians

In this section we shall show that every Lepagean two-form can be locally expressed in a certain canonical form—a kind of normal form of a two-form. This important property is closely related to the possibility of lowering the order of the corresponding Lagrangians and has many applications in integration theories.

4.5.1. Conventions. (1) For simplicity of notation, when speaking about a dynamical or locally variational form E *on* J^sY, we shall suppose that E is not projectable, i.e., at least one of its components is *nontrivially of order* s.

(2) In what follows we denote

$$s = 2c \qquad \text{if } s \text{ is even,}$$
$$s = 2c + 1 \qquad \text{if } s \text{ is odd.}$$

4.5.2. Theorem (Canonical form of Lepagean two-form). *Let E be a locally variational form of order s, $s \geqslant 1$, let α_E be its Lepagean equivalent. Then there is an open covering $\mathcal{O}$ of $J^{s-1}Y$ such that*

(1) *for each $W \in \mathcal{O}$ there is a fiber chart (V, ψ) on Y such that $W \subset V_{s-1}$,*

(2) *on each $W \in \mathcal{O}$ there are defined functions H, p^k_ν, $1 \leqslant \nu \leqslant m$, $0 \leqslant k \leqslant s-c-1$, such that the restriction of α_E to W is expressed in the form*

$$\alpha_E = -dH \wedge dt + \sum_{k=0}^{s-c-1} dp^k_\nu \wedge dq^\nu_k . \tag{4.32}$$

Proof. Since the form α_E is closed, there exists a covering $\mathcal{O}$ satisfying (1) such that on each $W \in \mathcal{O}$ it holds $\alpha_E = d\rho$ for a one-form ρ defined on W. Using Proposition 4.3.2 and the formula (2.11) we obtain (up to a projection)

$$\rho = A\alpha_E = \left(q^\sigma \int_0^1 (E_\sigma \circ \chi_s)\, du\right)dt + \sum_{k=0}^{s-1}\left(\sum_{j=0}^{s-1} 2q^\sigma_j \int_0^1 (F^{jk}_{\sigma\nu} \circ \chi_{s-1})\, u\, du\right)\omega^\nu_k . \tag{4.33}$$

We shall show that there are functions f, H, p^k_ν, $1 \leqslant \nu \leqslant m$, $0 \leqslant k \leqslant s-c-1$, on W such that (4.33) can be equivalently expressed in the form

$$\rho = -H\, dt + \sum_{k=0}^{s-c-1} p^k_\nu\, dq^\nu_k + df . \tag{4.34}$$

Consider the mapping $\chi_{s-1,s-c} : [0,1] \times W \to W$,

$$\chi_{s-1,s-c}(v, (t, q^\sigma, \dots, q^\sigma_{s-1})) = (t, q^\sigma, \dots, q^\sigma_{s-c-1}, vq^\sigma_{s-c}, \dots, vq^\sigma_{s-1}) . \tag{4.35}$$

Put

$$\begin{aligned} f = &\sum_{k=s-c}^{s-1} \sum_{j=0}^{s-1-k} 2q^\nu_k q^\sigma_j \int_0^1 \left(\int_0^1 (F^{jk}_{\sigma\nu} \circ \chi_{s-1})\, u\, du\right) \circ \chi_{s-1,s-c}\, dv \\ &+ \phi(t, q^\rho, \dots, q^\rho_{s-c-1}) , \end{aligned} \tag{4.36}$$

where ϕ is an arbitrary but fixed function, and define

$$p_\nu^k = \sum_{j=0}^{s-k-1} 2q_j^\sigma \int_0^1 (F_{\sigma\nu}^{jk} \circ \chi_{s-1})\, u\, du - \frac{\partial f}{\partial q_k^\nu}, \quad 1 \leqslant \nu \leqslant m, \quad 0 \leqslant k \leqslant s-c-1, \tag{4.37}$$

$$H = -q^\sigma \int_0^1 (E_\sigma \circ \chi_s)\, du + \sum_{k=0}^{s-1} \sum_{j=0}^{s-k-1} 2q_j^\sigma q_{k+1}^\nu \int_0^1 (F_{\sigma\nu}^{jk} \circ \chi_{s-1})\, u\, du + \frac{\partial f}{\partial t}. \tag{4.38}$$

Using the identity

$$\frac{\partial f}{\partial q_k^\nu} = \sum_{j=0}^{s-k-1} 2q_j^\sigma \int_0^1 (F_{\sigma\nu}^{jk} \circ \chi_{s-1})\, u\, du, \quad s-c \leqslant k \leqslant s-1, \tag{4.39}$$

we get

$$H = -q^\sigma \int_0^1 (E_\sigma \circ \chi_s)\, du + \sum_{k=0}^{s-c-1} p_\nu^k q_{k+1}^\nu + \frac{df}{dt}. \tag{4.40}$$

Substituting into (4.34) we obtain (4.33). This completes the proof. □

4.5.3. Definition. The expression of α_E in the form (4.32) will be called *canonical form* of the Lepagean two-form α_E, or, *canonical form* of the Lagrangean system α_E.

For any fixed function $\phi(t, q^\rho, \ldots, q_{s-c-1}^\rho)$, the functions H and p_ν^k, $1 \leqslant \nu \leqslant m$, $0 \leqslant k \leqslant s-c-1$, defined by (4.38) and (4.37) will be called the *Hamilton function* (or *Hamiltonian*) and *momenta* of the locally variational form E/of the Lagrangean system α_E.

By Theorem 4.5.2, Hamiltonian and momenta are functions related *directly* to given *variational equations* (to the given *Lagrangean system*), i.e., they refer to the whole class of equivalent Lagrangians. We stress that different choices of ϕ's in (4.36) lead to different sets of Hamiltonian and momenta for a given Lagrangean system, but all of them are of order $s-1$ (where s is the order of the equations).

4.5.4. Proposition. *Let E be a locally variational form on J^sY, $(p_\sigma, \ldots, p_\sigma^{s-c-1})$ momenta of E. Then for $0 \leqslant k \leqslant s-c-1$, $1 \leqslant \sigma \leqslant m$,*

$$\begin{aligned} \frac{\partial p_\nu^k}{\partial q_i^\sigma} - \frac{\partial p_\sigma^i}{\partial q_k^\nu} &= 2F_{\sigma\nu}^{ik}, \quad 0 \leqslant i \leqslant s-c-1, \\ \frac{\partial p_\nu^k}{\partial q_i^\sigma} &= 2F_{\sigma\nu}^{ik}, \quad s-c \leqslant i \leqslant s-1, \end{aligned} \tag{4.41}$$

where the $F_{\sigma\nu}^{ik}$'s are given by (4.10).

Proof. The formulas (4.41) are easily obtained from the definition of momenta (4.37) if one uses (4.16) and the following relation

$$F_{\sigma\nu}^{jk} = \int_0^1 d(u^2 F_{\sigma\nu}^{jk} \circ \chi_{s-1}) = 2\int_0^1 (F_{\sigma\nu}^{jk} \circ \chi_{s-1}) u\, du + \sum_{l=0}^{s-1} q_l^\rho \int_0^1 \Big(\frac{\partial F_{\sigma\nu}^{jk}}{\partial q_l^\rho} \circ \chi_{s-1}\Big) u^2 du.$$

Now, we shall prove an important property of variational equations.

Let λ_1 and λ_2 be two equivalent Lagrangians, λ_1 of order k and λ_2 of order r. If $k > r$, we shall say that the order of the Lagrangian λ_1 *can be reduced* to r.

4.5.5. Theorem (Order-reduction). *Every Lagrangian can be locally reduced to the lowest possible order, i.e., to the order $s/2$ if the order s of the Euler-Lagrange equations is even, and to $(s+1)/2$ if the order s of the Euler-Lagrange equations is odd.*

Proof. Since any Lagrangian is equivalent with a Vainberg-Tonti Lagrangian, it is sufficient to prove the order-reducibility property for Vainberg-Tonti Lagrangians.

Let λ be the Vainberg-Tonti Lagrangian of a locally variational form E on J^sY. By (4.6), λ is defined on an open subset of J^sY. Consider f defined by (4.36). Putting

$$\lambda_{\min} = \lambda - h\,df \tag{4.42}$$

one gets, for any choice of ϕ, a local Lagrangian equivalent with λ. It is an easy exercise to show that this Lagrangian is of order $c = s/2$ (resp. $c + 1 = (s+1)/2$) if s is even (resp. odd). □

In particular, in the frequently used case of $s = 2$ the formula (4.42) for the order-reduction of a Vainberg-Tonti Lagrangian to an equivalent first order Lagrangian takes the form

$$L_{\min} = q^\sigma \int_0^1 (E_\sigma \circ \chi_2)\, du - \frac{d}{dt}\Big(q^\sigma q_1^\nu \int_0^1 \Big(\int_0^1 (B_{\sigma\nu} \circ \chi_1)\, u\, du\Big) \circ \chi_{1,1}\, dv\Big) + \frac{d\phi}{dt}, \tag{4.43}$$

where ϕ is a function of (t, q^ν), and

$$\begin{aligned} \chi_2(u, t, q^\sigma, q_1^\sigma, q_2^\sigma) &= (t, uq^\sigma, uq_1^\sigma, uq_2^\sigma), \\ \chi_1(u, t, q^\sigma, q_1^\sigma) &= (t, uq^\sigma, uq_1^\sigma), \\ \chi_{1,1}(v, t, q^\sigma, q_1^\sigma) &= (t, q^\sigma, vq_1^\sigma). \end{aligned}$$

Lagrangians of the lowest possible order will be called *minimal-order Lagrangians*.

The problem on local order-reduction of a Lagrangian in higher-order mechanics has been solved by A. L. Vanderbauwhede [1], using Vainberg's potential operator methods. We have taken the proof from O. Krupková [2], where the theorem has been first obtained by means of methods of differential geometry. In field theory one does not have an analogous theorem; some results concerning local order-reducibility of Lagrangians can be found in D. Krupka [10].

Note that now it is easy to see the meaning of the free function ϕ in the formulas (4.36)–(4.38); namely, *ϕ is the gauge function parametrizing the class of equivalent minimal-order Lagrangians.*

4.5.6. Remark. If $s = 2c$ (even-order Euler-Lagrange equations/odd-order Lagrangean systems) then minimal-order Lagrangians are of the form $\lambda = L\,dt$, where L is a function of $(t, q^\sigma, \ldots, q_c^\sigma)$, satisfying the condition

$$\frac{\partial^2 L}{\partial q_c^\sigma \partial q_c^\nu} \neq 0 \quad \text{for at least one } \sigma,\ \nu\,. \tag{4.44}$$

If $s = 2c + 1$ (odd-order Euler-Lagrange equations/even-order Lagrangean systems) then minimal-order Lagrangians are of the form $\lambda = L(t, q^\sigma, \dots, q^\sigma_c, q^\sigma_{c+1})\, dt$, where L satisfies the conditions

$$\begin{aligned} \frac{\partial^2 L}{\partial q^\sigma_{c+1} \partial q^\nu_{c+1}} &= 0 \quad \text{for all } \sigma,\ \nu, \\ \frac{\partial^2 L}{\partial q^\sigma_c \partial q^\nu_{c+1}} - \frac{\partial^2 L}{\partial q^\sigma_{c+1} \partial q^\nu_c} &\neq 0 \quad \text{for at least one } \sigma,\ \nu. \end{aligned} \tag{4.45}$$

In other words, minimal-order Lagrangians of even-order Lagrangean systems are *affine* in the highest derivatives, i.e., of the form

$$L_{\min} = a + b_\sigma q^\sigma_{c+1}, \tag{4.46}$$

with a, b_σ not depending upon the q^ρ_{c+1}'s, and satisfying

$$\frac{\partial b_\nu}{\partial q^\sigma_c} - \frac{\partial b_\sigma}{\partial q^\nu_c} \neq 0.$$

4.5.7. Corollary. *Let E be a locally variational form on $J^s Y$, let λ be a (local) Lagrangian of E of order $r \leqslant s$. Then the Lepagean equivalent θ_λ of λ is projectable onto an open subset of $J^{s-1}Y$.*

Proof. If λ is a (local) Lagrangian of order $r \leqslant s$ then it holds $\lambda = \lambda_{\min} + h\, df$ where $\lambda_{\min}$ is a minimal-order Lagrangian (i.e., of order $s - c$), and f is a function of order $s - 1$. Hence, $\theta_\lambda = \theta_{\lambda_{\min}} + df$ which is of order $2(s - c) - 1$, i.e., of order $s - 1$ if $s = 2c$ and of order s if $s = 2c + 1$. Now, it remains to consider the latter case. By the above remark, $L_{\min}$ is of the form (4.46); this means that

$$\theta_{\lambda_{\min}} = (a + b_\sigma q^\sigma_{c+1})dt + \sum_{i=1}^{c-1} f^{i+1}_\sigma \omega^\sigma_i + b_\sigma \omega^\sigma_c = a\, dt + \sum_{i=1}^{c-1} f^{i+1}_\sigma \omega^\sigma_i + b_\sigma dq^\sigma_c\,,$$

and since $f^1_\sigma, \dots, f^c_\sigma$ are of order $2c = s - 1$, the $\theta_{\lambda_{\min}}$ is projectable onto $J^{s-1}Y$. Consequently, θ_λ is of order $s - 1$. □

We shall show an important property of Hamiltonians and momenta of locally variational forms; namely, that the formulas (4.38) and (4.37) can be interpreted in a way which relates them to (local) *minimal-order Lagrangians*, and that the "non-uniqueness" of Hamiltonians and momenta corresponds to the equivalence property of such "admissible" Lagrangians. On the one hand, this will make the practical computations more convenient, and on the other hand, this will clarify relations to standard definitions of Hamiltonians and momenta used in classical mechanics.

4.5.8. Theorem. *Let E be a locally variational form on $J^s Y$.*

(1) *Let H, p^k_ν, $1 \leqslant \nu \leqslant m$, $0 \leqslant k \leqslant s - c - 1$, be a Hamiltonian and momenta of E. Then there exists a minimal-order Lagrangian $\lambda_{\min}$ for E such that*

$$\theta_{\lambda_{\min}} = -H\, dt + \sum_{k=0}^{s-c-1} p^k_\nu\, dq^\nu_k\,. \tag{4.47}$$

(2) *Let $\lambda_{\min}$ be a minimal-order Lagrangian for E,*

$$\theta_{\lambda_{\min}} = L_{\min}\, dt + \sum_{k=0}^{s-c-1} (f_{\min})_{\nu}^{k+1} \, (dq_k^\nu - q_{k+1}^\nu dt)$$

its Lepagean equivalent. Then

$$\begin{aligned} H &= -L_{\min} + \sum_{k=0}^{s-c-1} (f_{\min})_{\nu}^{k+1} \, q_{k+1}^\nu \,, \\ p_\nu^k &= (f_{\min})_{\nu}^{k+1} \,, \quad 1 \leqslant \nu \leqslant m \,, \quad 0 \leqslant k \leqslant s-c-1 \end{aligned} \tag{4.48}$$

are a Hamiltonian and momenta of E.

Proof. (1) According to the proof of Theorem 4.5.2 we have

$$-H\, dt + \sum_{k=0}^{s-c-1} p_\nu^k \, dq_k^\nu = \rho - df \,,$$

where ρ is the Lepagean equivalent of the Vainberg-Tonti Lagrangian (cf. (4.33)), and f is defined by (4.36). This means that the above one-form is the Lepagean equivalent of the Lagrangian $h\rho - hdf$, which by Theorem 4.5.5, is a minimal-order Lagrangian for E.

(2) In the notation of the proof of Theorem 4.5.5, $\theta_{\lambda_{\min}} = \rho - df$, where ρ is the Lepagean equivalent of the Vainberg-Tonti Lagrangian of E. Now, using (4.34), we get (4.48). □

By the above Theorem, we get for momenta and the corresponding Hamiltonian of a Lagrangean system α the following formulas equivalent to (4.37)–(4.38):

$$\begin{aligned} p_\nu^k &= \sum_{j=0}^{s-c-k-1} (-1)^j \frac{d^j}{dt^j} \frac{\partial L_{\min}}{\partial q_{k+1+j}^\nu} \,, \quad 1 \leqslant \nu \leqslant m \,, \; 0 \leqslant k \leqslant s-c-1 \,, \\ H &= -L_{\min} + \sum_{k=0}^{s-c-1} p_\nu^k \, q_{k+1}^\nu \,, \end{aligned} \tag{4.49}$$

where $\lambda_{\min} = L_{\min} dt$ is a minimal-order Lagrangian for α.

4.5.9. Remark. We have seen that to a locally variational form of order s (to a Lagrangean system of order $s-1$) one can canonically assign certain systems of $m(s-c)+1$ functions of order $s-1$ (parametrized by a free gauge function ϕ of order $s-c-1$), each system consisting from one Hamiltonian and $m(s-c)$ momenta. Hamiltonians and momenta of a locally variational form are in one-to-one correspondence with its minimal-order Lagrangians. Moreover, it is easy to see that two systems (H, p_ν^k) and (H', p'^k_ν) of a Hamiltonian and momenta defined on an open subset of $J^{s-1}Y$ are *equivalent* (in the sense that they correspond to the same locally variational form E) if and only if

$$H' = H + \frac{\partial \phi}{\partial t} \,, \quad p'^k_\nu = p_\nu^k - \frac{\partial \phi}{\partial q_k^\nu} \tag{4.50}$$

for a gauge function $\phi(t, q^\rho, \ldots, q^\rho_{s-c-1})$. The corresponding minimal-order Lagrangians λ, λ' then evidently satisfy the relation $L' = L + d\phi/dt$.

Let us stop for a moment to discuss an important particular family of Lagrangean systems which are frequently considered in higher-order mechanics, namely, the *time-independent Lagrangean systems*.

Consider a fibered manifold $\pi : R \times M \to R$. Suppose that a Lagrangean system of order $s-1$ on π possesses local *time-independent minimal-order Lagrangians* (i.e., such that $\partial L_{\mathrm{min}}/\partial t = 0$, where t is the global coordinate on R); such a Lagrangean system is called *autonomous*, or *time-independent*. Denote by r the minimal order (recall that $r = s/2$ if s is even, and $r = (s+1)/2$ if s is odd). Then the class of time-independent minimal-order Lagrangians is of the form $L'_{\mathrm{min}} = L_{\mathrm{min}} + d\phi/dt$, where ϕ is an affine function of the time, $\phi = ct + g(q^\sigma_j)$, where $c = \mathrm{const}$ and $0 \leqslant j \leqslant r-1$. Now, from (4.50) we get that Hamiltonians corresponding to different equivalent time-independent minimal-order Lagrangians differ only by a constant. Consequently, using (4.38) we have

4.5.10. Proposition. *A time-independent Lagrangean system of order $s-1$, $s \geqslant 1$, on a fibered manifold $\pi : R \times M \to R$ possesses a unique up to a constant time-independent Hamiltonian.*

In fibered coordinates we have this Hamiltonian represented in the form

$$H = -q^\sigma \int_0^1 (E_\sigma \circ \chi_s)\, du + \sum_{k=0}^{s-1} \sum_{j=0}^{s-k-1} 2 q^\sigma_j \, q^\nu_{k+1} \int_0^1 (F^{jk}_{\sigma\nu} \circ \chi_{s-1})\, u\, du + c. \tag{4.51}$$

The Hamiltonian (4.51) will be called the *total energy* of the time-independent Lagrangean system α_E.

In other words, time-independent Lagrangean systems on $R \times T^{s-1}M$ are characterized by a time-independent Hamiltonian, which is *unique up to a constant*.

On the other hand, for *time-dependent* Lagrangean systems on $\pi : R \times M \to R$, or for (any) Lagrangean systems on a more general fibered manifold one has no distinguished Hamiltonians.

In this section, we have introduced Hamiltonian and momenta as functions defined on the phase space, associated directly to *variational equations*, or equivalently, to *minimal-order Lagrangians*. Notice that this understanding of Hamiltonian and momenta is in correspondence with the classical calculus of variations where Hamiltonian and momenta are functions related to a given *first-order Lagrangian* (which, being of the form $T - V$, clearly is a *minimal-order* Lagrangian). Unfortunately, higher-order generalizations of these classical concepts concentrated to a more or less heuristic generalization of the *formulas* $p = \partial L/\partial q$ and $H = -L + p\dot{q}$ to the case when L depends on higher derivatives. Within this approach, Hamiltonian and momenta were introduced as functions related to a *Lagrangian*. More precisely, for a Lagrangian L of order r, the Hamiltonian and momenta were defined by the following formulas, originally due to Ostrogradskii:

$$\begin{aligned} \hat{p}^k_\nu &= \sum_{j=0}^{r-k-1} (-1)^j \frac{d^j}{dt^j} \frac{\partial L}{\partial q^\nu_{k+1+j}}, \quad 1 \leqslant \nu \leqslant m,\ 0 \leqslant k \leqslant r-1, \\ \hat{H} &= -L + \sum_{k=0}^{r-1} \hat{p}^k_\nu q^\nu_{k+1} \end{aligned} \tag{4.52}$$

(see eg. Th. De Donder [1], P. Dedecker [7], M. de León and P. R. Rodrigues [1], D. Krupka and J. Musilová [1], M. Ferraris and M. Francaviglia [2], and others). Notice that *the order* and also *the number* of these functions depends on the order of the particular Lagrangian. Since a definition of Hamiltonian and momenta usually represents a key to Hamilton theory and the Hamilton-Jacobi theory, as well as to the possibility of quantization, this approach has strongly influenced on the research in this field in the last 40 years.

Let us compare now the Hamiltonians and momenta associated to variational equations with the Hamiltonians and momenta of particular Lagrangians (4.52). Let L be a Lagrangian of order r, defining a variational form E of order s. If $\hat{H}$ and $\hat{p}$ are the Ostrogradskii Hamiltonian and momenta of L, and H and p are Hamiltonian and momenta of E defined by (4.38) and (4.37) (or by (4.49)), we have

$$\hat{p}^k_\nu = \frac{\partial f}{\partial q^\nu_k}, \quad r-1 \geqslant k \geqslant s-c,$$

$$\hat{p}^k_\nu = p^k_\nu + \frac{\partial f}{\partial q^\nu_k}, \quad s-c-1 \geqslant k \geqslant 0,$$

$$\hat{H} = H - \frac{\partial f}{\partial t},$$

for a function $f(t, q^\rho, \ldots, q^\rho_{r-1})$. Clearly, $\hat{p}$'s and $\hat{H}$'s coincide with the p's and H's on the set of *minimal-order Lagrangians*. For the other Lagrangians the $\hat{H}$'s and $\hat{p}$'s represent more functions. Moreover, while (H, p) are of order $s-1$, the order of $(\hat{H}, \hat{p})$ depends on the order of the Lagrangian L and, if $r > s$, it is greater than $s-1$. Summarizing, (compared with our concept of Hamiltonian and momenta related to the class of equivalent Lagrangians) for Lagrangians of order greater than the minimal one, one gets from the Ostrogradskii definition (4.52) *more* functions, living generally *higher* than on the space where the dynamics proceeds; this can lead to problems and confusion if one wants to introduce the Hamilton formalism and to understand the concepts of regularity, Legendre transformation, or to study constrained systems.

To illustrate the basic differences on an example, let us consider the following *third order Lagrangians*:

$$L_1 = \tfrac{1}{2}\dot{q}^2 + \ddot{q}^2 + \dddot{q}\, q,$$

$$L_2 = \dot{q}^2(1+\ddot{q}) + t\dot{q}\ddot{q} + q\ddot{q}^2 + q\dot{q}\,\dddot{q}.$$

Computing the momenta and the Hamiltonian of these *Lagrangians* we get

$$\hat{p}_0 = \dot{q}, \quad \hat{p}_1 = \ddot{q}, \quad \hat{p}_2 = q, \quad \hat{H} = \tfrac{1}{2}\dot{q}^2$$

for the Lagrangian L_1, and

$$\hat{p}_0 = \dot{q}(1+\ddot{q}), \quad \hat{p}_1 = q\ddot{q} + t\dot{q}, \quad \hat{p}_2 = q\dot{q}, \quad \hat{H} = 0,$$

for the Lagrangian L_2. Note that in this way, we have assigned to any of these Lagrangians one Hamiltonian and *three* momenta, which are functions of order 2 (depending on *accelerations*). On the other hand, let us compute Hamiltonians and momenta of the *Euler-Lagrange form* corresponding to these Lagrangians. First note that the Lagrangians L_1, L_2 are *equivalent* and *reducible to the first order*. This can be seen at once e.g. if one computes the corresponding Lepagean two-forms: one gets

$$d\theta_{\lambda_1} = -\dot{q}\, d\dot{q} \wedge dt + d\dot{q} \wedge dq = d\theta_{\lambda_2}.$$

In other words, both the Lagrangians define the same *first order Lagrangean system*. This means, however, that any family of momenta of this Lagrangean system consists of *one* momentum only and that all Hamiltonians and momenta are of order *one* (depending on velocities, but not on the accelerations). Since

$$L_1 = L_0 + \frac{d}{dt}(\dot{q}\ddot{q}),$$
$$L_2 = L_0 + \frac{d}{dt}\left(\tfrac{1}{2}t\dot{q}^2 + q\dot{q}\ddot{q}\right),$$

where $L_0 = \dot{q}^2/2$, we get

$$p = \dot{q} + \frac{\partial \phi}{\partial q}, \quad H = \tfrac{1}{2}\dot{q}^2 - \frac{\partial \phi}{\partial t},$$

where ϕ is any function of (t, q).

In the following chapters we shall see that since Hamiltonians and momenta related to the class of equivalent Lagrangians/locally variational form live on the space where the dynamics takes place, they have a transparent geometrical meaning, and can be used to simplify integration of the Euler-Lagrange equations.

The meaning of the Ostrogradski momenta and Hamiltonian will be discussed in the next chapter.

4.6. Lower-order Lagrangean systems

We have seen that by Definition 4.3.12, the *order* of a Lagrangean system is defined as the order of the corresponding Lepagean two-form α (not as the order of a particular Lagrangian), and it can be easily derived from the order of the corresponding locally variational form $E = p_1\alpha$. Namely, if E is of order s then we have to do with a Lagrangean system of order $s - 1$.

Let us illustrate this property on two interesting particular cases of Lagrangians:

4.6.1. Lagrangians affine in the velocities: zero-order Lagrangean systems. By Definition 4.3.12, a Lagrangean system α is *of order zero* (i.e., defined on the total space Y of the fibered manifold π) if and only if the corresponding locally variational form $E = p_1\alpha$ is of order one. *Minimal-order Lagrangians* for a zero-order Lagrangean system are of order one; more precisely, it is easy to check that a first order Lagrangian λ leads to Euler-Lagrange equations of order *one* if and only if $\lambda = L\,dt$ where the function $L = L(t, q^\rho, q_1^\rho)$ satisfies the following *reducibility condition*

$$\frac{\partial^2 L}{\partial q_1^\sigma \partial q_1^\nu} = 0 \quad \text{for all} \quad \sigma, \nu. \tag{4.53}$$

Equivalently, L is *affine in the velocities*, i.e., of the form

$$L = a(t, q^\rho) + b_\sigma(t, q^\rho) q_1^\sigma. \tag{4.54}$$

Hamiltonians and momenta of a zero-order Lagrangean system defined by the Lagrangian (4.54) are according to (4.49) given by

$$p_\nu = b_\nu - \frac{\partial \phi}{\partial q_\nu}, \quad H = -a + \frac{\partial \phi}{\partial t}, \tag{4.55}$$

where $\phi(t, q^\sigma)$ is an arbitrary gauge function of order zero. Note that all the Hamiltonians and momenta are functions *of order zero* (living on Y).

It should be stressed that, according to Definition 4.3.12, the *phase space* of zero-order Lagrangean systems is the *total space* Y of the fibered manifold π.

Although in textbooks an emphasis is put on studying first and higher-order Lagrangean systems, zero-order Lagrangean systems often appear in applications, and, due to recent research, they seem to be even more fundamental than the first and higher-order systems. The reasons are the following: First, as we shall see in the next chapter, every Lagrangean system of any order can be represented as a zero-order Lagrangean system (Hamilton theory). Next, many mechanical systems (not necessarily Lagrangean) can be represented as zero-order *Lagrangean* systems (this will be shown in Section 6.7). Finally, for zero-order Lagrangean systems an effective and relatively easy quantization method has been developed (see R. Jackiw [1]). Last but not least there are *true* zero-order Lagrangean systems in physics, namely in field theory (e.g. the Dirac field).

4.6.2. Lagrangians affine in the accelerations: second and first-order Lagrangean systems. A general second order Lagrangian leads to Euler-Lagrange equations of order 4; the equations are of order 3 if and only if

$$\frac{\partial^2 L}{\partial q_2^\sigma \partial q_2^\nu} = 0 \quad \text{for all} \quad \sigma, \nu, \tag{4.56}$$

i.e., L is *affine in the accelerations*. Denote

$$L = f(t, q^\rho, q_1^\rho) + g_\sigma(t, q^\rho, q_1^\rho) q_2^\sigma. \tag{4.57}$$

If at least for one σ, ν,

$$\frac{\partial g_\nu}{\partial q_1^\sigma} - \frac{\partial g_\sigma}{\partial q_1^\nu} \neq 0 \tag{4.58}$$

then the Euler-Lagrange equations are nontrivially of order 3, i.e., L defines a *second-order Lagrangean system*. Clearly, in this case, L is a *minimal-order Lagrangian* (i.e., it is not reducible). For a Hamiltonian and momenta of this Lagrangean system we get

$$\begin{aligned} p_\nu^1 &= g_\nu, \\ p_\nu^0 &= \frac{\partial f}{\partial q_1^\nu} + \frac{\partial g_\sigma}{\partial q_1^\nu} q_2^\sigma - \frac{d g_\nu}{dt} = \frac{\partial f}{\partial q_1^\nu} - \frac{\partial g_\nu}{\partial t} - \frac{\partial g_\nu}{\partial q^\sigma} q_1^\sigma + \left(\frac{\partial g_\sigma}{\partial q_1^\nu} - \frac{\partial g_\nu}{\partial q_1^\sigma} \right) q_2^\sigma, \\ H &= -f + p_\nu^0 q_1^\nu \end{aligned} \tag{4.59}$$

(modulo derivatives of a gauge function of order *one*). Note that all the Hamiltonians and momenta are *second order functions*, affine in the accelerations. The phase space is J^2Y.

If for all σ, ν

$$\frac{\partial g_\nu}{\partial q_1^\sigma} - \frac{\partial g_\sigma}{\partial q_1^\nu} = 0, \tag{4.60}$$

and

$$-\frac{\partial^2 f}{\partial q_1^\sigma \partial q_1^\nu} + \frac{\partial g_\nu}{\partial q^\sigma} + \frac{\partial g_\sigma}{\partial q^\nu} + \frac{\partial^2 g_\sigma}{\partial t \partial q_1^\nu} + \frac{\partial^2 g_\sigma}{\partial q^\rho \partial q_1^\nu} q_1^\rho \neq 0 \tag{4.61}$$

then the Euler-Lagrange equations are of order 2, i.e., L defines a *first-order Lagrangean system*. Now, however, *L is not a minimal-order Lagrangian*. Compute a Hamiltonian and momenta of this Lagrangean system. If L_0 is a minimal order Lagrangian, equivalent with L, we get $L = L_0 + dF/dt$, where

$$g_\sigma = \frac{\partial F}{\partial q_1^\sigma}, \quad L_0 = f - \frac{\partial F}{\partial t} - \frac{\partial F}{\partial q^\nu} q_1^\nu. \tag{4.62}$$

Now,

$$\begin{aligned} p_\nu &= \frac{\partial L_0}{\partial q_1^\nu} = \frac{\partial f}{\partial q_1^\nu} - \frac{\partial g_\nu}{\partial t} - \frac{\partial g_\nu}{\partial q^\sigma} q_1^\sigma - \frac{\partial F}{\partial q^\nu}, \\ H &= -L_0 + \frac{\partial L_0}{\partial q_1^\nu} q_1^\nu = -f + \frac{\partial f}{\partial q_1^\nu} q_1^\nu - \left(\frac{\partial g_\nu}{\partial t} + \frac{\partial g_\nu}{\partial q^\sigma} q_1^\sigma\right) q_1^\nu + \frac{\partial F}{\partial t} \end{aligned} \tag{4.63}$$

(modulo derivatives of a gauge function of order *zero*). Note that Hamiltonians and momenta do not depend on accelerations. The phase space is J^1Y. It is interesting to note that in field theory, the Hilbert-Einstein Lagrangian is precisely of the above type (see D. Krupka and O. Štěpánková [1]); we shall return to this example in Chapter 6.

If finally the Euler-Lagrange equations of (4.57) are of order 1 then L is reducible to a Lagrangian affine in the velocities, the arising Lagrangean system is *of order zero*, and Hamiltonian and momenta of E are of the form (4.55).

4.7. Transformation properties of Lepagean forms

Let ϕ be a local diffeomorphism of J^rY. If ρ is a Lepagean one-form (resp. two-form) of order r then $\phi^*\rho$ need not be Lepagean. However, we shall show that the situation is different if one takes an *isomorphism of the fibered manifold* π.

4.7.1. Theorem. *Let ϕ be an isomorphism of $\pi : Y \to X$.*

(1) *If ρ is a Lepagean one-form on J^rY then $J^r\phi^*\rho$ is a Lepagean one-form; the corresponding Lagrangians $h\rho$ and $hJ^r\phi^*\rho$ satisfy the relation*

$$h\, J^r\phi^*\rho = J^{r+1}\phi^*\, h\rho.$$

(2) *If α is a Lepagean two-form on J^rY then $J^r\phi^*\alpha$ is a Lepagean two-form; the corresponding locally variational forms $p_1\alpha$ and $p_1 J^r\phi^*\alpha$ satisfy the relation*

$$p_1\, J^r\phi^*\alpha = J^{r+1}\phi^*\, p_1\alpha.$$

Proof. Since α is closed, i.e., locally, $\alpha = d\rho$, the form $J^r\phi^*\alpha = J^r\phi^*\, d\rho = d\, J^r\phi^*\rho$ is also closed. Hence, to prove that $J^r\phi^*\rho$ is a Lepagean one-form, and $J^r\phi^*\alpha$ is a Lepagean two-form, one has to prove that $p_1 J^r\phi^* d\rho = p_1 J^r\phi^*\alpha$ is a dynamical form. This, however, can be done by a direct computation. The remaining two formulas follow from Propositions 2.6.1 and 2.6.2 (notice that since $J^r\phi$ transfers one-contact one-forms into one-contact one-forms, it transfers also i-contact forms into i-contact forms, hence, in particular, commutes with p_1). □

4.7.2. Corollary. *Let ϕ be an isomorphism of $\pi : Y \to X$.*

(1) *Let λ be a Lagrangian on J^rY, θ_λ its Lepagean equivalent. Then the forms $J^{2r-1}\phi^*\theta_\lambda$ and $\theta_{J^r\phi^*\lambda}$ are defined on the same open subset of $J^{2r-1}Y$, and*

$$\theta_{J^r\phi^*\lambda} = J^{2r-1}\phi^*\,\theta_\lambda.$$

(2) *Let E be a locally variational form on J^sY, let α_E be its Lepagean equivalent. Then the forms $J^{s-1}\phi^*\alpha_E$ and $\alpha_{J^s\phi^*E}$ are defined on the same open subset of $J^{s-1}Y$, and*

$$\alpha_{J^s\phi^*E} = J^{s-1}\phi^*\,\alpha_E.$$

As a consequence of Theorem 4.7.1. we get an important relation between the transformed Euler-Lagrange form of a Lagrangian and the Euler-Lagrange form of the transformed Lagrangian:

4.7.3. Theorem. *Let ϕ be an isomorphism of the fibered manifold $\pi : Y \to X$, λ a Lagrangian on J^rY. Then the forms $J^{2r}\phi^*E_\lambda$ and $E_{J^r\phi^*\lambda}$ are defined on the same open subset of $J^{2r}Y$, and*

$$J^{2r}\phi^*E_\lambda = E_{J^r\phi^*\lambda}\,.$$

Proof. Denote θ_λ the Lepagean equivalent of the Lagrangian λ. Then $J^{2r-1}\phi^*\theta_\lambda$ is the Lepagean equivalent of the Lagrangian $J^r\phi^*\lambda$. Now, for the Euler-Lagrange form of this Lagrangian we get

$$\begin{aligned} E_{J^r\phi^*\lambda} &= E_{J^{2r}\phi^*h\theta_\lambda} = E_{hJ^{2r-1}\phi^*\theta_\lambda} \\ &= p_1\,dJ^{2r-1}\phi^*\theta_\lambda = p_1\,J^{2r-1}\phi^*d\theta_\lambda = J^{2r}\phi^*\,p_1 d\theta_\lambda = J^{2r}\phi^*E_\lambda. \end{aligned}$$

□

We have also an "infinitesimal version" of the above assertions:

4.7.4. Theorem. *Let ξ be a π-projectable vector field on Y.*

(1) *If ρ is a Lepagean one-form on J^rY then $\partial_{J^r\xi}\rho$ is a Lepagean one-form and*

$$h\,\partial_{J^r\xi}\rho = \partial_{J^{r+1}\xi}\,h\rho.$$

(2) *If α is a Lepagean two-form on $J^{s-1}Y$ then $\partial_{J^{s-1}\xi}\alpha$ is a Lepagean two-form and*

$$p_1\,\partial_{J^{s-1}\xi}\alpha = \partial_{J^s\xi}\,p_1\alpha.$$

Proof. Obviously, to prove that $\partial_{J^r\xi}\rho$ is a Lepagean one-form, and $\partial_{J^r\xi}\alpha$ is a Lepagean two-form, it is sufficient to show that for every π-projectable vector field ξ on Y, the form $p_1\,di_{J^{s-1}\xi}\alpha$ is dynamical. Expressing α in the form (4.7), computing $p_1\,di_{J^{s-1}\xi}\alpha$, and using the identities (4.12)–(4.15) and the prolongation formula (2.3) for the components of ξ, we get the result. These computations also give us

$$p_1\,di_{J^{s-1}\xi}\alpha = i_{J^s\xi}\,dE + di_{J^s\xi}E,$$

where $E = p_1\alpha$, proving that $p_1\, \partial_{J^{s-1}\xi}\alpha = \partial_{J^s\xi}\, p_1\alpha$.

Finally, the relation $h\, \partial_{J^r\xi}\rho = \partial_{J^{r+1}\xi}\, h\rho$ is nothing but the infinitesimal first variation formula (3.17).

4.7.5. Theorem. *Let ξ be a π-projectable vector field on Y, λ a Lagrangian on J^rY. Then*

$$\partial_{J^{2r}\xi} E_\lambda = E_{\partial_{J^r\xi}\lambda}\,.$$

Proof. Using the assertion (1) of Theorem 4.7.4, we get

$$\partial_{J^{2r-1}\xi}\alpha_{E_\lambda} = di_{J^{2r-1}\xi}\alpha_{E_\lambda} = di_{J^{2r-1}\xi}d\theta_\lambda = d(\partial_{J^{2r-1}\xi}\theta_\lambda) = d\theta_{\partial_{J^r\xi}\lambda} = \alpha_{E_{\partial_{J^r\xi}\lambda}},$$

proving that

$$p_1\, \partial_{J^{2r-1}\xi}\alpha_{E_\lambda} = E_{\partial_{J^r\xi}\lambda}.$$

Now, Theorem 4.7.5 follows from the assertion (2) of Theorem 4.7.4. □

4.7.6. Corollary. *Let ξ be a π-projectable vector field on Y, λ a Lagrangian on J^rY. Then*

$$\partial_{J^{2r-1}\xi}\theta_\lambda = \theta_{\partial_{J^r\xi}\lambda}, \qquad \partial_{J^{2r-1}\xi}\alpha_{E_\lambda} = \alpha_{\partial_{J^{2r}\xi}E_\lambda} = \alpha_{E_{\partial_{J^r\xi}\lambda}}\,.$$

Transformation properties of Lepagean equivalents of Lagrangians under isomorphisms of fibered manifolds have been studied by D. Krupka in [2,5], where also Theorems 4.7.1 (1), 4.7.3, and 4.7.5 were obtained.

4.8. Holonomic constraints

Let $\pi : Y \to X$ be a fibered manifold, $\dim X = 1$. A submanifold $M \subset Y$ is called a *holonomic constraint* if $\dim M < m$, and the projection π restricted to M, i.e., $\pi|_M : M \to X$, is a fibered submanfiold of π. Let $\operatorname{codim} M = \dim Y - \dim M = k$. Then the submanifold M can be locally represented by k equations

$$f^i(t, q^\nu) = 0, \qquad 1 \leqslant i \leqslant k$$

such that

$$\operatorname{rank}\left(\frac{\partial f^i}{\partial q^\sigma}\right) = k.$$

This means that at each point $y \in Y$ there is a fiber chart $(\bar V, \bar\psi)$ *adapted* to the submanifold M. Keeping the above notations, we can take $\bar\psi = (t, q^1, \dots, q^{m-k}, f^1, \dots, f^k)$.

Consider a Lagrangean system α of oder $s-1$, where $s \geqslant 1$, on π. The constraint M naturally prolongs to a constraint $J^{s-1}M \subset J^{s-1}Y$. The equations of $J^{s-1}M$ are of the form

$$f^i = 0, \quad \frac{df^i}{dt} = 0, \quad \dots, \quad \frac{d^{s-1}f^i}{dt^{s-1}} = 0$$

(the original equations $f^i = 0$ and their derivatives up to the order $s-1$). Denote by ι the canonical embedding of M into Y, and by $J^{s-1}\iota$ its $(s-1)$th prolongation.

We have the following important assertion.

4.8.1. Theorem. *If α is a Lepagean two form on $J^{s-1}Y$ and $M \subset Y$ is a holonomic constraint then $J^{s-1}\iota^*\alpha$ is a Lepagean two-form on $J^{s-1}M$.*

Proof. Obviously, $J^{s-1}\iota^*\alpha$ is a closed two-form on $J^{s-1}M$. Hence, if $s = 1$, we are done. For $s > 1$ it remains to show that $p_1 J^{s-1}\iota^*\alpha$ is a dynamical form on $J^{s-1}M$. We have

$$p_1\, J^{s-1}\iota^*\alpha = J^s\iota^*\, p_1\alpha,$$

and the form on the right-hand side apparently is a dynamical form. This completes the proof. □

4.8.2. Corollary. *Let α be a Lagrangean system of order $s-1$, $s \geqslant 1$, E its related dynamical form. Let $M \subset Y$ be a holonomic constraint, $\iota : M \to Y$ the canonical embedding. Then $J^{s-1}\iota^*\alpha$ is a Lagrangean system of order $s-1$ on $\pi|_M$, and its related dynamical form is $J^s\iota^*E$. If λ is a (local) Lagrangian of order r for α then $J^r\iota^*\lambda$ is a (local) Lagrangian for $J^{s-1}\iota^*\alpha$.*

Proof. Since $E = p_1\alpha$, we have by the above theorem for the constrained dynamical form,

$$E' = p_1\, J^{s-1}\iota^*\alpha = J^s\iota^*\, p_1\alpha = J^s\iota^*E.$$

If λ is a Lagrangian of order r for α and θ_λ is its Cartan form, then (up to projection) $d\theta_\lambda = \alpha$, hence

$$J^{s-1}\iota^*\alpha = J^{2r-1}\iota^*\, d\theta_\lambda = d\, J^{2r-1}\iota^*\theta_\lambda = d\theta_{J^r\iota^*\lambda}.$$

□

As a consequence of the above results one can see that Lagrangean systems subject to holonomic constraints can be equivalently considered as unconstrained Lagrangean systems on a fibered submanifold of the original fibered manifold. This property of Lagrangean systems has been well known in classical mechanics for a long time. Within the classical terminology, the number $\dim M - 1$ is referred to as the *number of degrees of freedom* of the given mechanical system; if (t, q^σ) arc local coordinates on M then the q^σ's are called *generalized coordinates*.

In mechanics, often more general kinds of constraints are considered. For example, classical mechanics works with constraints on J^1Y which are represented by the equations

$$a^i_\sigma(t, q^\nu)\dot{q}^\sigma + b^i(t, q^\nu) = 0, \quad 1 \leqslant i \leqslant k,$$

which are a particular case of

$$f^i(t, q^\nu, \dot{q}^\nu) = 0, \quad 1 \leqslant i \leqslant k.$$

Such constraints are called *non-holonomic*. Unfortunately, these constraints do not save the variationality property. More precisely, a Lagrangean system if subject to non-holonomic constraints, can become non-Lagrangean. This means that to deal with non-holonomic constraints, one needs a more general approach to mechanics than that we have introduced so far, covering all (not only variational) mechanical systems. It should be stressed, however, that a complete geometric theory of non-holonomic systems still remanis an open subject of research. For more details on recent developments in this field the reader can consult e.g. papers by C. M. Marle [2], G. Giachetta [1], J. Koiller [1], J. F. Cariñena and M. F. Rañada [3], M. F. Rañada [4], W. Sarlet, F. Cantrijn and D. J. Saunders [1], M. de León and D. M. de Diego [3], P. Libermann [3], D. J. Saunders, W. Sarlet and F. Cantrijn [1], W. Sarlet [8,9], M. de León, J. C. Marrero and D. M. de Diego [2,3], O. Krupková [13], and references therein.

Chapter 5.

HAMILTONIAN SYSTEMS

5.1. Introduction

We have seen in Chapter 1 that the Euler-Lagrange equations

$$\frac{\partial L}{\partial q^\sigma} - \frac{d}{dt}\frac{\partial L}{\partial \dot{q}^\sigma} = 0 \tag{5.1}$$

which are a system of m second-order ODE for sections γ of a fibered manifold $\pi : Y \to X$ can be represented by means of a system of $2m$ *first-order* equations for sections δ of a fibered manifold $\pi_1 : J^1Y \to X$, called the *Hamilton equations*. Namely, introducing momenta

$$p_\sigma = \frac{\partial L}{\partial \dot{q}^\sigma}$$

we get the Euler-Lagrange equations in the form

$$\frac{dp_\sigma}{dt} = \frac{\partial L}{\partial q^\sigma} \tag{5.2}$$

reminding of the Newton equations $dp/dt = F$. This form of the Euler-Lagrange equations suggests one the idea to represent the Euler-Lagrange equations by means of a *vector field*. Apparently this would be possible if the q^σ and p_σ could be considered as independent: then equations (5.2) turn to be a part of equations for sections $t \to (t, q^\sigma, p_\sigma)$ of the fibered manifold π_1. The remaining set of equations then reads

$$\frac{dq^\sigma}{dt} = \dot{q}^\sigma \tag{5.3}$$

where the right-hand side is considered to be a function of the coordinates (t, q^ν, p_ν). Equations (5.2), (5.3) then describe the desired representation of the Euler-Lagrange equations (5.1) by means of a vector field (which, however, is not defined on the configuration space Y, but on its first jet prolongation). Apparently, they are *equivalent* with the Euler-Lagrange equations (5.1) in the sense that the solutions of (5.1) are in *one-to-one* correspondence with those of (5.2), (5.3). Equations (5.2), (5.3) can take a more convenient form if one introduces the Hamilton function

$$H = -L + p_\sigma \dot{q}^\sigma .$$

Then (5.2) and (5.3) take the familiar form

$$\frac{dq^\sigma}{dt} = \frac{\partial H}{\partial p_\sigma}, \quad \frac{dp_\sigma}{dt} = -\frac{\partial H}{\partial q^\sigma}, \tag{5.4}$$

called the *canonical form* of the Hamilton equations. We stress an essential feature of the above construction—namely that the momenta p_σ could be considered a part of new *coordinates* on J^1Y. This means that

$$\det\left(\frac{\partial p_\sigma}{\partial \dot{q}^\nu}\right) = \det\left(\frac{\partial^2 L}{\partial \dot{q}^\sigma \partial \dot{q}^\nu}\right) \neq 0. \tag{5.5}$$

The geometric interpretation of Hamilton equations as *equations for sections of the fibered manifold* π_1 was first noticed by Goldschmidt and Sternberg. In their setting, if λ is a Lagrangian and θ_λ is its Cartan form, then the Euler-Lagrange equations read as follows: a section $\gamma : X \to Y$ is an extremal if for every vector field ξ on J^1Y,

$$J^1\gamma^* i_\xi d\theta_\lambda = 0, \tag{5.6}$$

while the Hamilton equations are the equations

$$\delta^* i_\xi d\theta_\lambda = 0, \tag{5.7}$$

for sections $\delta : X \to J^1Y$. The latter equations, in principle, need not be equivalent with the Euler-Lagrange equations since they admit *more* solutions (not every section δ of π_1 is *holonomic*, i.e., of the form $\delta = J^1\gamma$ for a section γ of π). Indeed, it turns out that (5.7) becomes equivalent with (5.6) if one makes an additional assumption on *regularity*, supposing that the Lagrangian λ satisfies the condition (5.5). On the other hand, equations (5.7) were obtained without any regularity assumption, i.e., they represent an intrinsic form of Hamilton equations for an *arbitrary* Lagrangian, describing generally an *extended dynamics* on J^1Y.

From the above considerations one can become aware of another important feature of the Hamilton equations, namely that, in fact, they refer to the class of all *equivalent Lagrangians* rather than to a particular Lagrangian (this can be seen e.g. from (5.7) if one takes into account that the form $d\theta_\lambda$ is "the same" for all equivalent Lagrangians). Having in mind the definition of a Lagrangean system as a *Lepagean two-form* (Chapter 4), representing the class of equivalent Lagrangians, we conclude that generalized Hamilton equations should be of the form

$$\delta^* i_\xi \alpha = 0, \tag{5.8}$$

where α is a Lepagean two-form of order $s-1$ ($s \geqslant 1$), and δ refers to sections of the fibered manifold π_{s-1} (i.e., with the range in the phase space $J^{s-1}Y$ where the form α lives).

Such viewpoint, however, has fundamental consequences: First, one has to reappraise the concept of *regularity*. Recall that conventionally regularity is defined to be a property of *Lagrangians*. (For example, the Lagrangian $L = m\dot{q}^2/2$ is regular, satisfying the condition (5.5), whereas $L' = -mq\ddot{q}/2$ is considered singular, being linear in the highest derivatives. However, these Lagrangians are equivalent differing by a total derivative, and consequently, they give the same Euler-Lagrange expressions $-m\ddot{q}$.) Thus, the concept of regularity naturally should refer to a *Lagrangean system* (equivalently, to a *locally variational form*, or a *class of equivalent Lagrangians*), not to a particular Lagrangian. Next, Hamiltonian and momenta should be associated with a Lepagean two-form rather than with a particular Lagrangian (this viewpoint is in accordance with the definition of Hamiltonian and momenta in Chapter 4). Finally, the generalized Hamilton equations (5.8) apparently have a direct interpretation in terms of exterior differential systems, as *equations for integral sections of the characteristic distribution of the closed two-form* α.

Conventional Hamilton theory related to particular Lagrangians leads to many ambiguities namely if one deals with higher-order Lagrangians, or with first-order Euler-Lagrange equations. These difficulties transfer to quantum mechanics if one needs to quantize equivalent Lagrangians, possessing the same dynamics. On the other hand, a Hamilton theory, unique for the class of equivalent Lagrangians (independent of the choice of a gauge function) could serve as an adequate mathematical tool to overcome the difficulties in quantization of equivalent Lagrangians. Indeed, due to recent research in this field it is known that equivalent Lagrangians lead to the same quantum mechanics (Y. Kaminaga [1], C. A. P. Galvão and N. A. Lemos [1]; in the latter paper some ideas close to O. Krupková [2] have been proposed).

The aim of this chapter is to introduce generalized Hamilton equations in an intrinsic way mentioned above. It should be pointed out that the presented Hamilton theory differs from a conventional one in many respects. First of all, it covers Lagrangean systems in general, i.e., without any restrictive assumptions on order, regularity, etc. The Hamilton equations refer not to a particular Lagrangian, as it is standard, but to a *Lagrangean system* in the sense of Definition 4.3.12 (i.e., related to a *locally variational form*, or, otherwise speaking, to the *class of equivalent Lagrangians*). Significantly, one gets global Hamilton equations even in case that there does not exist a global Lagrangian. We stress the *geometric viewpoint*, interpreting the Hamilton equations as equations for *integral sections of a distribution on the phase space*. Finally, the phase space is the manifold $J^{s-1}Y$ where the extended dynamics takes place, and where Hamiltonians and momenta live. Within this approach, important properties of Hamilton equations of classical mechanics are saved, namely, (1) the Hamilton equations are *first-order* ODE, which are *locally variational*, i.e., can be derived from (local) Lagrangians, (2) for regular problems they are *equivalent* with the Euler-Lagrange equations. However, as pointed above, Hamilton equations possess in general more solutions than the original Euler-Lagrange equations—the dynamics described by the Hamilton equations is an *extension* of the original dynamics described by the Euler-Lagrange equations. However, it is significant that even in the general case there is a transparent relation between these two dynamics, namely, extremals are in *one-to-one correspondence* with those integral sections of the Euler-Lagrange distribution which are prolongations of sections of $\pi : Y \to X$. Within this viewpont, the geometrical and practical meaning of the generalized Hamilton theory consists in the possibility to represent the dynamics by means of a *Pfaffian system*, while extremals (solutions of Euler-Lagrange equations) *cannot* be characterized by such an important property.

Certain Lagrangean systems admit a natural representation by means of a closed two-form on $R \times T^*M$. Within the range of symplectic geometry, such a representation is often referred to as a "*Hamiltonian formalism*". Such a construction, however, is not possible generally—it is limited to a particular family of Lagrangean systems. A discussion to this point is postponed to Chapter 10, where Lagrangean systems admitting "Hamiltonian formalism" will be considered in more detail.

The exposition in this section follows the results of O. Krupková [2,4,7,9].

5.2. Hamilton two-form and generalized Hamilton equations

Let $\pi : Y \to X$ be a fibered manifold, $\dim X = 1$. Let $s \geqslant 1$. In this section we shall consider the 1-jet prolongation of the fibered manifold $\pi_{s-1} : J^{s-1}Y \to X$, i.e., the fibered manifold $(\pi_{s-1})_1 : J^1(J^{s-1}Y) \to X$. To avoid confusion, we shall use the notation $\tilde{h}$ (resp. $\tilde{p}$, resp. $\tilde{p}_1$) for the horizontalization (resp. contactization, resp. 1-contactization) with respect to the projection π_{s-1}.

Let $s \geqslant 1$, let α be a Lepagean two-form on $J^{s-1}Y$, i.e., by Definition 4.3.12, a *Lagrangean system of order $s-1$ on the fibered manifold π*. Recall from Chapter 4 that in a fiber chart, α has one of the following equivalent representations

$$\pi^*_{s,s-1}\alpha_E = E_\sigma \, \omega^\sigma \wedge dt + \sum_{j,k=0}^{s-1} F^{jk}_{\sigma\nu} \, \omega^\sigma_j \wedge \omega^\nu_k,$$

or,

$$\alpha_E = A_\sigma \, \omega^\sigma \wedge dt + \sum_{j,k=0}^{s-2} F^{jk}_{\sigma\nu} \, \omega^\sigma_j \wedge \omega^\nu_k + B_{\sigma\nu} \, \omega^\sigma \wedge dq^\nu_{s-1},$$

(relevant for $s \geqslant 2$), or,

$$\alpha_E = \Big(E_\sigma - \sum_{k=0}^{s-1} 2F^{0k}_{\sigma\nu} q^\nu_{k+1}\Big) dq^\sigma \wedge dt - \sum_{j=1}^{s-1}\sum_{k=0}^{s-1-j} 2F^{jk}_{\sigma\nu} q^\nu_{k+1} \, dq^\sigma_j \wedge dt + \sum_{j,k=0}^{s-1} F^{jk}_{\sigma\nu} \, dq^\sigma_j \wedge dq^\nu_k,$$

where the $F^{jk}_{\sigma\nu}$'s are given by (4.10). By Theorem 4.5.2 α locally takes the canonical form

$$\alpha = -dH \wedge dt + \sum_{k=0}^{s-c-1} dp^k_\nu \wedge dq^\nu_k,$$

where the Hamiltonian and momenta are defined by means of the Euler-Lagrange expressions E_σ by (4.38) and (4.37), or, equivalently, by means of minimal-order Lagrangians by (4.49).

The corresponding dynamical form to α,

$$E = p_1 \alpha$$

on J^sY, is locally variational, and extremals of E are (local) sections γ of π which satisfy the Euler-Lagrange equations

$$J^{s-1}\gamma^* \, i_{J^{s-1}\xi}\alpha = 0$$

for every π-vertical vector field ξ on Y; these equations are ODE of order s for γ.

Since α is a closed two-form on $J^{s-1}Y$, it can be considered as a *Lagrangean system of order zero on the fibered manifold π_{s-1}*. Now, we get a dynamical form

$$\mathcal{H} = \tilde{p}_1 \alpha \tag{5.9}$$

of order 1 on $J^1(J^{s-1}Y)$ which is locally variational. Extremals of $\mathcal{H}$ are (local) sections δ of π_{s-1} satisfying the Euler-Lagrange equations

$$\delta^* \, i_\xi \alpha = 0 \tag{5.10}$$

for every π_{s-1}-vertical vector field ξ on $J^{s-1}Y$. Equations (5.10) are ODE of order *one* for δ. Notice that since α is uniquely determined by E, the form $\mathcal{H}$ and the equations (5.10) are also uniquely determined by the locally variational form E.

Using a fiber-chart expression of α, and denoting the associated coordinates on $J^1(J^{s-1}Y)$ by $(t, q^\sigma, \ldots, q^\sigma_{s-1}, q^\sigma_{,1}, \ldots, q^\sigma_{s-1,1})$, we get for $\mathcal{H}$

$$\mathcal{H} = \sum_{i=0}^{s-1} H^i_\sigma \, dq^\sigma_i \wedge dt \,, \quad H^i_\sigma = E_\sigma \, \delta^{0i} + \sum_{k=0}^{s-1-i} 2F^{ik}_{\sigma\nu}(q^\nu_{k,1} - q^\nu_{k+1}) \,. \tag{5.11}$$

5.2.1. Definition. The form $\mathcal{H}$ defined by (5.9) is called the *Hamilton form* of the Lagrangean system α on π (or the *Hamilton form* associated with the locally variational form $E = p_1\alpha$).

The Lepagean equivalent $\alpha_{\mathcal{H}}$ of $\mathcal{H}$, which is a closed two-form on $J^1(J^{s-1}Y)$, projectable onto $J^{s-1}Y$, will be referred to as the *Hamiltonian system* associated with the Lagrangean system α_E.

Extremals of $\mathcal{H}$ are called *Hamilton extremals* of the Lagrangean system α on π (or of the locally variational form $E = p_1\alpha$).

Equations for extremals of $\mathcal{H}$ (i.e., the Euler-Lagrange equations of $\mathcal{H}$) are called *generalized Hamilton equations* of the Lagrangean system α on π (or of the locally variational form E).

The concept of the Hamilton form (related to a Lagrangian) has been introduced by D. Krupka and O. Štěpánková [1]; in the paper D. Krupka and J. Musilová [1] it has been transferred to higher-order mechanics. In O. Krupková [2] this concept was related to the class of equivalent Lagrangians.

The geometric meaning of the Hamilton form is similar to that of the Euler-Lagrange form; namely, the components of the Hamilton form have the meaning of "left-hand sides of the Hamilton equations", similarly as the components of the Euler-Lagrange form have the meaning of the "left-hand sides of the Euler-Lagrange equations". The expression (5.11) represents these components in fibered coordinates. Other frequently used expressions (in Legendre coordinates, and in Darboux coordinates) will be written down later (Chapters 6, 7 and 9).

By the above considerations we have assigned to a Lagrangean system of order $s-1$ on π another Lagrangean system on π_{s-1}—the associated Hamiltonian system—which is *of order zero*. Let us study the relations between these Lagrangean systems in more detail.

Directly from the definition of the Hamilton form we obtain the following propositions:

5.2.2. Proposition. *Let E be a locally variational form on J^sY, and let $\mathcal{H}$ be a locally variational form on $J^1(J^{s-1}Y)$. $\mathcal{H}$ is the Hamilton form associated with E if and only if their Lepagean equivalents coincide, i.e.,*

$$\alpha_{\mathcal{H}} = \alpha_E \,.$$

5.2.3. Proposition. *Let E be a locally variational form on J^sY, α_E its Lepagean equivalent. There exists a unique two-form $\mathcal{H}$ on $J^1(J^{s-1}Y)$ such that for every π_{s-1}-*

vertical vector field ξ *on* $J^{s-1}Y$

$$i_{J^1\xi}\,\mathcal{H} = \tilde{h}\, i_\xi\, \alpha_E. \tag{5.12}$$

$\mathcal{H}$ *is the Hamilton form associated with* E.

By definition, *Hamilton form is locally variational.* Since it is of order one, this means that to every point x in $J^1(J^{s-1}Y)$ there exists a neighborhood U of x and a Lagrangian κ defined on U such that $\mathcal{H} = E_\kappa$; such Lagrangians are obviously minimal-order Lagrangians for $\mathcal{H}$.

5.2.4. Theorem. *Let* $\mathcal{H}$ *be a Hamilton form associated with a locally variational form* E *on* J^sY. *A* π_{s-1}*-horizontal* 1*-form* κ *is a (local) first-order Lagrangian of* $\mathcal{H}$ *if and only if*

$$\kappa = \tilde{h}\theta_\lambda \tag{5.13}$$

where λ *is a (local) Lagrangian of order* $r \leqslant s$ *of* E.

Proof. Denote by α the Lepagean equivalent of $\mathcal{H}$ and E.

Let κ be a (local) first-order Lagrangian of $\mathcal{H}$. Then $\kappa = \tilde{h}\rho$ where ρ is a one-form on an open subset of $J^{s-1}Y$ such that $d\rho = \alpha$. Since α is a Lepagean equivalent of E, the form $\lambda = h\rho$ is a (local) Lagrangian of E. This Lagrangian is obviously of order $\leqslant$ s.

Conversely, let λ be a (local) Lagrangian of order $r \leqslant s$ of E. Then, by Corollary 4.5.7, θ_λ is of order $s - 1$. Putting $\kappa = \tilde{h}\theta_\lambda$ we obtain a (local) Lagrangian of order 1 for the fibered manifold π_{s-1}. Since locally $d\theta_\lambda = \alpha$, we get $E_\kappa = \tilde{p}_1\, d\theta_\lambda = \mathcal{H}$, proving that κ is a Lagrangian of $\mathcal{H}$. □

Note that by the above theorem, minimal-order Lagrangians of $\mathcal{H}$ are in one-to-one correspondence with Lagrangians of order $r \leqslant s$ of the corresponding locally variational form E.

5.2.5. Definition. Let $\mathcal{H}$ be a Hamilton form associated with a locally variational form E on J^sY. Every (local) Lagrangian of $\mathcal{H}$ which is of the form (5.13) is called *extended Lagrangian* of E.

If λ is a Lagrangian of order $r \leqslant s$ for E, we shall usually use the notation $\tilde{\lambda}$ for the corresponding extended Lagrangian $\tilde{h}\theta_\lambda$.

Let us find the fiber-chart expression for extended Lagrangians: Let $\lambda = L\,dt$, and denote $\tilde{\lambda} = \tilde{h}\theta_\lambda$ by $\tilde{L}\,dt$. Then (in the notation of (3.11), (3.12), or (4.52)),

$$\tilde{L} = L + \sum_{i=0}^{s-1} f_\sigma^{i+1}\,(q_{i,1}^\sigma - q_{i+1}^\sigma) = \sum_{i=0}^{s-1} \hat{p}_\sigma^i q_{i,1}^\sigma - \hat{H}, \tag{5.14}$$

where $\hat{H}$, $\hat{p}$ are the Ostrogradskii Hamiltonian and momenta of the Lagrangian L.

Denoting by ι the canonical embedding of J^sY into $J^1(J^{s-1}Y)$ we can see that

$$\iota^*\tilde{\lambda} = \lambda. \tag{5.15}$$

Notice that the form $\theta_{\tilde{\lambda}}$ is projectable onto an open subset of $J^{s-1}Y$ and, up to this projection, it equals θ_λ.

By (5.14) and (4.47) we get that if $\lambda_{\min} = L_{\min}\,dt$ is a minimal-order Lagrangian for α, then the corresponding extended Lagrangian $\tilde{\lambda}_0 = \tilde{h}\theta_{\lambda_{\min}}$ is

$$L_0 = \sum_{i=0}^{s-c-1} p^i_\sigma q^\sigma_{i,1} - H.$$

The relation between $\tilde{L}$ and $\tilde{L}_0$ obviously reads $\tilde{L} = \tilde{L}_0 + \tilde{d} f/dt$, where f is a function of order $s-1$.

5.2.6. Proposition. *Let $\mathcal{H}$ be a Hamilton form associated with a locally variational form E. $\mathcal{H}$ is globally variational if and only if E is globally variational.*

Proof. By Theorem 5.2.4, if λ is a global Lagrangian of E then $\tilde{\lambda}$ is a global Lagrangian of $\mathcal{H}$.

Conversely, if $\mathcal{H}$ is globally variational then, by Corollary 4.3.10, there exists a global first-order Lagrangian for $\mathcal{H}$, i.e., by Theorem 5.2.4, a global Lagrangian $\tilde{\lambda} = \tilde{h}\theta_\lambda$. However, since the Lepagean equivalent of $\tilde{\lambda}$ is θ_λ and is defined on $J^{s-1}Y$, we get that $\lambda = h\theta_\lambda$ is a global Lagrangian of E. $\square$

Theorem 3.3.2 applied to the Hamilton form gives equivalent expressions for the generalized Hamilton equations:

5.2.7. Theorem. *Let α be a Lagrangean system of order $s-1 \geqslant 0$, let $E = p_1\alpha$ be the associated locally variational form on J^sY, and $\mathcal{H} = \tilde{p}_1\alpha$ the associated Hamilton form on $J^1(J^{s-1}Y)$. Let δ be a (local) section of π_{s-1}. The following conditions are equivalent:*

(1) *δ is a Hamilton extremal.*

(2) *The Hamilton form $\mathcal{H}$ vanishes along $J^1\delta$, i.e.,*

$$\mathcal{H} \circ J^1\delta = 0.$$

(3) *For every vector field ξ on $J^{s-1}Y$*

$$\delta^*\, i_\xi\, \alpha = 0.$$

(4) *For every π_{s-1}-projectable vector field ξ on $J^{s-1}Y$*

$$\delta^*\, i_\xi\, \alpha = 0.$$

(5) *For every π_{s-1}-vertical vector field ξ on $J^{s-1}Y$*

$$\delta^*\, i_\xi\, \alpha = 0.$$

(6) *In every fiber chart, δ satisfies the system of ms first-order ordinary differential equations*

$$H^i_\sigma \circ J^1\delta = 0, \quad 0 \leqslant i \leqslant s-1, \; 1 \leqslant \sigma \leqslant m,$$

where H^i_σ are given by (5.11).

We have assigned to a locally variational form E of order s another locally variational form $\mathcal{H}$ of order one. This means that now we have the possibility to study instead of the original dynamics, represented by extremals which are sections of π, a *new dynamics*, represented by Hamilton extremals which are sections of π_{s-1}. Our aim now is to clarify relations between these two dynamics, and develop techniques, which will enable us to learn the original dynamics with the help of the new one.

To understand the relation between a given Lagrangean system and its associated Hamiltonian system, it is necessary to clarify the relation between extremals and Hamilton extremals. First, note that by definition of Hamilton extremals, solutions of the generalized Hamilton equations are *not supposed* to be *holonomic* sections of π_{s-1} (this means a Hamilton extremal need not to be of the form of a *prolongation* of a section of π). Recall that a Hamilton extremal δ of a locally variational form E on J^sY is called *holonomic* if $\delta = J^{s-1}\gamma$ for an extremal γ of E.

Generally, extremals are not in one-to-one correspondence with Hamilton extremals, since among Hamilton extremals there may appear sections which do not correspond to any solution of the original variational equations. Let us illutrate this feature by an easy example.

5.2.8. Example. Consider the first-order Lagrangean system on the fibered manifold $\pi : R \times R^2 \to R$, arising from the minimal-order Lagrangian

$$L = \tfrac{1}{2}(\dot{q}^1 + \dot{q}^2)^2,$$

i.e., let

$$\alpha = d\dot{q}^1 \wedge \omega^1 + d\dot{q}^2 \wedge \omega^2 + d\dot{q}^1 \wedge \omega^2 + d\dot{q}^2 \wedge \omega^1.$$

We get the Hamilton equations

$$\delta^*(\omega^1 + \omega^2) = 0, \quad \delta^*(d\dot{q}^1 + d\dot{q}^2) = 0,$$

i.e.,

$$\frac{d}{dt}(\dot{q}^1 + \dot{q}^2) \circ \delta = 0, \quad \frac{d}{dt}(q^1 + q^2) \circ \delta = (\dot{q}^1 + \dot{q}^2) \circ \delta$$

and the Euler-Lagrange equations

$$\frac{d}{dt}(\dot{q}^1 + \dot{q}^2) \circ J^1\gamma = 0.$$

A general solution of the Hamilton equations is $\delta(t) = (t, q^1, c_1 t + c_2 - q^1, \dot{q}^1, c_1 - \dot{q}^1)$. Thus, for example, the section $\delta_0 : t \to (t, t, 0, t, 1 - t)$ is a Hamilton extremal, but is not an extremal of the Lagrangean system α.

The fact that the Hamilton equations for singular Lagrangians can have more solutions than their Euler-Lagrange equations is fairly well-known; for the first time it was pointed out by M. J. Gotay and J. M. Nester [1].

Taking into account theorems 5.2.7 and 3.3.2 we can see immediately that the set of Hamilton extremals *contains all* (prolonged) extremals, and more precisely that *extremals are in one-to-one correspondence with holonomic Hamilton extremals*:

5.2.9. Proposition. *Let E be a locally variational form on J^sY. If γ is an extremal of E then $J^{s-1}\gamma$ is a Hamilton extremal of E. Conversely, if for a (local) section γ of π, $J^{s-1}\gamma$ is a Hamilton extremal of E then γ is an extremal.*

Thus, we shall also say that the Euler-Lagrange equations describe the *proper dynamics* and the generalized Hamilton equations describe the *extended dynamics* of the given Lagrangean system.

5.2.10. Remark. Let $\alpha_{\mathcal{H}}$ be Hamiltonian system, i.e., a zero-order Lagrangean system for the fibered manifold π_{s-1}, $s \geqslant 1$, associated with an $(s-1)$-th order Lagrangean system α for π. It is worthwhile to notice that the *Hamilton equations* for $\alpha_{\mathcal{H}}$ coincide with the Hamilton equations for α. Computing a Hamiltonian and momenta of the Hamiltonian system $\alpha_{\mathcal{H}}$ (e.g. from Theorem 4.5.8 by means of a Lagrangian (5.14)), and denoting them by $\tilde{H}$ and $\tilde{p}^i_\sigma$, respectively, we get (in the notation of (5.14))

$$\tilde{p}^i_\sigma = \frac{\partial \tilde{L}}{\partial q^\sigma_i} = \hat{p}^i_\sigma, \quad \tilde{H} = -\tilde{L} + \sum \tilde{p}^i_\sigma q^\sigma_{i,1} = \hat{H}.$$

By these formulas we get an interpretation for the Ostrogradskii momenta and Hamiltonian related to a Lagrangian *of order* $r \leqslant s$ (formulas (4.52)), as the *momenta and Hamiltonian of the associated to* α (zero-order) *Hamiltonian system* $\alpha_{\mathcal{H}}$.

Hamiltonian systems recently became quite popular in physics due to a new quantization method proposed by Faddeev and Jackiw (see R. Jackiw [1]). This method is based on quantizing an *extended Lagrangian* rather than the original one.

5.3. The characteristic and the Euler-Lagrange distributions

According to the definition, generalized Hamilton equations are first-order variational equations associated to given s-order variational equations. They are of the form

$$\mathrm{F}\,\dot{x} = b\,,$$

where x stands for the components of a section δ of π_{s-1} (i.e., $x^\nu_k = q^\nu_k \circ \delta$), and $\mathrm{F} = (2F^{jk}_{\sigma\nu})$ where the $F^{jk}_{\sigma\nu}$'s are defined by (4.10). Since the matrix F generally is not regular, these equations cannot be interpreted as equations for integral sections of a vector field.

Let us turn to a deeper study of the geometric structure of generalized Hamilton equations to clarify the nature of the extended dynamics on the phase space.

Let $s \geqslant 1$, and let α be a Lagrangean system on $J^{s-1}Y$. Denote by E the corresponding locally variational form (defined by $E = p_1\alpha$), and by $\mathcal{D}$ the *characteristic distribution* of α. This distribution is generated by means of the *one-forms*

$$i_\xi\,\alpha\,, \quad \text{where } \xi \text{ runs over the set of } all \text{ vector fields on } J^{s-1}Y, \tag{5.16}$$

or equivalently, it is spanned by the *vector fields* ζ such that

$$i_\zeta\,\alpha = 0. \tag{5.17}$$

It is easy to see that (in the notation of chapters 2 and 4)

$$\mathcal{D} \approx \operatorname{span}\{\eta_0,\ \eta_\sigma,\ \eta^i_\sigma,\ 1 \leqslant i \leqslant s-1,\ 1 \leqslant \sigma \leqslant m\}, \tag{5.18}$$

where

$$\eta_0 = \Big(A_\sigma - \sum_{k=0}^{s-2} 2F^{0k}_{\sigma\nu} q^\nu_{k+1}\Big)\, dq^\sigma - \sum_{i=1}^{s-1}\sum_{k=0}^{s-1-i} 2F^{ik}_{\sigma\nu} q^\nu_{k+1}\, dq^\sigma_i,$$

$$\eta_\sigma = A_\sigma\, dt + \sum_{k=0}^{s-2} 2F^{0k}_{\sigma\nu}\omega^\nu_k + B_{\sigma\nu}\, dq^\nu_{s-1}, \tag{5.19}$$

$$\eta^i_\sigma = \sum_{k=0}^{s-1-i} 2F^{ik}_{\sigma\nu}\omega^\nu_k, \quad 1 \leqslant i \leqslant s-1, \quad 1 \leqslant \sigma \leqslant m.$$

The rank of the distribution $\mathcal{D}$ is generally nonconstant, and at each point $x \in J^{s-1}Y$,

$$\operatorname{corank} \mathcal{D}(x) = \operatorname{rank} \alpha(x).$$

We stress that since we suppose α be smooth, *the distribution $\mathcal{D}$ is defined by smooth one-forms, hence generally cannot be spanned by continuous vector fields*. Thus, for Lagrangean systems of a nonconstant rank it is more matural to use the description of $\mathcal{D}$ by the system of smooth Pfaffian forms (5.16). At the same time, to get the extended dynamics explicitly characterized by means of vector fields (5.17) one needs to apply a quite complicated procedure, called the "constraint algorithm" (see Chapter 7 for details).

If rank α is constant on $J^{s-1}Y$ then the characteristic distribution $\mathcal{D}$ has a constant rank, and is *completely integrable*. Hence, $\mathcal{D}$ can be spanned by a system of smooth vector fields. Notice that the number of vector fields generating $\mathcal{D}$ may be greater than one (although the Hamilton equations are *ordinary* differential equations)—this reflects the fact that Hamilton equations need not be solvable with respect to the first derivatives.

5.3.1. Example. Consider the Lagrangean system defined in Example 5.2.8. The characteristic distribution is

$$\mathcal{D} \approx \operatorname{span}\{d\dot q^1 + d\dot q^2,\ \omega^1 + \omega^2\},$$

and rank $\mathcal{D}$ = const = 3. Consequently, $\mathcal{D}$ can be locally spanned by 3 smooth vector fields,

$$\frac{\partial}{\partial t} + (\dot q^1 + \dot q^2)\frac{\partial}{\partial q^1}, \quad \frac{\partial}{\partial q^1} - \frac{\partial}{\partial q^2}, \quad \frac{\partial}{\partial \dot q^1} - \frac{\partial}{\partial \dot q^2}.$$

Thus, the extended dynamics is characterized by the above 3 fector fields on J^1Y. Obviously, the distribution spanned by these vector fields admits integral sections which are not holonomic. To exclude sections of this kind, one has to consider the subdistribution of rank 2, spanned by the vector fields

$$\frac{\partial}{\partial t} + \dot q^1 \frac{\partial}{\partial q^1} + \dot q^2 \frac{\partial}{\partial q^2} + g\Big(\frac{\partial}{\partial \dot q^1} - \frac{\partial}{\partial \dot q^2}\Big),$$

where g is an arbitrary function. By this subdistribution the proper dynamics is completely characterized (and we note that the characterization by means of a *unique* vector field is not possible). We shall return to this example in Chapter 7.

Summarizing, we can see that those Hamilton equations which cannot be solved with respect to the first derivatives cannot be locally described by a unique vector field on $J^{s-1}Y$, but rather are characterized by a system of smooth Pfaffian forms. Only in the particular case rank α = const there is an equivalent description by means of a system of smooth local vector fields on $J^{s-1}Y$. For the Euler-Lagrange equations the situation is even more complicated. We shall see that the condition rank α = const is not sufficient for the proper dynamics be characterized by a system of vector fields, and that more subtle considerations are needed. This will be our concern in the next section, and in detail in Chapter 7.

Before turning to deeper study of extended and proper dynamics, let us notice that we have also another significant distribution on the phase space, related with the closed two-form α:

5.3.2. Definition. Let α be a Lagrangean system of order $s-1$ on π. The distribution Δ defined on $J^{s-1}Y$ and generated by means of the one-forms $i_\xi \alpha$, where ξ runs over the set of π_{s-1}-*vertical* vector fields is called the *Euler-Lagrange distribution* of α. Vector fields belonging to the Euler-Lagrange distribution are called *Hamilton vector fields*.

The Euler-Lagrange distribution (associated to a particular Lagrangian) has been introduced by D. Krupka and J. Musilová [1]. In O. Krupková [2] this concept has been re-defined for Lepagean two-forms.

We can see that the characteristic distribution is a subdistribution of the Euler-Lagrange distribution, $\mathcal{D} \subset \Delta$. Note that

$$\Delta \approx \operatorname{span}\{\eta_\sigma,\ \eta^i_\sigma,\ 1 \leqslant i \leqslant s-1,\ 1 \leqslant \sigma \leqslant m\}. \tag{5.20}$$

Instead of the Pfaffian system (5.20) defining the Euler-Lagrange distribution we can also consider the annihilator of (5.20), i.e., the system of Hamilton vector fields. However, since the Euler-Lagrange distribution is in general of a nonconstant rank, Hamilton vector fields need not be continuous.

By Theorem 5.2.7 and Proposition 5.2.9 we can see the geometric meaning of the distributions $\mathcal{D}$ and Δ:

5.3.3. Proposition.

(1) *A section δ of π_{s-1} is an integral section of Δ if and only if it is an integral section of $\mathcal{D}$.*

(2) *Integral sections of both the distributions Δ and $\mathcal{D}$ coincide with Hamilton extremals. Holonomic integral sections of both Δ and $\mathcal{D}$ coincide with the $(s-1)$-th prolongations of extremals.*

Roughly we can say that, though generally different, the distributions $\mathcal{D}$ and Δ "intersect on Hamilton extremals". We shall see in the next section that both the distributions $\mathcal{D}$ and Δ are usefull for studies of the dynamics of Lagrangean systems, and that especially the difference between them will be helpfull in understanding singular systems.

5.3.4. Remark. Note that for *zero-order Lagrangean systems* (=*first-order variational equations*) the Hamilton theory *coincides* with the Lagrange theory: the phase space = the configuration space, extended dynamics = proper dynamics, Hamilton equations = Euler-Lagrange equations, Hamilton extremals = extremals. This means that in this case, studying the Euler-Lagrange distribution we get directly the *structure of extremals*.

5.4. Structure theorems

By Proposition 5.3.3, the generalized Hamilton equations can be interpreted as equations for integral *sections* of either the characteristic distribution or the Euler-Lagrange distribution. These distributions, though generally different, have *common integral sections*, and both they are suitable for studying the extended dynamics. In this section we shall see that the difference between them has a clear geometric meaning which will help us to understand the dynamics of singular Lagrangean systems.

First of all, notice that it may happen that through a point in the phase space there passes an integral curve (or even a maximal integral manifold) but *no* integral *section*, which means that the corresponding Hamilton equations have at that point no solution (for concrete examples of such Lagrangean systems we refer to Chapter 7).

On the other hand, if there is a Hamilton extremal passing through a point $x \in J^{s-1}Y$ then at this point there must exist a non-vertical vector belonging to $\mathcal{D}(x)$ (and hence to $\Delta(x)$). In other words, both the distributions must be *weakly horizontal* at x. Thus, the points of the phase space where $\mathcal{D}$ and Δ are not weakly horizontal are not allowed to become *initial points* for the Hamiltonian motion.

Let us turn to study *conditions under which the distributions $\mathcal{D}$ and Δ are weakly horizontal.*

For convenience, we shall use the following notation: If α is a Lagrangean system on $J^{s-1}Y$, we shall use its fiber chart representation in the form as given by Proposition 4.3.2, and for the locally variational form $E = p_1\alpha$ we shall write

$$E = (A_\sigma + B_{\sigma\nu} q_s^\nu)\, dq^\sigma \wedge dt$$

where the A_σ's and $B_{\sigma\nu}$'s do not depend upon the q_s^ρ's. Further we shall consider the matrices

$$\mathrm{B} = (B_{\sigma\nu}), \qquad (\mathrm{B} \mid \mathrm{A}) = (B_{\sigma\nu}, A_\sigma), \tag{5.21}$$

where the left (resp. right) index numbers rows (resp. columns), and

$$\mathrm{F} = (2F_{\sigma\nu}^{ik}), \qquad (\mathrm{F} \mid \mathrm{A}) = \begin{pmatrix} 2F_{\sigma\nu}^{0k} & A_\sigma \\ 2F_{\sigma\nu}^{ik} & 0 \end{pmatrix}. \tag{5.22}$$

where the left indices $\sigma, j, \leqslant \sigma \leqslant m, 0 \leqslant j \leqslant s-1$, number rows, and the right indices ν, k label columns, $1 \leqslant \nu \leqslant m, 0 \leqslant k \leqslant s-1$. Note that since by (4.27) and (4.10)

$$B_{\sigma\nu} = \frac{\partial E_\sigma}{\partial q_s^\nu} = 2F_{\sigma\nu}^{0,s-1} = -2F_{\sigma\nu}^{1,s-2} = \cdots = (-1)^{s-1} 2F_{\sigma\nu}^{s-1,0}, \tag{5.23}$$

the matrix F is of the form

$$\mathrm{F} = 2\begin{pmatrix} F_{\sigma\nu}^{0,0} & F_{\sigma\nu}^{0,1} & \cdots & F_{\sigma\nu}^{0,s-1} \\ F_{\sigma\nu}^{1,0} & \cdots & F_{\sigma\nu}^{1,s-2} & 0 \\ \vdots & & & \\ F_{\sigma\nu}^{s-1,0} & 0 & \cdots & 0 \end{pmatrix} = \begin{pmatrix} 2F_{\sigma\nu}^{0,0} & 2F_{\sigma\nu}^{0,1} & \cdots & B_{\sigma\nu} \\ 2F_{\sigma\nu}^{1,0} & \cdots & -B_{\sigma\nu} & 0 \\ \vdots & & & \\ (-1)^{s-1}B_{\sigma\nu} & 0 & \cdots & 0 \end{pmatrix}.$$

Now, we can prove a fundamental theorem for learning the structure of Hamilton extremals (see O. Krupková [4,9]).

5.4.1. Theorem. *Let α be a Lagrangean system on $J^{s-1}Y$, let $x \in J^{s-1}Y$ be a point. The following five conditions are equivalent:*

(1) *The characteristic distribution $\mathcal{D}$ of α is weakly horizontal at x.*

(2) *The Euler-Lagrange distribution* Δ *is weakly horizontal at* x.
(3) $\Delta(x) = \mathcal{D}(x)$.
(4) *At* x, *the equation* $i_\zeta \alpha = 0$ *has a nonvertical solution* ζ_x.
(5) $\operatorname{rank} \mathrm{F}(x) = \operatorname{rank}(\mathrm{F} \mid \mathrm{A})(x)$.

If the condition $\operatorname{rank} \mathrm{B}(x) = \operatorname{rank}(\mathrm{B} \mid \mathrm{A})(x)$ *holds then any of the assertions* (1)–(5) *is satisfied.*

Proof. (1) $\Rightarrow$ (2), since $\mathcal{D} \subset \Delta$.

The assertions (2) and (4) are obviously equivalent.

(2) $\Leftrightarrow$ (5): Let ξ be a (not necessarily continuous) vector field belonging to Δ, defined in a neighborhood U of x, and such that $T\pi_{s-1}.\xi \neq 0$ at x, let

$$\xi = \xi^0 \frac{\partial}{\partial t} + \sum_{j=0}^{s-1} \xi_j^\sigma \frac{\partial}{\partial q_j^\sigma}$$

be a fiber-chart expression of ξ. By assumption, ξ satisfies the equations

$$\begin{gathered}\Big(A_\sigma - \sum_{k=0}^{s-2} 2F_{\sigma\nu}^{0k} q_{k+1}^\nu\Big)\xi^0 + \sum_{k=0}^{s-2} 2F_{\sigma\nu}^{0k} \xi_k^\nu + B_{\sigma\nu} \xi_{s-1}^\nu = 0, \\ -\sum_{k=0}^{s-1-i} 2F_{\sigma\nu}^{ik} q_{k+1}^\nu \xi^0 + \sum_{k=0}^{s-1-i} 2F_{\sigma\nu}^{ik} \xi_k^\nu = 0, \quad 1 \leqslant \sigma \leqslant m,\ 1 \leqslant i \leqslant s-1.\end{gathered} \tag{5.24}$$

Without the loss of generality we can assume $\xi^0(x) = -1$. Hence, at x we obtain a system of ms linear non-homogeneous algebraic equations for ms unknowns $\xi_j^\sigma(x)$. Since the matrix of the system is equivalent to the matrix $(\mathrm{F} \mid \mathrm{A})(x)$, we get from the Frobenius theorem on the existence of solutions of algebraic equations that $\operatorname{rank}\mathrm{F} = \operatorname{rank}(\mathrm{F} \mid \mathrm{A})$ at x. Conversely, if (5) holds, then by Frobenius theorem at the point x there exists a solution $\xi(x)$ to (5.24) satisfying $\xi^0(x) = -1$. Hence, $\xi(x) \in \Delta(x)$ proving (2).

(3) $\Leftrightarrow$ (5): Suppose (3). Then $\eta_0(x)$ is a linear combination of the 1-forms $\eta_\sigma^i(x)$, $0 \leqslant i \leqslant s-1$, $1 \leqslant \sigma \leqslant m$. Since the matrix of the generators of $\mathcal{D}$ is equivalent to the matrix

$$\begin{pmatrix} \mathrm{F} & \mathrm{A} \\ \mathrm{A} & 0 \end{pmatrix} = \begin{pmatrix} & & & B_{\sigma\nu} & A_\sigma \\ & & -B_{\sigma\nu} & 0 & 0 \\ & \cdots & & & \\ (-1)^{s-1} B_{\sigma\nu} & 0 & \cdots & 0 & 0 \\ A_\nu & 0 & \cdots & 0 & 0 \end{pmatrix}, \tag{5.25}$$

where the submatrix $(\mathrm{F} \mid \mathrm{A})$ is the matrix of generators of Δ, we get that the last row has to be a linear combination of the other rows (at x). Consequently, considering the condition $F_{\sigma\nu}^{ik} = -F_{\nu\sigma}^{ki}$ which holds for F, we get that (at x) the last column of the matrix (5.25) is a linear combination of its other columns. This means that at x

$$\operatorname{rank}\begin{pmatrix} \mathrm{F} & \mathrm{A} \\ \mathrm{A} & 0 \end{pmatrix} = \operatorname{rank}(\mathrm{F} \mid \mathrm{A}) = \operatorname{rank}\begin{pmatrix} \mathrm{F} \\ \mathrm{A} \end{pmatrix} = \operatorname{rank}\mathrm{F}.$$

Conversely, if the condition $\operatorname{rank}\mathrm{F}(x) = \operatorname{rank}(\mathrm{F} \mid \mathrm{A})(x)$ holds then, by the same arguments as above, $\mathcal{D}(x)$ is spanned by the forms $\eta_\sigma^i(x)$, $0 \leqslant i \leqslant s-1$, $1 \leqslant \sigma \leqslant m$. Hence $\mathcal{D}(x) = \Delta(x)$.

It remains to show (2) $\Rightarrow$ (1), which, however, is now trivial.

Finally, if $\operatorname{rank} B(x) = \operatorname{rank}(B \mid A)(x)$ then obviously $\operatorname{rank} F(x) = \operatorname{rank}(F \mid A)(x)$. This completes the proof. □

Theorem 5.4.1 brings a geometric characterization of the weak horizontality conditions of the characteristic and Euler-Lagrange distribution. Equivalently, it provides us with an explicit algebraic condition (condition (5)) which can be easily applied in every concrete situation to exclude the points of the phase space where the generalized Hamilton equations have no solution.

Having in mind the definition of a Lagrangean system as a Lepagean two-form α, the distribution $\mathcal{D}$ (as the charactericstic distribution of α), compared with Δ, seems at a first sight to be a more natural object describing the Hamiltonian system associated with α. However, according to Theorem 5.4.1, both the distributions $\mathcal{D}$ and Δ are usefull from the theoretical and practical point of view. Indeed, if the Lagrangean system is weakly horizontal then (5.16) leads to consider superfluous (not independent) 1-forms. In such a situation, working with the distribution Δ is more simple. Also, there is a theoretical argument for taking into account both the distributions $\mathcal{D}$ and Δ, namely the condition (3) of Theorem 5.4.1, which leads to a geometric interpretation of the set of points of the phase space where Hamilton extremals a priori are not allowed to pass through as the set of points where the distributions $\mathcal{D}$ and Δ are different.

We have seen that Lagrangean systems can possess Hamilton extremals which do not correspond to extremals. The above theorem helps us to find extended dynamics, however, to study proper dynamics, we have to find additional conditions excluding the nonholonomic Hamilton extremals. Clearly, while Hamilton extremals are connected with non-vertical vector fields, holonomic Hamilton extremals are connected with semisprays. We have the following assertions which will help us to learn the *structure of holonomic Hamilton extremals* (= prolonged extremals).

5.4.2. Theorem. *Let α be a Lagrangean system of order $s-1$, Δ its Euler-Lagrange distribution. Let $x \in J^{s-1}Y$ be a point. The following conditions are equivalent:*

(1) *In a neighborhood of x there exists a (possibly non-continuous) semispray ξ such that $\xi(x) \in \Delta(x)$.*

(2) $\operatorname{rank} B(x) = \operatorname{rank}(B \mid A)(x)$.

Proof. Let ξ be a semispray such that $\xi(x) \in \Delta(x)$. Then $\xi(x)$ satisfies the equations (5.24), and since $\xi^0 = 1$, $\xi_i^\sigma = q_{i+1}^\sigma$, $1 \leqslant \sigma \leqslant m$, $0 \leqslant i \leqslant s-2$, the equations (5.24) simplify to

$$(A_\sigma + B_{\sigma\nu}\xi_{s-1}^\nu)(x) = 0. \tag{5.26}$$

This system is solvable with respect to the $\xi_{s-1}^\nu(x)$'s, i.e., the condition (2) is satisfied.

Conversely, suppose $\operatorname{rank} B(x) = \operatorname{rank}(B \mid A)(x)$, and let β_x^ν, $1 \leqslant \nu \leqslant m$ be a solution of the equation $A_\sigma(x) + B_{\sigma\nu}(x)\beta_x^\nu = 0$. Putting

$$\xi = \frac{\partial}{\partial t} + \sum_{i=0}^{s-2} q_{i+1}^\sigma \frac{\partial}{\partial q_i^\sigma} + \xi_{s-1}^\sigma \frac{\partial}{\partial q_{s-1}^\sigma}$$

where $\xi_{s-1}^\sigma(x) = \beta_x^\sigma$, $1 \leqslant \sigma \leqslant m$, and $\xi_{s-1}^\sigma(y) = 0$ for all $y \neq x$ we get a semispray satisfying the condition (1). □

Taking into account the equations (5.26) we get

5.4.3. Theorem. *Let α be a Lagrangean system of order $s-1$, Δ its Euler-Lagrange distribution, let r be an integer, $1 \leqslant r \leqslant m$. Let $x \in J^{s-1}Y$ be a point. The following two conditions are equivalent:*

(1) *In a neighborhood of x there exist semisprays $\xi_1, \dots, \xi_r$ (not necessarily continuous) possessing the following properties:*

(1_a) *$\xi_1, \dots, \xi_r$ are linearly independent at x,*

(1_b) *$\xi_j(x) \in \Delta(x)$, $1 \leqslant j \leqslant r$,*

(1_c) *if ξ is another semispray in a neighborhood of x such that $\xi(x) \in \Delta(x)$ then the vectors $\xi_1(x), \dots, \xi_r(x), \xi(x)$ are linearly dependent.*

(2) $\operatorname{rank} \mathrm{B}(x) = \operatorname{rank}(\mathrm{B} \mid \mathrm{A})(x) = m + 1 - r$.

Proof. Suppose (1) and denote for $i = 1, \dots, r$ by $\beta_i = (\beta_i^\sigma)$ the components of ξ_i at $\partial/\partial q_{s-1}^\sigma$, $1 \leqslant \sigma \leqslant m$ at the point x. The β_i's satisfy the equations $\mathrm{A}(x) + \mathrm{B}(x)\beta_i = 0$, which means that $\operatorname{rank} \mathrm{B}(x) = \operatorname{rank}(\mathrm{B} \mid \mathrm{A})(x)$, and it holds $\beta_2 = \beta_1 + \bar{\beta}_2, \dots, \beta_r = \beta_1 + \bar{\beta}_r$ where $\bar{\beta}_2, \dots, \bar{\beta}_r$ are solutions of the homogeneous equations $\mathrm{B}(x)\bar{\beta} = 0$. If $\sum_{l=2}^r a_l \bar{\beta}_l = 0$ then $\sum_{l=2}^r a_l(\beta_l - \beta_1) = -(\sum_l a_l)\beta_1 + \sum_l a_l \beta_l = 0$, i.e., by assumption, $a_2 = \cdots = a_r = 0$, proving the linear independence of the $\bar{\beta}_i$'s. Let ξ be another semispray such that $\xi(x) \in \Delta(x)$. Then $\xi(x) = \sum_{j=1}^r b_j \xi_j(x)$ for some constants $b_1, \dots, b_r$ such that $b_1 + \cdots + b_r = 1$. Hence $\beta = b_1\beta_1 + \sum_{l=2}^r b_l(\beta_1 + \bar{\beta}_l) = \beta_1 + \sum_{l=2}^r \bar{\beta}_l$, proving that $\bar{\beta}_2, \dots, \bar{\beta}_r$ form a basis of solutions of the system of the homogeneous equations with the matrix $\mathrm{B}(x)$. This means that $r - 1 = m - \operatorname{rank} \mathrm{B}(x)$.

Conversely, taking a basis $\bar{\beta}_2, \dots, \bar{\beta}_r$ of solutions of $\mathrm{B}(x)\bar{\beta} = 0$ and a particular solution β_1 of $\mathrm{A}(x) + \mathrm{B}(x)\beta = 0$, and putting $\beta_2 = \beta_1 + \bar{\beta}_2, \dots, \beta_r = \beta_1 + \bar{\beta}_r$ we get similarly as in the proof of the preceding theorem a system of semisprays $\xi_1, \dots, \xi_r$ satisfying (1_b). Now, if $\sum b_j \xi_j(x) = 0$ for some constants b_j, $1 \leqslant j \leqslant r$, then $\sum b_j = 0$ and $\sum b_j \beta_j = 0$, i.e., $(\sum b_j)\beta_1 + \sum_{l=2}^r b_l \bar{\beta}_l = \sum_{l=2}^r b_l \bar{\beta}_l = 0$, which means that $b_l = 0$ for $2 \leqslant l \leqslant r$, and consequently also $b_1 = 0$, proving (1_a). If ξ is another semispray such that $\xi(x) \in \Delta(x)$ then $\beta = \beta_1 + \sum b_l \bar{\beta}_l = \beta_1 + \sum b_l(\beta_l - \beta_1)$; now, $\xi(x) = (1 - \sum_{l=2}^r b_l)\xi_1(x) + \sum_{l=2}^r b_l \xi_l(x)$, proving ($1_c$). $\square$

We shall use the following terminology suggested by theorems 5.4.1 and 5.4.2.

5.4.4. Definition. The subset of the phase space

$$\tilde{\mathcal{P}} = \{x \in J^{s-1}Y \mid \operatorname{rank} \mathrm{F} = \operatorname{rank}(\mathrm{F} \mid \mathrm{A})\}$$

will be called *primary constraint set*.

The set $\mathcal{P} \subset \tilde{\mathcal{P}}$ defined by

$$\mathcal{P} = \{x \in J^{s-1}Y \mid \operatorname{rank} \mathrm{B} = \operatorname{rank}(\mathrm{B} \mid \mathrm{A})\}$$

will be called *primary semispray-constraint set.*

Due to relations to physics, Theorems 5.4.1 and 5.4.2 can be called *Theorem on primary dynamical constraints*, and *Theorem on primary semispray constraints*, respectively. The reason for such a terminology becomes apparent in Chapter 7 when we shall deal in detail with singular Lagrangean systems.

5.5. The integration problem for variational ODE

Within the standard theory of distributions of a constant rank, the integration problem has two steps:

(1) to clarify the *structure of solutions* of the distribution in question (namely, if it is completely integrable then the maximal integral manifolds form a *foliation* of the given manifold),

(2) to *find the maximal integral manifolds*, which, in case of involutive distributions, practically means to find adapted charts to the foliation (= complete sets of independent first integrals).

For involutive distributions, the first problem is completely solved by the Frobenius theorem; the second problem is rather complicated and means to apply some of the known integration methods based on studying first integrals. If the distribution happens not to be involutive then no general theory, clarifying the structure of solutions and providing methods for finding them explicitly, is known.

We have shown that variational ODE can be interpreted as equations for *holonomic integral sections of the Euler-Lagrange distribution*. Consequently, the problem of integration of such equations identifies with the problem of *integration of a distribution*. However, *it is highly nontrivial, since only in a very particular case it is reduced to the known case of finding maximal integral manifolds of an involutive distribution of a constant rank* (we shall see that the only Lagrangean systems possessing this convenient property are the so called *regular* Lagrangean systems).

To be able to deal with all Lagrangean systems, we have to clarify the aspects of the integration problem in the general situation. Again, we can divide the integration problem into two steps:

(1) to clarify the *structure of solutions* of the corresponding Euler-Lagrange distribution,

(2) to *find the solutions explicitly*.

The second item clearly means to develop suitable integration methods, at least for some types of Lagrangean systems. This, however, is not possible without knowing the structure of solutions; we postpone this question to Chapter 9.

To discover the structure of solutions, i.e., the "dynamical picture", notice that if the order of the variational equations is $\geqslant 2$, one has to distinguish *two levels* of the integration problem:

(1) finding the extended dynamics = Hamilton extremals,

(2) finding the proper dynamics = prolonged extremals.

The first level of the problem means to solve the Euler-Lagrange distribution, more precisely, to find *integral sections* of this distribution.

The second level means to pick up only those solutions which are *holonomic* (i.e., of the form of a prolongation). Obviously, every extremal defines a one-dimensional (immersed) submanifold M of the phase space and a *semispray* ξ *along* M such that for every $y \in M$, $\xi(y)$ belongs to the Euler-Lagrange distribution Δ at y. Conversely, every vector field ξ belonging to the Euler-Lagrange distribution Δ, i.e., satisfying the Hamilton equation $i_\xi \alpha = 0$, and such that ξ is *a semispray along an at least one-dimensional immersed submanifold* M *of the phase space*, defines a solution of the Euler-Lagrange equations.

It should be stressed that ξ *need not be a semispray* everywhere on the phase space. This means that for finding the proper dynamics generally it is *not sufficient* to find the subdistribution of the Euler-Lagrange distribution, spanned by semisprays.

Within the range of the presymplectic mechanics, the problem of distinguishing solutions of the Euler-Lagrange equations in the set of solutions of the Hamilton equations is often called "the SODE (= second-order differential equation) problem". Here (in the context of our geometric study of higher-order Lagrangean systems) we shall speak about the *higher-order semispray problem.*

Since different Lagrangean systems can be described by Euler-Lagrange distributions with essentially different properties, the basic step in clarifying the dynamics of singular systems is their *geometric classification* according to the properties of their Euler-Lagrange distributions. The second step then consists in finding an *algorithm* for solving the problem of the structure of Hamilton extremals, and of the higher-order semispray problem. We shall deal with these questions in Chapter 7.

Chapter 6.

REGULAR LAGRANGEAN SYSTEMS

6.1. Introduction

By the standard definition of regularity (see e.g. P. Dedecker [7], M. de León and P. R. Rodrigues [1], and others), a Lagrangian L of order r is called *regular* if

$$\det\left(\frac{\partial^2 L}{\partial q_r^\sigma \partial q_r^\nu}\right) \neq 0. \tag{6.1}$$

This definition depends on a particular Lagrangian, hence, regularity is not a property of the Euler-Lagrange equations. This means that by this definition, a regular Lagrangian can be equivalent to a non-regular one. An easy example is provided by the Lagrangians

$$L_1 = \tfrac{1}{2}\dot{q}^2, \quad L_2 = \ddot{q} + \tfrac{1}{2}\dot{q}^2 = L_1 + \frac{d\dot{q}}{dt}$$

(which both define the classical free particle). Since the concept of regularity is in the standard approach closely connected with the Hamilton theory, this leads to Hamilton equations which depend on the particular Lagrangians and can be different by nature: taking two equivalent Lagrangians such that one is regular and the other is singular (as e.g. the above Lagrangians L_1, L_2), one is lead to consider the symplectic Hamilton equations (having a uniquely determined solution) for the first Lagrangian, and the presymplectic Hamilton equations for the second one; thus *the same dynamics* is deterministic (i.e., uniquely determined by the initial conditions) if the first Lagrangian is used, and possesses the properties of a constrained system in the second case. This inconsistency becomes even more transparent if the considered system cannot be represented by a global Lagrangian.

In the previous chapters we have defined a Lagrangean system as a Lepagean two-form (representing the class of *equivalent Lagrangians*), and we have shown that the dynamics of a Lagrangean system is described by a distribution on the phase space (the Euler-Lagrange distribution, or the characteristic distribution). Evidently, the most simple Lagrangean systems are those for which this distribution is of a *constant rank equal to one*; equivalently, such Lagrangean systems are locally represented by a *vector field* on the phase space. It is natural to call such Lagrangean systems *regular*.

Now, we shall be interested in this important class of Lagrangean systems in detail. We shall study the geometric structure of Hamilton extremals and prolonged extremals, which in this case turn out to coincide, and we shall also deal with the concept of *Legendre transformation*, which will be a generalization of the well-known Legendre transformation of classical mechanics.

The above new concept of regularity is fundamental especially for the Dirac theory of constraints and for quantum mechanics. The concept of a "regular Lagrangian" is enlarged to a wider class of Lagrangians than is that covered by the conventional definition of regularity (6.1). Moreover, regularity becomes the property of the Lepagean two-form, i.e., the *class of equivalent Lagrangians* (not of a particular Lagrangian, as in the case of definition (6.1)). Consequently, many problems which conventionally have to be treated by means of the Dirac theory, turn out to be quite simple: the Dirac theory for them is needed not at all (when applied, it even represents an additional complication which often leads to confusion—cf. e.g. the discussion in Y. Kaminaga [1]).

The theory of regular Lepagean two-forms also provides us with a new insight into the local inverse problem of the calculus of variations, including the problem of "variational integrating factors". We shall be interested in this topic in the last section of the present chapter.

We keep notations introduced in the previous chapters.

6.2. Regularity as a geometrical concept

6.2.1. Definition. A Lagrangean system (or, a locally variational form) will be called *regular* if its Euler-Lagrange distribution has a constant rank equal to one. A Lagrangian will be called *regular* if its Euler-Lagrange form is regular.

This concept of regularity has been proposed by O. Krupková [2]. A generalization of the standard definition of regularity similar to the above has been recently considered by D. J. Saunders [5].

Applying the results of the previous chapter to regular Lagrangean systems we get

6.2.2. Theorem. *Let α be a Lagrangean system on $J^{s-1}Y$, $s \geqslant 1$, let Δ (resp. $\mathcal{D}$) be its Euler-Lagrange (resp. characteristic) distribution. The following conditions are equivalent:*

(1) *α is regular.*

(2) rank *α is constant and maximal on $J^{s-1}Y$.*

(3) *The matrix $\mathrm{F} = (F^{jk}_{\sigma\nu})$ is regular on the phase space.*

(4) *At each point of $J^{s-1}Y$*

$$\det\left(\frac{\partial E_\sigma}{\partial q^\nu_s}\right) = \det(B_{\sigma\nu}) \neq 0\,. \tag{6.2}$$

(5) *Every point in the phase space has a neighborhood U such that $U \subset V_{s-1}$ for a fiber chart (V, ψ), $\psi = (t, q^\sigma)$ on Y, and the equation*

$$i_\zeta \alpha = 0 \tag{6.3}$$

has a unique solution ζ on U, satisfying the "scale condition"

$$T\pi_{s-1}.\zeta = \frac{\partial}{\partial t}. \tag{6.4}$$

(6) *Δ is a semispray connection.*

(7) *Hamilton equations and Euler-Lagrange equations of α are equivalent, i.e., every Hamilton extremal is holonomic.*

(8) $\Delta = \mathcal{D}$, *and it is a one-dimensional subdistribution of the Cartan distribution on the phase space.*

By the above theorem, regular Lagrangean systems of order $s-1$ (i.e., regular variational equations of order s) can be locally represented *equivalently* either by a semispray of order $s-1$ which is of the form

$$\zeta = \frac{\partial}{\partial t} + \sum_{i=0}^{s-2} q^{\sigma}_{i+1} \frac{\partial}{\partial q^{\sigma}_{i}} - B^{\sigma\nu} A_{\nu} \frac{\partial}{\partial q^{\sigma}_{s-1}}, \tag{6.5}$$

where $(B^{\sigma\nu})$ is the inverse matrix to $(B_{\sigma\nu})$, or by a Pfaffian system

$$A_{\sigma}\, dt + B_{\sigma\nu}\, dq^{\nu}_{s-1}, \quad \omega^{\sigma}, \quad \omega^{\sigma}_{1}, \quad \ldots, \quad \omega^{\sigma}_{s-2}, \quad 1 \leqslant \sigma \leqslant m. \tag{6.6}$$

The condition (6.2) is evidently independent of the choice of a fiber chart on $J^{s-1}Y$; we shall call it a *regularity condition.*

Regular Lagrangean systems on $J^{s-1}Y$ are characterized by the following dynamical picture:

The primary constraint set coincides with the primary semispray-constraint set, and $\tilde{\mathcal{P}} = \mathcal{P} = J^{s-1}Y$.

Hamilton extremals coincide with prolonged extremals.

Through every point of the phase space there passes a unique maximal Hamilton extremal (= prolonged extremal).

This means however, that regular Lagrangean systems obey the *Newtonian determinism* (the motion is uniquely determined by the initial conditions).

Note that the above concept of regularity completely corresponds to that in analytical mechanics: Indeed, within the range of classical analytical mechanics, where *only first-order Lagrangians* leading to *second-order equations* are considered, regularity, represented by the familiar condition

$$\det\left(\frac{\partial^2 L}{\partial \dot{q}^{\sigma}\, \partial \dot{q}^{\nu}}\right) \neq 0, \tag{6.7}$$

means in fact that:

– the Euler-Lagrange equations are solvable with respect to the second derivatives,

– the Euler-Lagrange equations can be represented by a vector field on the phase space,

– the Euler-Lagrange equations and the Hamilton equations are equivalent,

– Lagrangians equivalent to a regular Lagrangian are all regular (since classical analytical mechanics restricted itself to first-order Lagrangians).

Problems or even confusion appeared later, when people have started to study *nonclassical* Lagrangians (as e.g. Lagrangians linear in the velocities, or Lagrangians of order greater than one), but the concept of regularity have understood *heuristically* ((6.7) as a *definition* of regularity for *all* first-order Lagrangians, (6.1) as a *definition* of regularity for higher-order Lagrangians).

Among regular Lagrangean systems an important role is played by *symplectic* and *cosymplectic* systems. Later, when we shall study Lagrangean systems on a fibered manifold $\pi : R \times M \to R$, we shall see that there is a close correspondence between the symplectic (and cosymplectic) approach and the geometry of Lepagean two-forms (cf. Chapter 10).

6.2.3. Remark. By the variationality conditions (4.9), for *even-order Lagrangean systems* (i.e., odd-order variational equations) the matrix $\mathrm{B} = (B_{\sigma\nu})$ is *antisymmetric*. This means that *regular even*-order Lagrangean systems can exist only on such fibered manifolds for which $m = \dim Y - 1$ is *even*; in other words, *if m is odd then every even-order Lagrangean system is singular*.

In correspondence with classical mechanics we have the following definition.

6.2.4. Definition. Let $s \geqslant 1$, let α be a regular Lagrangean system of order $s - 1$. Let ζ be a Hamilton vector field of α defined on an open subset W of the phase space. If f is a function on W then the function

$$\partial_\zeta f \tag{6.8}$$

will be called *the evolution of f*.

Notice that if

$$\partial_\zeta f = 0, \tag{6.9}$$

i.e., if $i_\zeta df = 0$, then df belongs to the Euler-Lagrange distribution of α, and for every Hamilton extremal δ (= prolonged extremal), $\delta^* df = 0$. This means that f is a constant along prolonged extremals, i.e., f is a *first integral* of the Euler-Lagrange distribution of α.

6.2.5. Remark (Regularity of Hamilton equations). By Theorem 6.2.2, the Hamilton form is regular (as a first-order locally variational form) iff the matrix F is regular. Equivalently, *the Hamiltoninan system associated with a Lagrangean system is regular if and only if the corresponding Lagrangean system is regular*. In other words, Hamilton equations are regular iff the corresponding Euler-Lagrange equations are regular.

We remind the reader that the Hamiltonian system associated with any Lagrangean system α is of order *zero*, and that its minimal-order Lagrangians (i.e., extended Lagrangians of α) are *affine* in the first derivatives (cf. (5.14)).

6.3. Regularity conditions for Lagrangians

Usually, regularity is understood as a property of Lagrangians, and for equivalent Lagrangians the concept depends on the choice of a gauge function. On the other hand, Definition 6.2.1 has a clear geometrical meaning, and redefines regularity to be a property of the *class of equivalent Lagrangians*. Of course, one can rewrite this definition by means of particular Lagrangians and obtain in this way various *regularity conditions for Lagrangians*. Since it is useful to see the explicit form of such conditions, let us do this work at least for a few types of Lagrangians.

6.3.1. Regular first-order Lagrangians. *A first-order Lagrangian $L(t, q^\rho, q_1^\rho)$ is regular if and only if one of the following two conditions is satisfied:*

$$\det\Big(\frac{\partial^2 L}{\partial q_1^\sigma \partial q_1^\nu}\Big) \neq 0 \tag{6.10}$$

$$\frac{\partial^2 L}{\partial q_1^\sigma \partial q_1^\nu} = 0 \quad \forall \sigma, \nu, \quad \textit{and} \quad \det\Big(\frac{\partial^2 L}{\partial q^\sigma \partial q_1^\nu} - \frac{\partial^2 L}{\partial q_1^\sigma \partial q^\nu}\Big) \neq 0. \tag{6.11}$$

The familiar condition (6.10) describes all regular first-order Lagrangians the Euler-Lagrange equations of which are nontrivially of *second* order. In other words, Lagrangians satisfying the condition (6.10) are *minimal-order Lagrangians for regular first-order Lagrangean systems*; the phase space in this case is the manifold J^1Y.

The condition (6.11) describes all *regular* first-order Lagrangians which are *affine in the velocities*, i.e., of the form

$$L = f(t, q^\rho) + g_\nu(t, q^\rho) q_1^\nu,$$

and it says that a Lagrangian of this kind is regular if and only if

$$\det\Big(\frac{\partial g_\nu}{\partial q^\sigma} - \frac{\partial g_\sigma}{\partial q^\nu}\Big) \neq 0.$$

Note that for $m = 1$ (one-dimensional fibres) every Lagrangian affine in the velocities is singular, which is in correspondence with Remark 6.2.3.

Euler-Lagrange equations of Lagrangians affine in the velocities are clearly *first-order* ODE. This means that Lagrangians satisfying the condition (6.11) are *minimal-order Lagrangians for regular zero-order Lagrangean systems*. Notice that in the zero-order case the phase space coincides with the configuration space Y.

In particular, every extended Lagrangian is of this form, and we remind the reader that $\tilde{L}$ is regular if and only if L is regular. It is worthwhile to note at this place that according to a quantization method proposed by Faddeev and Jackiw (R. Jackiw [1]), to obtain the quantum mechanical description for a Lagrangian, the corresponding extended Lagrangian can be utilized.

6.3.2. Regular second-order Lagrangians. *A second-order Lagrangian $L(t, q^\rho, q_1^\rho, q_2^\rho)$ is regular if and only if one of the following four conditions is satisfied:*

$$\det\Big(\frac{\partial^2 L}{\partial q_2^\sigma \partial q_2^\nu}\Big) \neq 0, \tag{6.12}$$

$$\frac{\partial^2 L}{\partial q_2^\sigma \partial q_2^\nu} = 0 \quad \forall \sigma, \nu, \quad \textit{and} \quad \det\Big(\frac{\partial^2 L}{\partial q_1^\sigma \partial q_2^\nu} - \frac{\partial^2 L}{\partial q_2^\sigma \partial q_1^\nu}\Big) \neq 0, \tag{6.13}$$

$$\begin{gathered}\frac{\partial^2 L}{\partial q_2^\sigma \partial q_2^\nu} = 0, \quad \frac{\partial^2 L}{\partial q_1^\sigma \partial q_2^\nu} - \frac{\partial^2 L}{\partial q_2^\sigma \partial q_1^\nu} = 0 \quad \forall \sigma, \nu, \quad \textit{and} \\ \det\Big(\frac{\partial^2 L}{\partial q_1^\sigma \partial q_1^\nu} - \frac{\partial^2 L}{\partial q_2^\sigma \partial q^\nu} - \frac{\partial^2 L}{\partial q^\sigma \partial q_2^\nu} - \frac{d}{dt}\frac{\partial^2 L}{\partial q_1^\sigma \partial q_2^\nu}\Big) \neq 0.\end{gathered} \tag{6.14}$$

$$\frac{\partial^2 L}{\partial q_2^\sigma \partial q_2^\nu} = 0 , \quad \frac{\partial^2 L}{\partial q_1^\sigma \partial q_2^\nu} - \frac{\partial^2 L}{\partial q_2^\sigma \partial q_1^\nu} = 0 ,$$

$$\frac{\partial^2 L}{\partial q_1^\sigma \partial q_1^\nu} - \frac{\partial^2 L}{\partial q_2^\sigma \partial q^\nu} - \frac{\partial^2 L}{\partial q^\sigma \partial q_2^\nu} - \frac{d}{dt} \frac{\partial^2 L}{\partial q_1^\sigma \partial q_2^\nu} = 0 \quad \forall \sigma, \nu, \quad \textit{and} \tag{6.15}$$

$$\det\Big(\frac{\partial^2 L}{\partial q^\sigma \partial q_1^\nu} - \frac{\partial^2 L}{\partial q_1^\sigma \partial q^\nu} + \frac{d}{dt}\Big(\frac{\partial^2 L}{\partial q_2^\sigma \partial q^\nu} - \frac{\partial^2 L}{\partial q^\sigma \partial q_2^\nu}\Big)\Big) \neq 0 .$$

(6.12) is the regularity condition for second order Lagrangians the Euler-Lagrange equations of which are nontrivially of order *four*. Otherwise speaking, Lagrangians satisfying (6.12) are *minimal-order Lagrangians for regular third-order Lagrangean systems*. For such Lagrangean systems the phase space is J^3Y.

(6.13), (6.14) and (6.15) are regularity conditions for second order Lagrangians *affine in accelerations*, i.e., of the form

$$L = f(t, q^\rho, q_1^\rho) + g_\nu(t, q^\rho, q_1^\rho) q_2^\nu . \tag{6.16}$$

In this notation, the condition (6.13) reads

$$\det\Big(\frac{\partial g_\nu}{\partial q_1^\sigma} - \frac{\partial g_\sigma}{\partial q_1^\nu}\Big) \neq 0. \tag{6.17}$$

Lagrangians of this kind are *minimal-order Lagrangians for regular second-order Lagrangean systems*; their Euler-Lagrange equations are of order *three*, the phase space is J^2Y.

Similarly, with help of (6.16), conditions (6.14) read

$$\frac{\partial g_\nu}{\partial q_1^\sigma} = \frac{\partial g_\sigma}{\partial q_1^\nu} , \quad \det\Big(\frac{\partial^2 f}{\partial q_1^\sigma \partial q_1^\nu} - \frac{\partial g_\sigma}{\partial q^\nu} - \frac{\partial g_\nu}{\partial q^\sigma} - \frac{\partial^2 g_\nu}{\partial t \partial q_1^\sigma} - \frac{\partial^2 g_\nu}{\partial q^\rho \partial q_1^\sigma} q_1^\rho\Big) \neq 0 . \tag{6.18}$$

Lagrangians of this kind define *regular first-order Lagrangean systems*, i.e., their Euler-Lagrange equations are of *second* order and the phase space is J^1Y. Notice that Lagrangians of this kind are obviously reducible to first-order Lagrangians, the reducibility conditions being the first set of (6.18).

Finally, conditions (6.15) take the form

$$\frac{\partial g_\nu}{\partial q_1^\sigma} = \frac{\partial g_\sigma}{\partial q_1^\nu}, \quad \frac{\partial^2 f}{\partial q_1^\sigma \partial q_1^\nu} - \frac{\partial g_\sigma}{\partial q^\nu} - \frac{\partial g_\nu}{\partial q^\sigma} - \frac{\partial^2 g_\nu}{\partial t \partial q_1^\sigma} - \frac{\partial^2 g_\nu}{\partial q^\rho \partial q_1^\sigma} q_1^\rho = 0 , \quad \text{and}$$

$$\det\Big(\frac{\partial^2 f}{\partial q^\sigma \partial q_1^\nu} - \frac{\partial^2 f}{\partial q_1^\sigma q^\nu} + \frac{\partial}{\partial t}\Big(\frac{\partial g_\sigma}{\partial q^\nu} - \frac{\partial g_\nu}{\partial q^\sigma}\Big) + \frac{\partial}{\partial q^\rho}\Big(\frac{\partial g_\sigma}{\partial q^\nu} - \frac{\partial g_\nu}{\partial q^\sigma}\Big) q_1^\rho\Big) \neq 0. \tag{6.19}$$

Such Lagrangians obviously define *regular zero-order Lagrangean systems*; their Euler-Lagrange equations are of order *one* and the phase space is the manifold Y. Again, the first two sets of (6.19) are the reducibility conditions, while the last set of (6.19) is a true regularity condition.

6.3.3. Regularity conditions for minimal-order Lagrangians.

(1) *A minimal-order Lagrangian $\lambda_{\min}$ on $W \subset J^cY$ for an even-order locally variational form E on $J^{2c}Y$ is regular if and only if*

$$\det\Big(\frac{\partial^2 L}{\partial q_c^\sigma \partial q_c^\nu}\Big) \neq 0. \tag{6.20}$$

(2) *A minimal-order Lagrangian $\lambda_{\min}$ on $W \subset J^{c+1}Y$ for an odd-order locally variational form E on $J^{2c+1}Y$ is regular if and only if*

$$\det\left(\frac{\partial^2 L}{\partial q^\sigma_{c+1}\partial q^\nu_c} - \frac{\partial^2 L}{\partial q^\sigma_c \partial q^\nu_{c+1}}\right) \neq 0. \tag{6.21}$$

Similarly as in the above examples, one could obtain regularity conditions for any other types of Lagrangians. However, from the point of view of practical computations there is no need to search for such conditions, since one can easily work with the regularity condition in the form (6.2).

We have seen that the standard formula (6.1) is not appropriate as a *definition* of regularity. However, it can be viewed as one of the regularity *conditions*. Its meaning clearly is the following: if λ satisfies (6.1) then it is a minimal-order Lagrangian for a regular odd-order Lagrangean system; conversely, if an odd-order Lagrangean system is regular then every minimal-order Lagrangian satisfies (6.1).

The heuristic concept of regularity represented by formula (6.1) does not cover many important regular Lagrangians. Note that, in particular, within the standard approach, *no* Lagrangian leading to *odd-order Euler-Lagrange equations* (i.e., no even-order Lagrangean system) is regular. Consequently, this means that *all* Lagrangean systems of this kind conventionally have to be treated as singular with help of the Dirac theory of constrained systems. Thus, in many cases, the dynamics which is in fact locally representable by means of a semispray and is uniquelly determined by the initial conditions, is studied by means of complicated tools leading often to not very transparent results (cf. e.g. J. Barcelos-Neto and N. R. F. Braga [1], C. Batlle, J. Gomis, J. M. Pons and N. Román-Roy [1,2,3], D. Chinea, M. de León and J. C. Marrero [1], P. A. M. Dirac [1], M. de León and D. M. de Diego [1], M. de León, D. M. de Diego and P. Pitanga [1], V. V. Nesterenko [1], E. C. G. Sudarshan and N. Mukunda [1], and others).

6.4. Legendre transformation

The geometric concept of regularity leads to an understanding of Legendre transformations as of distinguished coordinate transformations on the phase space which are related directly with regular *Euler-Lagrange forms* (not with Lagrangians as it is standard). Consequently, a generalization of the conventional definition of Legendre transformation is obtained. We shall see that, on the contrary to the standard approach, *not every Lagrangian is appropriate to define a Legendre transformation*: in case of odd-order Lagrangean systems such "admissible Lagrangians" are just the minimal-order Lagrangians; in case of even-order Lagrangean systems the family of "admissible Lagrangians" is even closer, being only a subfamily of minimal-order Lagrangians.

We have shown in Chapter 4 that a Lagrangean system (or equivalently, a locally variational form) can be locally represented in a canonical form if one makes use of momenta and the related Hamiltonian (which are functions defined on an open subset of the phase space). An important property of *regular* Lagrangean systems is the possibility to use momenta as a part of *local coordinates* on the phase space. Namely, we have

6.4.1. Theorem. *Let α be a regular Lagrangean system of order $s-1 \geqslant 0$.*

(1) *Let $s = 2c$. Then for any family of momenta $(p_\nu, p^1_\nu, \dots, p^{c-1}_\nu)$ of α defined on an open subset W of the phase space, such that $W \subset V_{s-1}$ where (V, ψ), $\psi = (t, q^\sigma)$ is a fiber chart on Y, the mapping*

$$(t, q^\sigma, \dots, q^\sigma_{c-1}, q^\sigma_c, \dots, q^\sigma_{2c-1}) \to (t, q^\sigma, \dots, q^\sigma_{c-1}, p^{c-1}_\sigma, \dots, p_\sigma) \tag{6.22}$$

is a coordinate transformation on W.

(2) *Let $s = 2c+1$. Let x be a point in the phase space, let (V, ψ), $\psi = (t, q^\sigma)$ be a fiber chart at $\pi_{s-1,0}(x)$. There is a neighborhood W of x and a family of momenta $(p_\nu, p^1_\nu, \dots, p^c_\nu)$ of α defined on W such that $W \subset V_{s-1}$ and*

$$(t, q^\sigma, \dots, q^\sigma_{c-1}, q^\sigma_c, \dots, q^\sigma_{2c}) \to (t, q^\sigma, \dots, q^\sigma_{c-1}, p^c_\sigma, \dots, p_\sigma) \tag{6.23}$$

is a coordinate transformation on W.

Proof. If $(p_\sigma, \dots, p^{s-c-1}_\sigma)$ is a family of momenta of α, consider the square matrix

$$\Pi = \left(\frac{\partial p^k_\nu}{\partial q^\sigma_i} \right), \tag{6.24}$$

where ν, k label rows and σ, i label columns, $1 \leqslant \nu \leqslant m$, $0 \leqslant k \leqslant s-c-1$ and $1 \leqslant \sigma \leqslant m$, $c \leqslant i \leqslant s-1$.

(1) Let $s = 2c$. It is sufficient to show that $\det \Pi \neq 0$. By (4.41) we have

$$\Pi = (2F^{ik}_{\sigma\nu}), \quad c \leqslant i \leqslant 2c-1, \quad 0 \leqslant k \leqslant c-1. \tag{6.25}$$

Hence, the absolute value of $\det \Pi$ satisfies

$$|\det \Pi| = |\det \mathrm{B}|^c .$$

Now, if α is regular then the matrix Π is regular.

(2) Let $s = 2c+1$. Using (4.41) we get

$$\Pi = \left(\frac{\partial p^k_\nu}{\partial q^\sigma_c} \quad 2F^{ik}_{\sigma\nu} \right), \quad c+1 \leqslant i \leqslant 2c, \quad 0 \leqslant k \leqslant c. \tag{6.26}$$

For the absolute value of $\det \Pi$ we get

$$|\det \Pi| = |\det \mathrm{B}|^c \cdot |\det \mathrm{P}|,$$

where

$$\mathrm{P} = \left(\frac{\partial p^c_\nu}{\partial q^\sigma_c} \right) \tag{6.27}$$

If $\det \mathrm{P} \neq 0$, we are done, since by assumption $\det \mathrm{B} \neq 0$. Hence, suppose that $\det \mathrm{P} = 0$. Let x be a point. We shall show that there is a neighborhood W of x and a family of momenta $\{\bar{p}_\nu, \dots, \bar{p}^c_\nu\}$ on W such that

$$\det \left(\frac{\partial \bar{p}^c_\nu}{\partial q^\sigma_c} \right) \neq 0 \tag{6.28}$$

at each point of W. Consider the singular matrix $\mathrm{P}(x)$. Choose a symmetric $(m \times m)$-matrix $\Phi = (\Phi_{\sigma\nu})$ in such a way that

$$\det(\mathrm{P}(x) + \Phi) \neq 0.$$

Let $\phi(t, q^\sigma, \ldots, q^\sigma_c)$ be a function defined in a neighborhood of x and such that

$$\Phi_{\sigma\nu} = \frac{\partial^2\phi}{\partial q^\sigma_c \partial q^\nu_c}(x).$$

By the continuity property of the determinant we can see that there is a neighborhood W of x such that

$$\det\left(\frac{\partial p^c_\nu}{\partial q^\sigma_c} + \frac{\partial^2\phi}{\partial q^\sigma_c \partial q^\nu_c}\right) \neq 0$$

on W. Now, putting

$$\bar{p}^k_\nu = p^k_\nu + \frac{\partial\phi}{\partial q^\sigma_k}, \qquad 0 \leqslant k \leqslant c,$$

we get a family of momenta which satisfies (6.28), proving that for a regular Lagrangean system the mapping $(t, q^\sigma, \ldots, q^\sigma_{2c}) \to (t, q^\sigma, \ldots, q^\sigma_{c-1}, \bar{p}_\sigma, \ldots, \bar{p}^c_\sigma)$ is a coordinate transformation on W. □

Note that we have proved

6.4.2. Proposition. *Let α be a regular Lagrangean system of order $s-1 = 2c$, $c \geqslant 0$. Let x be a point in the phase space, let (V, ψ), $\psi = (t, q^\sigma)$ be a fiber chart at $\pi_{2c,0}(x)$. There exist a neighborhood W of x, $W \subset V_{2c}$, and a minimal-order Lagrangian $\lambda_{\min}$ on $\pi_{2c,c+1}(W)$ such that*

$$\det\left(\frac{\partial^2 L_{\min}}{\partial q^\sigma_{c+1} \partial q^\nu_c}\right) \neq 0. \tag{6.29}$$

In particular, if α is a regular Lagrangean system of order *zero*, then in a neighborhood of every point in Y, there is a family of momenta (p_σ) such that $(t, q^\sigma) \to (t, p_\sigma)$ is a coordinate transformation.

6.4.3. Definition. Let α be a regular Lagrangean system on $J^{s-1}Y$, $s \geqslant 1$. A chart (W, φ), $\varphi = (t, q^\sigma, \ldots, q^\sigma_{c-1}, p_\sigma, \ldots, p^{s-c-1}_\sigma)$ on the phase space satisfying the conditions of Theorem 6.4.1 will be called a *Legendre chart*, and the corresponding coordinates will be called *Legendre coordinates*. The transformation from fibered to Legendre coordinates will be called *Legendre transformation*.

Taking into account Theorem 4.5.8 we can compute Legendre coordinates directly from appropriate Lagrangians.

6.4.4. Corollary. *Let α be a regular Lagrangean system on $J^{s-1}Y$, $s \geqslant 1$.*

(1) *Let $s = 2c$. Then for any minimal-order Lagrangian $\lambda_{\min}$ of α, the functions $(t, q^\sigma, \ldots, q^\sigma_{c-1}, (f_{\min})^1_\sigma, \ldots, (f_{\min})^c_\sigma)$ are Legendre coordinates of α.*

(2) *Let $s = 1$. Then for any minimal-order Lagrangian $\lambda_{\min}$ of α satisfying the condition*

$$\det\left(\frac{\partial^2 L_{\min}}{\partial \dot{q}^\sigma \partial q^\nu}\right) \neq 0,$$

the $(t, (f_{\min})^1_\sigma)$ are Legendre coordinates of α.

(3) *If $s = 2c+1$, $c > 0$, then for any minimal-order Lagrangian satisfying (6.29) the $(t, q^\sigma, \ldots, q^\sigma_{c-1}, (f_{\min})^1_\sigma, \ldots, (f_{\min})^{c+1}_\sigma)$ are Legendre coordinates of α.*

6.4.5. Remark. Notice that regular even-order Lagrangean systems (i.e., odd-order variational equations) substantially differ from regular odd-order Lagrangean systems (described by even-order variational equations). Namely,

– their Legendre coordinates consist from more p's that q's,

– not every set of momenta can be used to define Legendre coordinates (in other words, for getting Legendre coordinates, not every minimal-order Lagrangian is admissible).

In particular, all the above concerns the *Hamiltonian systems*, i.e., the zero-order systems associated to Euler-Lagrange equations (cf. Remark 6.2.5).

The *inverse to the Legendre transformation* is obtained for $s = 1$ in an obvious way. For $s > 1$ we get

6.4.6. Proposition. *Let $s \geqslant 2$ and α be a regular Lagrangean system on $J^{s-1}Y$. Let (W, φ), $\varphi = (t, q^\sigma, \dots, q^\sigma_{c-1}, p_\sigma, \dots, p_\sigma^{s-c-1})$ be a Legendre chart of α, let H be the associated Hamiltonian. Then the mapping $(t, q^\sigma, \dots, q^\sigma_{c-1}, p_\sigma, \dots, p_\sigma^{s-c-1}) \to (\bar{t}, \bar{q}^\sigma, \dots, \bar{q}^\sigma_{s-1})$ defined by*

$$\begin{aligned} \bar{t} &= t, \\ \bar{q}^\sigma &= q^\sigma, \\ \bar{q}^\sigma_i &= \frac{\partial H}{\partial p^\sigma_{i-1}}, \quad 1 \leqslant i \leqslant c, \\ \bar{q}^\sigma_{c+k} &= \frac{d^k}{dt^k}\frac{\partial H}{\partial p^{c-1}_\nu}, \quad 1 \leqslant k \leqslant s-c-1 \end{aligned} \tag{6.30}$$

is the inverse to the Legendre transformation.

Proof. It is sufficient to prove that $\bar{q}^\sigma_i = q^\sigma_i$, $1 \leqslant i \leqslant c$.

Let $s = 2c+1$. By Theorem 4.5.8 and Proposition 6.4.2 there exists a minimal-order Lagrangian $L_{\min}$ such that (6.29) holds and

$$H = -L_{\min} + \sum_{i=0}^{c-2} p^i_\sigma q^\sigma_{i+1} + p^{c-1}_\sigma q^\sigma_c + p^\sigma_c q^\sigma_{c+1} \,. \tag{6.31}$$

From the definition of Legendre transformation by means of minimal-order Lagrangians we can see that

$$\frac{\partial q^\sigma_c}{\partial p^k_\nu} = 0, \quad 0 \leqslant k \leqslant c-1, \tag{6.32}$$

and that the matrix $(\partial q^\sigma_c/\partial p^c_\nu)$ is regular and inverse to $(\partial p^c_\sigma/\partial q^\nu_c)$. Using (6.30) we get

$$\bar{q}^\sigma_{i+1} = \frac{\partial H}{\partial p^i_\sigma} = q^\sigma_{i+1}, \quad 0 \leqslant i \leqslant c-1 .$$

Note that for the matrices

$$\mathrm{P} = \left(\frac{\partial p^c_\sigma}{\partial q^\nu_c}\right) = \left(\frac{\partial^2 L}{\partial q^\sigma_{c+1}\partial q^\nu_c}\right), \quad \mathrm{Q} = \left(\frac{\partial q^\sigma_c}{\partial p^c_\nu}\right) = \left(\frac{\partial^2 H}{\partial p^{c-1}_\sigma \partial p^c_\nu}\right) \tag{6.33}$$

we have $\mathrm{Q} = \mathrm{P}^{-1}$.

For $s = 2c$ the proof is similar. □

6.5. Legendre chart expressions

Consider a regular Lagrangean system α on $J^{s-1}Y$. Expressing the two-form α, the Hamilton form, Hamilton equations, extended Lagrangians or generators of the Euler-Lagrange distribution in Legendre coordinates we get them in the so called *canonical form*.

Denote by (W, φ), $\varphi = (t, q^\sigma, \ldots, q^\sigma_{c-1}, p_\sigma, \ldots, p^{s-c-1}_\sigma)$ a Legendre chart of α on $J^{s-1}Y$, and by $((\pi_{s-1})^{-1}_{1,0}W, \varphi_{1,0})$, $\varphi_{1,0} = (t, q^\sigma_i, p^j_\sigma, q^\sigma_{i,1}, p^j_{\sigma,1})$, $0 \leqslant i \leqslant c-1$, $0 \leqslant j \leqslant s-c-1$, the associated chart on $J^1(J^{s-1}Y)$. Further denote

$$\begin{aligned} \tilde{p}\, dq^\sigma_i &\equiv \tilde{\omega}^\sigma_i = dq^\sigma_i - q^\sigma_{i,1}\, dt, \\ \tilde{p}\, dp^j_\sigma &\equiv \tilde{\pi}^j_\sigma = dp^j_\sigma - p^j_{\sigma,1}\, dt, \end{aligned} \tag{6.34}$$

for $0 \leqslant i \leqslant c-1,\ , 0 \leqslant j \leqslant s-c-1$.

Now, if $s = 2c$, we have

$$\alpha = -dH \wedge dt + \sum_{k=0}^{c-1} dp^k_\nu \wedge dq^\nu_k, \tag{6.35}$$

hence

$$(\pi_{s-1})^*_{1,0}\alpha = -\sum_{i=0}^{c-1}\left(\frac{\partial H}{\partial q^\sigma_i} + p^i_{\sigma,1}\right)\tilde{\omega}^\sigma_i \wedge dt - \sum_{i=0}^{c-1}\left(\frac{\partial H}{\partial p^i_\sigma} - q^\sigma_{i,1}\right)\tilde{\pi}^i_\sigma \wedge dt + \sum_{i=0}^{c-1}\tilde{\pi}^i_\sigma \wedge \tilde{\omega}^\sigma_i,$$

and we get

$$\mathcal{H} = \tilde{p}_1\alpha = -\sum_{i=0}^{c-1}\left(\frac{\partial H}{\partial q^\sigma_i} + p^i_{\sigma,1}\right)\tilde{\omega}^\sigma_i \wedge dt - \sum_{i=0}^{c-1}\left(\frac{\partial H}{\partial p^i_\sigma} - q^\sigma_{i,1}\right)\tilde{\pi}^i_\sigma \wedge dt \tag{6.36}$$

for the *Hamilton form*,

$$\begin{gathered} -\frac{\partial H}{\partial q^\sigma_i} - \frac{d}{dt}(p^i_\sigma \circ \delta) = 0, \quad -\frac{\partial H}{\partial p^i_\sigma} + \frac{d}{dt}(q^\sigma_i \circ \delta) = 0, \\ 1 \leqslant \sigma \leqslant m, \quad 0 \leqslant i \leqslant c-1, \end{gathered} \tag{6.37}$$

for *Hamilton equations*, and

$$\begin{aligned} \mathcal{D} = \Delta &= \operatorname{span}\left\{\frac{\partial}{\partial t} + \sum_{k=0}^{c-1}\frac{\partial H}{\partial p^k_\nu}\frac{\partial}{\partial q^\nu_k} - \sum_{k=0}^{c-1}\frac{\partial H}{\partial q^\nu_k}\frac{\partial}{\partial p^k_\nu}\right\} \\ &\approx \operatorname{span}\left\{-\frac{\partial H}{\partial q^\nu_k}dt - dp^k_\nu,\ -\frac{\partial H}{\partial p^k_\nu}dt + dq^\nu_k,\ 0 \leqslant k \leqslant c-1,\ 1 \leqslant \nu \leqslant m\right\} \end{aligned} \tag{6.38}$$

for the *characteristic* and the *Euler-Lagrange distribution*. Now, the evolution of a function f reads

$$\partial_\zeta f = \frac{\partial f}{\partial t} + \sum_{k=0}^{c-1}\frac{\partial H}{\partial p^k_\nu}\frac{\partial f}{\partial q^\nu_k} - \sum_{k=0}^{c-1}\frac{\partial H}{\partial q^\nu_k}\frac{\partial f}{\partial p^k_\nu}. \tag{6.39}$$

If $\lambda_{\min} = L\, dt$ is a minimal-order Lagrangian then

$$(\pi_{s-1})^*_{1,0}\theta_{\lambda_{\min}} = \left(-H + \sum_{i=0}^{c-1} p^i_\sigma q^\sigma_{i,1}\right)dt + \sum_{i=0}^{c-1} p^i_\sigma \tilde{\omega}^\sigma_i,$$

hence,

$$\tilde{L} = -H + \sum_{i=0}^{c-1} p^i_\sigma q^\sigma_{i,1} \tag{6.40}$$

is the corresponding *extended Lagrangian*.

If $s = 2c+1$, we get the following Legendre chart expressions:

$$\alpha = -dH \wedge dt + \sum_{k=0}^{c-1} dp^k_\nu \wedge dq^\nu_k + dp^c_\nu \wedge dq^\nu_c, \tag{6.41}$$

$$\begin{aligned}\mathcal{H} = \tilde{p}_1\alpha = &-\sum_{i=0}^{c-1}\Big(\frac{\partial H}{\partial q^\sigma_i} + p^i_{\sigma,1} + p^c_{\nu,1}\frac{\partial q^\nu_c}{\partial q^\sigma_i}\Big)\tilde{\omega}^\sigma_i \wedge dt\\ &-\sum_{i=0}^{c-1}\Big(\frac{\partial H}{\partial p^i_\sigma} - q^\sigma_{i,1}\Big)\tilde{\pi}^i_\sigma \wedge dt - \Big(\frac{\partial H}{\partial p^c_\sigma} - \frac{\tilde{d}q^\sigma_c}{dt} + p^c_{\nu,1}\frac{\partial q^\nu_c}{\partial p^c_\sigma}\Big)\tilde{\pi}^c_\sigma \wedge dt\end{aligned} \tag{6.42}$$

for the *Hamilton form*, and

$$\begin{aligned}&-\frac{\partial H}{\partial q^\sigma_i} - \frac{d}{dt}(p^i_\sigma \circ \delta) - \frac{\partial q^\nu_c}{\partial q^\sigma_i}\frac{d}{dt}(p^c_\nu \circ \delta) = 0, \quad -\frac{\partial H}{\partial p^i_\sigma} + \frac{d}{dt}(q^\sigma_i \circ \delta) = 0,\\ &-\frac{\partial H}{\partial p^c_\sigma} + \frac{\partial q^\sigma_c}{\partial t} + \sum_{k=0}^{c-1}\frac{\partial q^\sigma_c}{\partial q^\nu_k}\frac{d}{dt}(q^\nu_k \circ \delta) + \Big(\frac{\partial q^\sigma_c}{\partial p^c_\nu} - \frac{\partial q^\nu_c}{\partial p^c_\sigma}\Big)\frac{d}{dt}(p^c_\nu \circ \delta) = 0,\\ &\quad 1 \leqslant \sigma \leqslant m, \quad 0 \leqslant i \leqslant c-1,\end{aligned} \tag{6.43}$$

for the canonical *Hamilton equations*. To express Hamilton vector fields, let us consider the matrices P, Q, Q = P^{-1}, defined by (6.33), and denote

$$\begin{aligned}N_{\sigma\nu} &= \frac{\partial p^c_\sigma}{\partial q^\nu_c} - \frac{\partial p^c_\nu}{\partial q^\sigma_c} = \frac{\partial^2 L}{\partial q^\sigma_{c+1}\partial q^\nu_c} - \frac{\partial^2 L}{\partial q^\sigma_c \partial q^\nu_{c+1}}\\ M^{\sigma\nu} &= \frac{\partial q^\sigma_c}{\partial p^c_\nu} - \frac{\partial q^\nu_c}{\partial p^c_\sigma} = \frac{\partial^2 H}{\partial p^{c-1}_\sigma \partial p^c_\nu} - \frac{\partial^2 H}{\partial p^{c-1}_\nu \partial p^c_\sigma};\end{aligned} \tag{6.44}$$

both the matrices N = $(N_{\sigma\nu})$ and M = $(M^{\sigma\nu})$ are regular, since N is regular by assumption, and

$$\mathrm{M} = -\mathrm{Q}\,\mathrm{N}\,\mathrm{Q}^T, \tag{6.45}$$

where Q^T is the transpose of Q. Denote by $\mathrm{M}^{-1} = (M_{\sigma\nu})$ the inverse matrix to M. Now, we get

$$\begin{aligned}\mathcal{D} = \Delta = \operatorname{span}&\Big\{\frac{\partial}{\partial t} + \sum_{k=0}^{c-1}\frac{\partial H}{\partial p^k_\nu}\frac{\partial}{\partial q^\nu_k} - \sum_{k=0}^{c-1}\Big(\frac{\partial H}{\partial q^\nu_k} + \frac{\partial q^\sigma_c}{\partial q^\nu_k}\zeta^c_\sigma\Big)\frac{\partial}{\partial p^k_\nu} + \zeta^c_\nu\frac{\partial}{\partial p^c_\nu}\Big\}\\ \approx \operatorname{span}&\Big\{-\frac{\partial H}{\partial q^\nu_k}dt - dp^k_\nu - \frac{\partial q^\sigma_c}{\partial q^\nu_k}dp^c_\sigma, -\frac{\partial H}{\partial p^k_\nu}dt + dq^\nu_k,\\ &-\frac{\partial H}{\partial p^c_\nu}dt + dq^\nu_c - \frac{\partial q^\sigma_c}{\partial p^c_\nu}dp^c_\sigma, \ 0 \leqslant k \leqslant c-1, \ 1 \leqslant \nu \leqslant m\Big\},\end{aligned} \tag{6.46}$$

for the *characteristic* and the *Euler-Lagrange distribution*, and

$$\partial_\zeta f = \frac{\partial f}{\partial t} + \sum_{k=0}^{c-1} \frac{\partial H}{\partial p_\nu^k} \frac{\partial f}{\partial q_k^\nu} - \sum_{k=0}^{c-1} \left(\frac{\partial H}{\partial q_k^\nu} + \frac{\partial q_c^\sigma}{\partial q_k^\nu} \zeta_\sigma^c \right) \frac{\partial f}{\partial p_\nu^k} + \zeta_\nu^c \frac{\partial f}{\partial p_\nu^c}, \tag{6.47}$$

for the *evolution of* f, where

$$\zeta_\sigma^c = M_{\sigma\rho} \Big(\frac{\partial H}{\partial p_\rho^c} - \frac{\partial q_c^\rho}{\partial t} - \sum_{i=0}^{c-1} \frac{\partial q_c^\rho}{\partial q_i^\nu} \frac{\partial H}{\partial p_\nu^i} \Big). \tag{6.48}$$

Finally, *extended Lagrangians* are of the form

$$\tilde{L} = -H + \sum_{i=0}^{c-1} p_\sigma^i q_{i,1}^\sigma + p_\sigma^c \frac{\tilde{d} q_c^\sigma}{dt}. \tag{6.49}$$

In the above formulas, $\tilde{d}/dt$ refers to the total derivative related with the fibered projection π_{s-1}, the q_c^σ's are considered as functions of the Legendre coordinates, $q_c^\sigma = \partial H / \partial p_\sigma^{c-1}$, and the relations (6.32) are used.

Notice that the extended Lagrangians are *regular* first-order Lagrangians, *affine in the first derivatives* (which is in correspondence with the fact that they define regular Lagrangean systems of order zero).

Although *zero-order Lagrangean systems* are a particular case of even-order Lagrangean systems, it is worth to write down explicit formulas for them.

We have seen in Section 4.6 that minimal-order Lagrangians for zero-order Lagrangean systems coincide with the first-order Lagrangians affine in the velocities, i.e., of the form

$$L = f(t, q^\rho) + g_\sigma(t, q^\rho) \dot{q}^\sigma. \tag{6.50}$$

Hamiltonians and momenta of a zero-order Lagrangean system are by (4.55)

$$p_\nu = g_\nu - \frac{\partial \phi}{\partial q_\nu}, \qquad H = -f + \frac{\partial \phi}{\partial t}, \tag{6.51}$$

where $\phi(t, q^\sigma)$ is an arbitrary gauge function of order zero. Note that all the Hamiltonians and momenta are functions *of order zero*.

Now, suppose that our Lagrangean system α is regular, i.e., every its minimal-order Lagrangian L satisfies the condition

$$\det \Big(\frac{\partial g_\nu}{\partial q^\sigma} - \frac{\partial g_\sigma}{\partial q^\nu} \Big) \neq 0 \tag{6.52}$$

(cf. (6.11)). Then we can choose ϕ in such a way that the mapping $(t, q^\sigma) \to (t, p_\sigma)$ is a Legendre transformation on an open subset of Y. In the Legendre coordinates, the Lagrangean system defined by L is of the form

$$\alpha = -dH \wedge dt + dp_\sigma \wedge dq^\sigma = \Big(\frac{\partial q^\sigma}{\partial t} - \frac{\partial H}{\partial p_\sigma} \Big) dp_\sigma \wedge dt + \frac{1}{2} \Big(\frac{\partial q^\sigma}{\partial p_\nu} - \frac{\partial q^\nu}{\partial p_\sigma} \Big) dp_\sigma \wedge dp_\nu,$$

and the Hamilton canonical equations are

$$\Big(\frac{\partial q^\sigma}{\partial p_\nu} - \frac{\partial q^\nu}{\partial p_\sigma} \Big) \frac{d}{dt} (p_\nu \circ \delta) = \frac{\partial H}{\partial p_\sigma} - \frac{\partial q^\sigma}{\partial t}. \tag{6.53}$$

The Euler-Lagrange distribution is locally spanned by the Hamilton vector field

$$\zeta = \frac{\partial}{\partial t} + M_{\sigma\nu}\Big(\frac{\partial H}{\partial p_\nu} - \frac{\partial q^\nu}{\partial t}\Big)\frac{\partial}{\partial p_\sigma}, \tag{6.54}$$

where $M = (M_{\sigma\nu})$ is the inverse matrix to

$$N = \Big(\frac{\partial q^\sigma}{\partial p_\nu} - \frac{\partial q^\nu}{\partial p_\sigma}\Big),$$

or, equivalently, by the one-forms

$$\Big(\frac{\partial q^\sigma}{\partial t} - \frac{\partial H}{\partial p_\sigma}\Big)\, dt + \Big(\frac{\partial q^\sigma}{\partial p_\nu} - \frac{\partial q^\nu}{\partial p_\sigma}\Big)\, dp_\nu. \tag{6.55}$$

6.6. Examples

6.6.1. Consider the first-order Lagrangian

$$L = \tfrac{1}{2}(q^1\dot{q}^2 - q^2\dot{q}^1 - (q^1)^2 - (q^2)^2) \tag{6.56}$$

(O. Krupková [9]). This Lagrangian defines the Lagrangean system

$$\alpha = -(q^1\, dq^1 + q^2\, dq^2) \wedge dt + dq^1 \wedge dq^2,$$

which is of *order zero*. The phase space is the total space of the given fibered manifold. The Euler-Lagrange distribution coincides with the characteristic distribution, and

$$\mathcal{D} = \Delta = \operatorname{span}\Big\{\frac{\partial}{\partial t} - q^2\frac{\partial}{\partial q^1} + q^1\frac{\partial}{\partial q^2}\Big\} \approx \operatorname{span}\{dq^1 + q^2\, dt,\ dq^2 - q^1\, dt\}.$$

The Lagrangean system is *regular*. Since it is of order zero, its *extremals* are described by a one-dimensional foliation of the configuration space, corresponding to the distribution Δ.

For illustration, let us write down the Hamilton equations also in Legendre coordinates. Since (6.56) is a minimal-order Lagrangian which satisfies (6.29), it is admissible for defining a Legendre transformation, and we can put

$$p_1 = \frac{\partial L}{\partial q^1} = -\tfrac{1}{2}q^2, \quad p_2 = \frac{\partial L}{\partial q^2} = \tfrac{1}{2}q^1, \quad H = -L + p_1\dot{q}^1 + p_2\dot{q}^2 = \tfrac{1}{2}\big((q^1)^2 + (q^2)^2\big),$$

and consider the Legendre transformation $(t, q^1, q^2) \to (t, p_1, p_2)$. In the Legendre coordinates we get for the Hamiltonian

$$H = 2(p_1^2 + p_2^2),$$

and for the Euler-Lagrange distribution

$$\mathcal{D} = \Delta = \operatorname{span}\Big\{\frac{\partial}{\partial t} - p_2\frac{\partial}{\partial p_1} + p_1\frac{\partial}{\partial p_2}\Big\} \approx \operatorname{span}\{dp_2 - p_1\, dt,\ dp_1 + p_2\, dt\}.$$

The Hamilton canonical equations read

$$\frac{d}{dt}(p_1 \circ \delta) = -p_2 \circ \delta, \quad \frac{d}{dt}(p_2 \circ \delta) = p_1 \circ \delta.$$

6.6.2. Let

$$L = (q^2 + q^3)\,\dot{q}^1 + q^4\dot{q}^3 + \tfrac{1}{2}\big((q^4)^2 - 2q^2q^3 - (q^3)^2\big) \tag{6.57}$$

(U. Kulshreshtha [1], see also J. Barcelos-Neto and N. R. F. Braga [1], and M. de León, D. Martín de Diego and P. Pitanga [1]). This is a first-order Lagrangian affine in the velocities. Checking the condition (6.11), we get that it is *regular*. For the corresponding momenta and Hamiltonian we have

$$\begin{aligned} p_1 &= q^2 + q^3, \quad p_2 = 0, \quad p_3 = q^4, \quad p_4 = 0, \\ H &= -\tfrac{1}{2}\left((q^4)^2 - 2q^2q^3 - (q^3)^2\right). \end{aligned}$$

Hence, the Lagrangian (6.57) defines a *regular zero-order Lagrangean system*

$$\begin{aligned} \alpha &= -dH \wedge dt + \sum_{\sigma=1}^{4} dp_\sigma \wedge dq^\sigma \\ &= (-q^3\,dq^2 - (q^2+q^3)\,dq^3 + q^4\,dq^4) \wedge dt + d(q^2+q^3) \wedge dq^1 + dq^4 \wedge dq^3. \end{aligned}$$

The Euler-Lagrange distribution is

$$\begin{aligned} \Delta = \mathcal{D} &= \operatorname{span}\left\{\frac{\partial}{\partial t} + q^3\frac{\partial}{\partial q^1} + q^4\frac{\partial}{\partial q_2} - q^4\frac{\partial}{\partial q^3} - q^2\frac{\partial}{\partial q^4}\right\} \\ &\approx \operatorname{span}\{d(q^2+q^3),\ -q^3\,dt + dq^1,\ -(q^2+q^3)\,dt + dq^1 - dq^4,\ q^4\,dt + dq^3\}. \end{aligned}$$

The dynamics proceeds on the *configuration space* and is *deterministic*. Extremals form a one-dimensional foliation of the configuration space, corresponding to the distribution Δ. Notice that the function

$$f = q^2 + q^3$$

is a *first integral* of the distribution Δ.

It is instructive to find a Legendre transformation for this Lagrangean system. Notice that (6.57) is a minimal-order Lagrangian, but it does not satisfy the condition (6.29). This means that its momenta cannot be used to define a Legendre transformation. To get a Legendre transformation, we must take a first-order Lagrangian which is equivalent with (6.57) and satisfies (6.29). We can take e.g. the following Lagrangian:

$$\begin{aligned} L' &= L + \frac{d}{dt}\left(q^1q^3 + \tfrac{1}{2}(q^2)^2 + \tfrac{1}{2}(q^4)^2\right) \\ &= (q^2 + 2q^3)\dot{q}^1 + q^2\dot{q}^2 + (q^1 + q^4)\dot{q}^3 + q^4\dot{q}^4 \\ &\quad + \tfrac{1}{2}\left((q^4)^2 - 2q^2q^3 - (q^3)^2\right) \end{aligned} \tag{6.58}$$

Then we get on the configuration space the Legendre transformation $(t, q^1, q^2, q^3, q^4) \to (t, p'_1, p'_2, p'_3, p'_4)$, where

$$p'_1 = q^2 + 2q^3, \quad p'_2 = q^2, \quad p'_3 = q^1 + q^4, \quad p'_4 = q^4.$$

The Hamiltonian reads

$$H = -\tfrac{1}{2}\left((q^4)^2 - 2q^2q^3 - (q^3)^2\right) = \tfrac{1}{8}(p'_1)^2 + \tfrac{1}{4}p'_1p'_2 - \tfrac{3}{8}(p'_2)^2 - \tfrac{1}{2}(p'_4)^2.$$

Now, the Hamilton canonical equations are

$$\begin{aligned} &\frac{d}{dt}(p'_1 \circ \delta) = -p'_4, \quad \frac{d}{dt}(p'_1 \circ \delta) + \frac{d}{dt}(p'_2 \circ \delta) = 0, \\ &\frac{d}{dt}(p'_3 \circ \delta) = \tfrac{1}{2}p'_1 - \tfrac{3}{2}p'_2, \quad \frac{1}{2}\frac{d}{dt}(p'_3 \circ \delta) - \frac{d}{dt}(p'_4 \circ \delta) = \tfrac{1}{4}(p'_1 + p'_2), \end{aligned}$$

and for the Euler-Lagrange distribution we get

$$\Delta = \mathcal{D} = \operatorname{span}\left\{\frac{\partial}{\partial t} - p'_4\frac{\partial}{\partial p'_1} + p'_4\frac{\partial}{\partial p'_2} + (\tfrac{1}{2}p'_1 - \tfrac{3}{2}p'_2)\frac{\partial}{\partial p'_3} - p'_2\frac{\partial}{\partial p'_4}\right\}.$$

6.6.3. Consider the second order Lagrangian (depending on accelerations)

$$L = -\tfrac{1}{2}(q^1\ddot{q}^1 + q^2\ddot{q}^2). \tag{6.59}$$

Computing α we get

$$\alpha = d\dot{q}^1 \wedge (dq^1 - \dot{q}^1\,dt) + d\dot{q}^2 \wedge (dq^2 - \dot{q}^2\,dt),$$

i.e., the above Lagrangian defines a *first-order* Lagrangean system. Since

$$\Delta = \mathcal{D} = \operatorname{span}\left\{\frac{\partial}{\partial t} + \dot{q}^1\frac{\partial}{\partial q^1} + \dot{q}^2\frac{\partial}{\partial q^2}\right\} \approx \operatorname{span}\{dq^1 - \dot{q}^1\,dt,\ dq^2 - \dot{q}^2\,dt,\ d\dot{q}^1,\ d\dot{q}^2\},$$

this Lagrangean system is *regular*, i.e., the extended motion coincides with the proper motion, and through every point of the phase space (which is the first jet of the given fibered manifold) there passes exactly one maximal prolonged extremal.

The Lagrangian (6.59) is not a minimal-order Lagrangian for the given Lagrangean system, hence it cannot be used to define a Legendre transformation. If we would like to express the Euler-Lagrange distribution (or the Hamilton equations) in Legendre coordinates, we could use the formulas (4.37), (4.38), or, more conveniently, we could take the momenta of any first-order Lagrangian equivalent with (6.59).

Note that since for the Lagrangian (6.59) we have

$$L = \tfrac{1}{2}\left((\dot{q}^1)^2 + (\dot{q}^2)^2\right) - \frac{1}{2}\frac{d}{dt}(q^1\dot{q}^1 + q^2\dot{q}^2),$$

the considered Lagrangean system is the classical free particle in two dimensions. Thus, Hamiltonians and momenta for (6.59) are defined by the familiar formulas

$$H = \tfrac{1}{2}\left((\dot{q}^1)^2 + (\dot{q}^2)^2\right) + \frac{\partial\phi}{\partial t} \qquad p_i = \dot{q}^i - \frac{\partial\phi}{\partial q^i}, \qquad i = 1, 2,$$

where ϕ is an arbitrary function depending upon (t, q^ν).

6.6.4. Let

$$L = \tfrac{1}{2}(\dot{q}^1\ddot{q}^2 - \dot{q}^2\ddot{q}^1) \tag{6.60}$$

(O. Krupková [9]). Since

$$\alpha = d\ddot{q}^2 \wedge (dq^1 - \dot{q}^1\,dt) - d\ddot{q}^1 \wedge (dq^2 - \dot{q}^2\,dt) + (d\dot{q}^1 - \ddot{q}^1\,dt) \wedge (d\dot{q}^2 - \ddot{q}^2\,dt),$$

this second-order Lagrangian defines a *second order* Lagrangean system (the phase space is the second jet of the given fibered manifold). For the Euler-Lagrange distribution we get

$$\mathcal{D} = \Delta = \operatorname{span}\left\{\frac{\partial}{\partial t} + \dot{q}^1\frac{\partial}{\partial q^1} + \dot{q}^2\frac{\partial}{\partial q^2} + \ddot{q}^1\frac{\partial}{\partial \dot{q}^1} + \ddot{q}^2\frac{\partial}{\partial \dot{q}^2}\right\}$$
$$\approx \operatorname{span}\{dq^1 - \dot{q}^1\,dt,\ dq^2 - \dot{q}^2\,dt,\ d\dot{q}^1 - \ddot{q}^1\,dt,\ d\dot{q}^2 - \ddot{q}^2\,dt,\ d\ddot{q}^1,\ d\ddot{q}^2\}.$$

We can see that this Lagrangean system is *regular*: through every point of the phase space there passes exactly one maximal prolonged extremal.

Similarly as in the above examples, we could express Δ, Hamilton equations, etc., in Legendre coordinates. Notice that to this purpose we could use the Lagrangian (6.60) since it is a minimal-order Lagrangian, satisfying the condition (6.29).

6.6.5. Let us list some other examples of regular Lagrangean systems—concrete computations of the Lepagean two-forms, characteristic distributions, Hamiltonians, momenta, Hamilton equations, etc. for these Lagrangians are left to the reader as an excercise.

First, let

$$\begin{aligned} L_1 &= \tfrac{1}{2}\dot{q}^2 + \ddot{q}, \\ L_2 &= -q\ddot{q}, \\ L_3 &= \tfrac{1}{2}\dot{q}^2 + \ddot{q}^2 + \dddot{q}\,\dot{q}, \\ L_4 &= \dot{q}^2(1+\ddot{q}) + t\dot{q}\ddot{q} + q\ddot{q}^2 + q\dot{q}\,\dddot{q}. \end{aligned} \tag{6.61}$$

All the Lagrangians (6.61) define (the same) *regular first-order Lagrangean system*; indeed, they are equivalent to the kinetic energy Lagrangian of the classical free particle.

Similarly, the Lagrangian

$$L = -\tfrac{1}{2}mx\ddot{x} - \tfrac{1}{2}kx^2 \tag{6.62}$$

(C. F. Hayes and J. M. Jankowski [1], C. A. P. Galvão and N. A. Lemos [1]), being equivalent to the classical harmonic oscillator Lagrangian, defines a *regular first-order* Lagrangean system.

Next, the Lagrangian

$$L = \tfrac{1}{2}m_1\dot{x}^2 + \tfrac{1}{2}m_2\dot{y}^2 + \tfrac{1}{2}(\dot{x}\ddot{y} - \dot{y}\ddot{x}) \tag{6.63}$$

(O. Krupková [2]), defines a *regular second-order* Lagrangean system (notice that the Euler-Lagrange equations are of order 3).

Finally, the Lagrangian

$$L = \tfrac{1}{2}\,(\ddot{q}^2 - \dot{q}^2) \tag{6.64}$$

(L. Klapka [1]), defines a *regular third-order* Lagrangean system.

Lagrangean systems arising from the Lagrangians (6.63) and (6.64) will be studied in Chapter 9.

Notice that all the examples of Lagrangean systems listed so far in this section but (6.64) are *singular in the standard sense*. This means that within the usual approach, they have to be studied with the tools of the Dirac theory of constrained systems. This, however, leads to complicated calculations and to not very transparent interpretation of results on dynamics of these Lagrangians, and consequently, often to confusion when quantization is applied.

6.6.6. The Hilbert-Einstein Lagrangian. We shall finish this section by touching an important example of Lagrangian, namely the Hilbert-Einstein Lagrangian of the general relativity theory. Although with this example we go behind the scope of this book, we want to mention briefly at least the main points to motivate the reader's interest in this subject.

It is well-known that the Hilbert-Einstein Lagrangian is a *second-order* Lagrangian (containing second order derivatives of the metric field), leading however to *second-order* Euler-Lagrange equations. Also, it is known that the Poincaré-Cartan form θ_λ of the Hilbert-Einstein Lagrangian is of order *one* (D. Krupka [1], W. Szczyrba [1], J. Kijowski [1], J. Novotný [2]). Hence, applying Definition 4.3.12, we conclude that this Lagrangian defines a *first-order Lagrangean system*. Moreover, we shall see below that it is *regular* (though traditionally considered degenerate). To this result we refer to D. Krupka and O. Štěpánková [1], where the reader can find a detailed exposition.

Consider a fibered manifold $\pi : Y \to X$, $\dim X = n$, let (x^i, y^σ) be local fibered coordinates on Y and $(x^i, y^\sigma, y^\sigma_j, y^\sigma_{jk})$ the associated coordinates on J^2Y. Denote

$$\omega_0 = dx^1 \wedge \cdots \wedge dx^n, \quad \omega_j = i_{\partial/\partial x^j}\, \omega_0.$$

We shall be interested in second order Lagrangians $\lambda = L\,\omega_0$ such that the function L is of the form

$$L = L_0(x^i, y^\sigma) + g^{jk}_\nu(x^i, y^\sigma) y^\nu_{jk}. \tag{6.65}$$

It can be checked by a direct calculation that the family of Lagrangians (6.65) is well-defined, i.e., the expression of L in the form (6.65) is saved with respect to different fibered charts. Computing the Poincaré-Cartan form of λ,

$$\theta_\lambda = L\,\omega_0 + \left(\left(\frac{\partial L}{\partial y^\sigma_j} - d_k \frac{\partial L}{\partial y^\sigma_{jk}}\right)\omega^\sigma + \frac{\partial L}{\partial y^\sigma_{jk}}\omega^\sigma_k\right) \wedge \omega_j, \tag{6.66}$$

where d_k denotes the total derivative operator with respect to x^k, we get that the θ_λ is projectable onto J^1Y:

$$\theta_\lambda = \left(L_0 + \frac{\partial g^{jk}_\sigma}{\partial y^\nu} y^\nu_k y^\sigma_j\right)\omega_0 - \left(\frac{\partial g^{ij}_\sigma}{\partial x^j} + \left(\frac{\partial g^{ij}_\sigma}{\partial y^\nu} + \frac{\partial g^{ij}_\nu}{\partial y^\sigma}\right) y^\nu_j\right) dy^\sigma \wedge \omega_i + d(g^{ij}_\sigma y^\sigma_j\, \omega_i). \tag{6.67}$$

Now, consider the Euler-Lagrange equations and the Hamilton equations of (6.65), namely,

$$J^1\gamma^*\, i_{J^1\xi} d\theta_\lambda = 0 \tag{6.68}$$

for every π-vertical vector field ξ on Y, and

$$\delta^*\, i_\xi d\theta_\lambda = 0 \tag{6.69}$$

for every π_1-vertical vector field ξ on J^1Y, respectively. Consider the matrix

$$Q^{ik}_{\sigma\nu} \equiv \frac{\partial^2 L_0}{\partial y^\rho_k \partial y^\sigma_i} - \frac{\partial g^{ik}_\sigma}{\partial y^\rho} - \frac{\partial g^{ik}_\rho}{\partial y^\sigma}, \tag{6.70}$$

where (σ, i) labels the rows and (ρ, k) the columns. One can show that under the hypothesis that the matrix (6.70) is *regular*, i.e., that

$$\det\left(\frac{\partial^2 L_0}{\partial y^\rho_k \partial y^\sigma_i} - \frac{\partial g^{ik}_\sigma}{\partial y^\rho} - \frac{\partial g^{ik}_\rho}{\partial y^\sigma}\right) \neq 0, \tag{6.71}$$

then the mapping $\gamma \to J^1\gamma$ of the set of extremals into the set of Hamilton extremals is *bijective*, i.e., *the Euler-Lagrange equations* (6.68) *are equivalent with the Hamilton equations* (6.69).

Put

$$H = -L_0 - \frac{\partial g_\sigma^{jk}}{\partial y^\nu}\, y_k^\nu y_j^\sigma \tag{6.72}$$

and

$$p_\sigma^i = -\frac{\partial g_\sigma^{ij}}{\partial x^j} - \left(\frac{\partial g_\sigma^{ij}}{\partial y^\nu} + \frac{\partial g_\nu^{ij}}{\partial y^\sigma}\right) y_j^\nu. \tag{6.73}$$

Then the form θ_λ takes the following *canonical form*

$$\theta_\lambda = -H\,\omega_0 + p_\sigma^i\, dy^\sigma \wedge \omega_i + d(g_\sigma^{ij} y_j^\sigma\, \omega_i). \tag{6.74}$$

We have the following Theorem:

If the matrix (6.70) *is regular then* $(x^i, y^\sigma, p_\sigma^j)$ *are local fibered coordinates on* J^1Y.

The transformation $(x^i, y^\sigma, y_j^\sigma) \to (x^i, y^\sigma, p_\sigma^j)$ with the p_σ^j defined by (6.73) is called *Legendre transformation.*

Notice that conventionally, Lagrangians (6.65) are considered singular, since they do not obey the standard definition of regularity,

$$\det\left(\frac{\partial^2 L}{\partial y_{jk}^\sigma \partial y_{il}^\nu}\right) \neq 0.$$

Moreover, within the standard approach, to L are assigned the momenta

$$\hat{p}_\sigma^i = \frac{\partial L}{\partial y_i^\sigma} - d_k \frac{\partial L}{\partial y_{ik}^\sigma} = -d_k\, g_\sigma^{ik} = p_\sigma^i + \frac{\partial g_\nu^{ij}}{\partial y^\sigma} y_j^\nu, \quad \hat{p}_\sigma^{ik} = \frac{\partial L}{\partial y_{ik}^\sigma} = g_\sigma^{ik},$$

and the Hamiltonian

$$\hat{H} = -L + p_\sigma^j y_j^\sigma + p_\sigma^{jk} y_{jk}^\sigma = -L_0 - y_j^\sigma\, d_k g_\sigma^{jk} = H - \frac{\partial g_\sigma^{jk}}{\partial x^k}\, y_j^\sigma.$$

The momenta $\hat{p}_\sigma^i$, $\hat{p}_\sigma^{ik}$ do not define a Legendre transformation on J^2Y, and the Lagrangian (6.65) is treated as a constrained system (in the sense of the Dirac theory of constrained systems) on J^2Y.

On the contrary, we have obtained the Lagrangean system (6.65) as a *regular* system on J^1Y (i.e., no constraints are present). Writing down the Hamilton equations (6.69) in the Legendre coordinates (defined by (6.73)) we get

$$\frac{\partial H}{\partial y^\nu} + \frac{\partial(p_\nu^i \circ \delta)}{\partial x^i} = 0, \qquad -\frac{\partial H}{\partial p_\nu^i} + \frac{\partial(y^\nu \circ \delta)}{\partial x^i} = 0. \tag{6.75}$$

The above Hamilton theory can be applied directly to the Einstien-Hilbert Lagrangian (= scalar curvature Lagrangian) of the general relativity theory. In this case we have the fibered manifold $\tau_0 : Y \to X$ where X is a 4-dimensional orientable manifold, and τ_0 is the restriction of the vector bundle $\tau : T_2^0 X \to X$ of symmetric tensors of type (0,2) over X to the open subset $T_{\mathrm{met}}X \subset T_2^0 X$ of *regular* tensors (τ_0 is the bundle of metrics over X). Denote by (x^i, g_{jk}), $1 \leqslant i, j, k \leqslant n$, $j \leqslant k$, local fibered coordinates on $T_{\mathrm{met}}X$. The Einstein-Hilbert Lagrangian is the following Lagrangian defined on $J^2 T_{\mathrm{met}}X$:

$$\lambda = R\sqrt{|g|}\,\omega_0, \tag{6.76}$$

where R is the scalar curvature of the metric g,

$$R = g^{ik} g^{jp} R_{ijkp},$$

where

$$R_{ijks} = \tfrac{1}{2}(g_{is,jk} + g_{jk,is} - g_{ik,js} - g_{js,ik}) + g_{pq}(\Gamma^p_{jk}\Gamma^q_{is} - \Gamma^p_{js}\Gamma^q_{ik}),$$
$$\Gamma^i_{jk} = \tfrac{1}{2} g^{is}(g_{sj,k} + g_{sk,j} - g_{jk,s}),$$

and $|g| = |\det(g_{jk})|$. It is easy to see that the Lagrangian (6.76) is of type (6.65). The Poincaré-Cartan form of (6.76) reads (see e.g. J. Novotný [2])

$$\theta_\lambda = \sqrt{|g|}\, g^{ip}(\Gamma^j_{ip}\Gamma^k_{jk} - \Gamma^j_{ik}\Gamma^k_{jp})\,\omega_0 + (g^{jp}g^{iq} - g^{pq}g^{ij})\,(dg_{pq,j} + \Gamma^k_{pq}\, dg_{jk}) \wedge \omega_i.$$

It can be verified by a dirct calculation that the Lagrangian (6.76) is *regular*, satisfying the regularity condition (6.71). Now, using (6.72), (6.73), the Hamiltonian H, momenta $P^{rs,i}$, and the Hamilton equations

$$\frac{\partial H}{\partial g_{rs}} + \tilde{d}_j P^{rs,j} = 0, \quad -\frac{\partial H}{\partial P^{rs,j}} + \tilde{d}_j g^{rs} = 0, \tag{6.77}$$

which are the equations for Hamilton extremals $(x^i) \to (x^i, g_{jk}(x^i), P^{jk,q}(x^i))$, may be obtained by a routine calculation. We remind the reader that the Hamilton equations (6.77) are first-order equations, *equivalent* with the well-known Einstein equations for vacuum. Thus, we have shown that general relativity, traditionally approached to as a second order constraint theory, can be viewed as a *regular first-order theory*.

Notice that for the case $\dim X = 1$ the formulas (6.71) (regularity condition) and (6.72), (6.73) (Hamiltonian and momenta) reduce to (6.14), and (4.63) where $F = g_\nu \dot{q}^\nu$.

6.7. Equivalence of dynamical forms and the inverse problem of the calculus of variations

We have already touched the inverse problem of the calculus of variations in Chapters 1 and 4. Now, let us return to this problem to discuss its more general formulations (cf. O. Krupková [2]).

Let E be a dynamical form of order $s \geqslant 1$ on a fibered manifold $\pi : Y \to X$, $\dim X = 1$. Denote $E = E_\sigma\, dq^\sigma \wedge dt$.

Recall from Section 4.2 that a (local) section γ of π is an *integral section* of E if

$$E \circ J^s\gamma = 0. \tag{6.78}$$

The above equation represents a system of m s-th order ODE for the components of γ.

We shall suppose that the functions E_σ are *affine in the highest derivatives*, i.e., that

$$E_\sigma = A_\sigma + B_{\sigma\nu} q^\nu_s, \tag{6.79}$$

where the functions A_σ and $B_{\sigma\nu}$ do not depend on the q^ρ_s's. Notice that

– the matrix $(B_{\sigma\nu})$ need not be symmetric.

– The property that the E_σ's are affine in the highest derivatives is independent of a choice of a fiber chart. Indeed, if (V, ψ), $\psi = (t, q^\sigma)$ and $(\bar{V}, \bar{\psi})$, $\bar{\psi} = (\bar{t}, \bar{q}^\sigma)$ are two overlaping charts, and $E = E_\sigma\, dq^\sigma \wedge dt = \bar{E}_\sigma\, d\bar{q}^\sigma \wedge d\bar{t}$ with the E_σ of the form (6.79), we get by a direct computation that $\bar{E}_\sigma = \bar{A}_\sigma + \bar{B}_{\sigma\nu}\bar{q}_s^\nu$ with the $\bar{A}_\sigma$ and $\bar{B}_{\sigma\nu}$ independent of the $\bar{q}_s^\rho$. It is easy to see that

$$\bar{B}_{\sigma\nu} = \frac{\partial q^\sigma}{\partial \bar{q}^\nu} B_{\sigma\nu}, \tag{6.80}$$

while the transformation rule for the A_σ is a little bit more complicated.

– The property that the E_σ's are affine in the highest derivatives does *not* represent a restrictive assumption, since *any* dynamical form can be considered to be of type (6.79). Indeed, if E is a dynamical form of order s, $E = E_\sigma\, dq^\sigma \wedge dt$, and the E_σ are not of the form (6.79), then one can consider $\pi^*_{s+1,s}E$ instead of E. This is, however, a dynamical form of ordr $s+1$ which *is* affine in the highest derivatives q^ρ_{s+1} (with $A_\sigma = E_\sigma$ and $B_{\sigma\nu} = 0$).

Thus, in what follows we shall suppose, without the loss of generality, that E is of type (6.79).

A dynamical form E of order $s \geqslant 1$ will be referred to as a *mechanical system*.

A dynamical form/mechanical system E will be called *regular* if the matrix $(B_{\sigma\nu})$ is regular, i.e., if

$$\det(B_{\sigma\nu}) = \det\left(\frac{\partial E_\sigma}{\partial q_s^\nu}\right) \neq 0. \tag{6.81}$$

Again, by (6.80), regularity is a well-defined property.

Notice that if E is locally variational then the concept of regularity (6.81) coincides with that for Lagrangean systems (cf. Theorem 6.2.2).

Let E be a *regular* dynamical form of order s. Let $w : J^{s-1}Y \to J^sY$ be a semispray connection (Sec. 2.7). We say that E and w are *related* if

$$w^*E = 0. \tag{6.82}$$

In fibered coordinates the condition (6.82) reads

$$w^\sigma = -B^{\sigma\nu}A_\nu, \tag{6.83}$$

where $(B^{\sigma\nu})$ is the inverse matrix to $(B_{\sigma\nu})$. Thus,

$$E_\sigma = B_{\sigma\nu}(q_s^\nu - w^\nu),$$

and for the semispray distribution Δ_w of w we get,

$$\begin{aligned}\Delta_w &= \operatorname{span}\left\{\frac{\partial}{\partial t} + \sum_{i=0}^{s-2} q^\sigma_{i+1}\frac{\partial}{\partial q^\sigma_i} - B^{\sigma\nu}A_\nu\frac{\partial}{\partial q^\sigma_{s-1}}\right\}\\ &\approx \operatorname{span}\{\omega^\sigma_i,\ dq^\sigma_{s-1} + B^{\sigma\nu}A_\nu\,dt,\ 1 \leqslant \sigma \leqslant m,\ 0 \leqslant i \leqslant s-1\}.\end{aligned}$$

Let E, E' be two regular dynamical forms of order s. We say that E, E' are *equivalent* if their related semispry distributions coincide. We easily get

6.7.1. Proposition.

(1) *E and E' are equivalent if and only if they have the same set of integral sections.*

(2) *E and E' are equivalent if and only if there exists an everywhere regular matrix (G^ν_σ) of order $s-1$ such that*

$$E'_\sigma = G^\nu_\sigma E_\nu.$$

Proof. The first assertion is obvious. Let us prove (2).

If E and E' are equivalent then by (6.83)

$$B^{\sigma\nu}A_\nu = B'^{\sigma\nu}A'_\nu.$$

Thus, $A'_\rho = B'_{\rho\sigma}B^{\sigma\nu}A_\nu$, and

$$E'_\sigma = A'_\sigma + B'_{\sigma\nu}q^\nu_s = B'_{\sigma\rho}B^{\rho\nu}A_\nu + B'_{\sigma\rho}q^\rho_s = B'_{\sigma\rho}B^{\rho\nu}(A_\nu + B_{\nu\mu}q^\mu_s) = B'_{\sigma\rho}B^{\rho\nu}E_\nu,$$

and putting

$$G^\nu_\sigma = B'_{\sigma\rho}B^{\rho\nu} \tag{6.84}$$

we are done. The converse is obvious. □

The matrix $G = (G^\nu_\sigma)$ defined in the preceding proposition will be referred to as a *multiplier*, or *integrating factor* for a dynamical form E.

Notice that every dynamical form E related with w is equivalent with the dynamical form E^w,

$$E^w_\sigma = \delta_{\sigma\nu}(q^\nu_s - w^\nu), \tag{6.85}$$

"explicitly solved with respect to the highest derivatives". Thus, the equivalence class of forms related with w is completely determined by E^w and the set of everywhere regular $(m \times m)$-matrices of order $s - 1$.

Summarizing, we have seen that to every regular dynamical form E there exists a *unique* related semispray connection (defined by (6.82)). Conversely, to every semispray connection w there is related an equivalence class of regular dynamical forms, $\{E^w\}$, differing by a multiplier G.

Now, we are prepared to discuss the local inverse problem of the calculus of variations. This problem can be posed in several different levels:

(1) Given a dynamical form E, one can ask if it is *locally variational*, i.e., if there exists (locally) a Lagrangian λ such that (possibly up to a projection) $E = E_\lambda$. The complete solution of this problem is known in the most general situation (higher-order field theory), and we have presented its solution for the case of higher-order mechanics in Chapter 4. Here we only remind the reader that the necessary and sufficient conditions for a dynamical form be locally variational, generalizing the *Helmholtz conditions* (1.8) resp. (1.15), are of the form

$$\frac{\partial E_\sigma}{\partial q^\nu_l} - (-1)^l \frac{\partial E_\nu}{\partial q^\sigma_l} - \sum_{k=l+1}^{s} (-1)^k \binom{k}{l} \frac{d^{k-l}}{dt^{k-l}} \frac{\partial E_\nu}{\partial q^\sigma_k} = 0,$$

where $0 \leqslant l \leqslant s$ and $1 \leqslant \sigma, \nu \leqslant m$ (see Corollary 4.3.9), and the corresponding equivalence class of Lagrangians is obtained from the Vainberg-Tonti Lagrangian (4.6) by adding the total derivative of an arbitrary function.

Using the Helmholtz conditions we obtain an assertion on the *structure of second-order Euler-Lagrange expressions* and the corresponding first-order Lagrangians (see O. Krupková [1]).

6.7.2. Theorem. *Let $U \subset R$ be an open interval, and $V \subset R^m$ an open ball with the center at the origin $0 \in R^m$, let E be a dynamical form on $J^2(U \times V)$. Denote*

$E = E_\sigma \, dq^\sigma \wedge dt$. Consider the mapping $\bar{\chi} : [0,1] \times J^1(U \times V) \to J^1(U \times V)$, defined by

$$\bar{\chi}(v, t, q^\nu, \dot{q}^\nu) = (t, q^\nu, v\dot{q}^\nu). \tag{6.86}$$

The following conditions are equivalent:

(1) *$E_\sigma(t, q^\nu, \dot{q}^\nu, \ddot{q}^\nu)$, $1 \leqslant \sigma \leqslant m$, are variational.*

(2) *It holds*

$$E_\sigma = A_\sigma + B_{\sigma\nu}\ddot{q}^\nu,$$

where $B_{\sigma\nu}$ are functions on $J^1(U \times V)$ satisfying

$$B_{\sigma\nu} = B_{\nu\sigma}, \qquad \frac{\partial B_{\sigma\nu}}{\partial \dot{q}^\rho} = \frac{\partial B_{\sigma\rho}}{\partial \dot{q}^\nu}, \tag{6.87}$$

and

$$A_\sigma = \Gamma_{\sigma\nu\rho}\dot{q}^\nu \dot{q}^\rho + \dot{q}^\nu \int_0^1 \left(\frac{\partial B_{\sigma\nu}}{\partial t} \circ \bar{\chi} \right) dv + \alpha_{\sigma\nu}\dot{q}^\nu + \beta_\sigma, \tag{6.88}$$

where

$$\Gamma_{\sigma\nu\rho} = \frac{1}{2} \int_0^1 \left(\frac{\partial B_{\sigma\nu}}{\partial q^\rho} + \frac{\partial B_{\sigma\rho}}{\partial q^\nu} - 2\frac{\partial B_{\nu\rho}}{\partial q^\sigma} \right) \circ \bar{\chi} \, dv + \int_0^1 \left(\frac{\partial B_{\nu\rho}}{\partial q^\sigma} \circ \bar{\chi} \right) v \, dv. \tag{6.89}$$

and

$$\alpha_{\sigma\nu} = -\alpha_{\nu\sigma}, \qquad \frac{\partial \alpha_{\sigma\nu}}{\partial q^\rho} + \frac{\partial \alpha_{\rho\sigma}}{\partial q^\nu} + \frac{\partial \alpha_{\nu\rho}}{\partial q^\sigma} = 0, \qquad \frac{\partial \beta_\sigma}{\partial q^\nu} - \frac{\partial \beta_\nu}{\partial q^\sigma} = \frac{\partial \alpha_{\sigma\nu}}{\partial t}. \tag{6.90}$$

To prove the theorem we shall need the following lemma.

6.7.3. Lemma. *Consider the mappings $\chi : [0,1] \times J^2(U \times V) \to J^2(U \times V)$, defined by*

$$\chi(u, t, q^\nu, \dot{q}^\nu, \ddot{q}^\nu) = (t, uq^\nu, u\dot{q}^\nu, u\ddot{q}^\nu), \tag{6.91}$$

and $\bar{\chi}$ defined by (6.86). Let F be a smooth function defined on $J^1(U \times V)$. Then it holds

$$F = \int_0^1 (F \circ \chi) \, du + q^\sigma \int_0^1 \left(\frac{\partial F}{\partial q^\sigma} \circ \chi \right) u \, du + \dot{q}^\sigma \int_0^1 \left(\frac{\partial F}{\partial \dot{q}^\sigma} \circ \chi \right) u \, du, \tag{6.92}$$

$$F = n \int_0^1 (F \circ \bar{\chi}) \, v^{n-1} dv + \dot{q}^\sigma \int_0^1 \left(\frac{\partial F}{\partial \dot{q}^\sigma} \circ \bar{\chi} \right) v^n \, dv, \quad n = 1, 2, 3, \ldots, \tag{6.93}$$

$$\int_0^1 \left(\int_0^1 (F \circ \bar{\chi}) \, dv \right) \circ \bar{\chi} \, v \, dv = \int_0^1 (F \circ \bar{\chi}) \, dv - \int_0^1 (F \circ \bar{\chi}) \, v \, dv, \tag{6.94}$$

$$\dot{q}^\sigma \int_0^1 \left(\frac{\partial F}{\partial \dot{q}^\sigma} \circ \bar{\chi} \right) dv - \dot{q}^\sigma \int_0^1 \left(\frac{\partial F}{\partial \dot{q}^\sigma} \circ \bar{\chi} \right) v \, dv - \dot{q}^\sigma \int_0^1 \left(\int_0^1 \left(\frac{\partial F}{\partial \dot{q}^\sigma} \circ \bar{\chi} \right) dv \right) \circ \bar{\chi} \, v \, dv = 0, \tag{6.95}$$

and there exists a function ϕ on $U \times V$ such that

$$F = \dot{q}^\sigma \int_0^1 \left(\frac{\partial F}{\partial \dot{q}^\sigma} \circ \bar{\chi} \right) dv + \phi(t, q^\nu). \tag{6.96}$$

Proof. (6.92) follows immediately from the identity

$$F = \int_0^1 d(u(F \circ \chi)).$$

Similarly, (6.93) is obtained from the identity

$$F = \int_0^1 d(v^n(F \circ \bar{\chi})), \quad n = 1, 2, 3, \dots.$$

The relation (6.94) is obtained from (6.93). Namely, writing down (6.93) for $n = 1$ and applying this relation to $\int_0^1 (F \circ \bar{\chi})\, v\, dv$, and similarly, applying (6.93) for $n = 2$ to the function $\int_0^1 (F \circ \bar{\chi})\, dv$, the identity (6.94) follows. Next, since

$$\dot{q}^\sigma \int_0^1 \Big(\frac{\partial F}{\partial \dot{q}^\sigma} \circ \bar{\chi}\Big)\, dv = \int_0^1 d(F \circ \bar{\chi}) = F - F(t, q^\nu, 0),$$

we get (6.96). Finally, using the above relation and (6.93) for $n = 1$, i.e.,

$$\dot{q}^\sigma \int_0^1 \Big(\frac{\partial F}{\partial \dot{q}^\sigma} \circ \bar{\chi}\Big)\, v\, dv = F - \int_0^1 (F \circ \bar{\chi})\, dv,$$

and taking into account that

$$\begin{aligned} &\dot{q}^\sigma \int_0^1 \Big(\int_0^1 \Big(\frac{\partial F}{\partial \dot{q}^\sigma} \circ \bar{\chi}\Big)\, dv\Big) \circ \bar{\chi}\, v\, dv \\ &= \int_0^1 \Big(\dot{q}^\sigma \int_0^1 \Big(\frac{\partial F}{\partial \dot{q}^\sigma} \circ \bar{\chi}\Big)\, dv\Big) \circ \bar{\chi}\, dv = \int_0^1 (F \circ \bar{\chi})\, dv - F(t, q^\nu, 0), \end{aligned}$$

we get (6.95). □

Proof of Theorem 6.7.2. Let A_σ, $1 \leqslant \sigma \leqslant m$, be functions defined on $J^1(U \times V)$. Since E_σ are variational, they obey the Helmholtz conditions, which we take in the form

$$\begin{aligned} &B_{\sigma\nu} = B_{\nu\sigma}, \quad \frac{\partial B_{\sigma\nu}}{\partial \dot{q}^\rho} = \frac{\partial B_{\sigma\rho}}{\partial \dot{q}^\nu}, \\ &\frac{\partial A_\sigma}{\partial \dot{q}^\nu} + \frac{\partial A_\nu}{\partial \dot{q}^\sigma} = 2\Big(\frac{\partial B_{\sigma\nu}}{\partial t} + \dot{q}^\rho \frac{\partial B_{\sigma\nu}}{\partial q^\rho}\Big), \\ &\frac{\partial B_{\sigma\rho}}{\partial q^\nu} - \frac{\partial B_{\nu\rho}}{\partial q^\sigma} = \frac{1}{2} \frac{\partial}{\partial \dot{q}^\rho} \Big(\frac{\partial A_\sigma}{\partial \dot{q}^\nu} - \frac{\partial A_\nu}{\partial \dot{q}^\sigma}\Big), \\ &\frac{\partial A_\sigma}{\partial q^\nu} - \frac{\partial A_\nu}{\partial q^\sigma} = \frac{1}{2}\Big(\frac{\partial}{\partial t}\Big(\frac{\partial A_\sigma}{\partial \dot{q}^\nu} - \frac{\partial A_\nu}{\partial \dot{q}^\sigma}\Big) + \dot{q}^\rho \frac{\partial}{\partial q^\rho}\Big(\frac{\partial A_\sigma}{\partial \dot{q}^\nu} - \frac{\partial A_\nu}{\partial \dot{q}^\sigma}\Big)\Big) \end{aligned} \tag{6.97}$$

(cf. (1.8)). Recall from Chapter 1 that the fourth of the above identites is not independent.

Similarly as in Chapter 1 denote

$$\phi_{\sigma\nu} = \frac{1}{2}\Big(\frac{\partial A_\sigma}{\partial \dot{q}^\nu} - \frac{\partial A_\nu}{\partial \dot{q}^\sigma}\Big), \quad \psi_{\sigma\nu} = \frac{1}{2}\Big(\frac{\partial A_\sigma}{\partial \dot{q}^\nu} + \frac{\partial A_\nu}{\partial \dot{q}^\sigma}\Big).$$

Then using (6.96) we get

$$\begin{aligned}A_\sigma &= \dot q^\rho \int_0^1 (\psi_{\sigma\rho}\circ\bar\chi)\,dv + \dot q^\rho \int_0^1 (\phi_{\sigma\rho}\circ\bar\chi)\,dv + \beta_\sigma \\ &= \dot q^\rho \int_0^1 (\psi_{\sigma\rho}\circ\bar\chi)\,dv + \dot q^\rho \int_0^1 \Big(\dot q^\kappa \int_0^1 \Big(\frac{\partial \phi_{\sigma\rho}}{\partial \dot q^\kappa}\circ\bar\chi\Big)\,dv + \alpha_{\sigma\rho}\Big)\circ\bar\chi\,dv + \beta_\sigma \\ &= \dot q^\rho \int_0^1 (\psi_{\sigma\rho}\circ\bar\chi)\,dv + \tfrac{1}{2}\dot q^\rho \dot q^\kappa \int_0^1 \Big(\int_0^1 \Big(\frac{\partial \phi_{\sigma\rho}}{\partial \dot q^\kappa} + \frac{\partial \phi_{\sigma\kappa}}{\partial \dot q^\rho}\Big)\circ\bar\chi\,dv\Big)\circ\bar\chi\,v\,dv \\ &\quad + \alpha_{\sigma\rho}\dot q^\rho + \beta_\sigma\end{aligned} \tag{6.98}$$

where $\alpha_{\sigma\nu}$, β_σ are functions defined on $U \times V$, and

$$\alpha_{\sigma\nu} = -\alpha_{\nu\sigma}.$$

Suppose (1). Then the functions E_σ satisfy the Helmholtz conditions, i.e., the $B_{\sigma\nu}$ obey (6.87), and, according to (6.98) and the third and fourth identities of (6.97),

$$\begin{aligned}A_\sigma &= \dot q^\rho \int_0^1 \Big(\frac{\partial B_{\sigma\rho}}{\partial t} + \dot q^\nu \frac{\partial B_{\sigma\rho}}{\partial q^\nu}\Big)\circ\bar\chi\,dv \\ &\quad + \tfrac{1}{2}\dot q^\rho \dot q^\kappa \int_0^1 \Big(\int_0^1 \Big(\frac{\partial B_{\kappa\sigma}}{\partial q^\rho} + \frac{\partial B_{\rho\sigma}}{\partial q^\kappa} - 2\frac{\partial B_{\kappa\rho}}{\partial q^\sigma}\Big)\circ\bar\chi\,dv\Big)\circ\bar\chi\,v\,dv + \alpha_{\sigma\rho}\dot q^\rho + \beta_\sigma.\end{aligned} \tag{6.99}$$

From this expression of A_σ we easily obtain (6.88) and (6.89) if we use (6.94). Now, since the functions (6.99) have to satisfy the last identity of (6.97), we get after some computation, using the fourth identity of (6.97) and applying (6.95),

$$\Big(\frac{\partial \alpha_{\sigma\rho}}{\partial q^\nu} + \frac{\partial \alpha_{\nu\sigma}}{\partial q^\rho} + \frac{\partial \alpha_{\rho\nu}}{\partial q^\sigma}\Big)\dot q^\rho + \frac{\partial \beta_\sigma}{\partial q^\nu} - \frac{\partial \beta_\nu}{\partial q^\sigma} - \frac{\partial \alpha_{\sigma\nu}}{\partial t} = 0.$$

Now, since the functions $\alpha_{\sigma\nu}$ and β_σ do not depend on $\dot q^\rho$, the (6.90) follow.

Suppose the converse. Then we have to show that the E_σ in the form (6.87)–(6.90) satisfy the Helmholtz conditions (1.8). This, however, is obtained after a straightforward computation with the help of the identities (6.93) and (6.95) of Lemma 6.7.3. □

Notice that if the $B_{\sigma\nu}$'s depend upon the q^ρ's only then (6.99) take the form

$$A_\sigma = \Gamma_{\sigma\nu\rho}\dot q^\nu \dot q^\rho + \alpha_{\sigma\rho}\dot q^\rho + \beta_\sigma,$$

where

$$\begin{aligned}\Gamma_{\sigma\nu\rho} &= \frac{1}{4}\Big(\frac{\partial B_{\sigma\rho}}{\partial q^\nu} + \frac{\partial B_{\sigma\nu}}{\partial q^\rho}\Big) + \frac{1}{4}\Big(\frac{\partial B_{\sigma\nu}}{\partial q^\rho} + \frac{\partial B_{\sigma\rho}}{\partial q^\nu} - 2\frac{\partial B_{\nu\rho}}{\partial q^\sigma}\Big) \\ &= \frac{1}{2}\Big(\frac{\partial B_{\sigma\nu}}{\partial q^\rho} + \frac{\partial B_{\sigma\rho}}{\partial q^\nu} - \frac{\partial B_{\nu\rho}}{\partial q^\sigma}\Big),\end{aligned}$$

are the *Christoffel symbols* of the $B_{\sigma\nu}(q^\rho)$'s.

As an easy example of application of the above theorem, let us aks the question *under what conditions the Newton equations for a classical (resp. relativistic) particle in a force field F*,

$$m\ddot q^\sigma = F^\sigma \quad \text{resp.} \quad \frac{m_0}{\sqrt{1-v^2}}\,\ddot q^\sigma = F^\sigma,$$

are variational or, in another formulation, under what conditions the above force field F is *potential* (J. Novotný [1], E. Engels and W. Sarlet [1]). Substituting $B_{\sigma\nu} = m$, resp., $B_{\sigma\nu} = \delta_{\sigma\nu}m_0/\sqrt{1-v^2}$ into (6.87)–(6.90), we get that, in both cases, the (6.87) are satisfied identically, and $\Gamma_{\sigma\nu\rho} = 0$, i.e.,

$$-F_\sigma = (\beta_\sigma + \alpha_{\sigma\nu})dq^\nu \wedge dt,$$

where $\beta_\sigma, \alpha_{\sigma\nu}$ satisfy (6.90). If $m = 3$, the above formulas give us that the force F is a *Lorentz-type force*: namely, putting

$$E^i = \delta^{ij}\beta_j, \quad \alpha_{ij} = -\varepsilon_{ijk}H^k,$$

where δ^{ij} is the Kronecker symbol and ε_{ijk} is the Levi-Civita symbol, we obtain

$$F = v \times H + E,$$

where

$$H = \operatorname{rot} A, \quad E = -\frac{\partial A}{\partial t} + \operatorname{grad}\phi$$

(cf. Chapter 1).

Notice that by (6.88) and (6.89) the expression for the functions A_σ is the sum of two terms: one uniquelly determined by means of the functions $B_{\sigma\nu}$ (having the meaning of a local generalized metric) and the second one which is a Lorentz force-type term.

Accordingly, also the corresponding first-order Lagrangians take the form of a sum of a kinetic energy-type term (depending only on a local "generalized metric" $B = (B_{\sigma\nu})$), and a potential energy term (depending on arbitrary functions). More precisely, one has

6.7.4. Corollary. *Every first-order Lagrangian for second order Euler-Lagrange expressions $E_\sigma(t, q^\nu, \dot{q}^\nu, \ddot{q}^\nu)$, $1 \leqslant \sigma \leqslant m$, is of the form*

$$L = T - V,$$

where

$$T = -\dot{q}^\sigma\dot{q}^\nu \int_0^1 \left(\int_0^1 (B_{\sigma\nu} \circ \bar{\chi})\, dv\right) \circ \bar{\chi}\, v\, dv, \tag{6.100}$$

and

$$V = -f_\sigma \dot{q}^\sigma + \varphi + \frac{d\phi}{dt}, \tag{6.101}$$

where

$$f_\sigma = q^\nu \int_0^1 (\alpha_{\sigma\nu} \circ \chi)\, u\, du, \quad \varphi = -q^\sigma \int_0^1 (\beta_\sigma \circ \chi)\, du, \tag{6.102}$$

and $\phi(t, q^p)$ is an arbitrary function.

Proof. The assertion can be obtained by a direct computation from the formula (4.43) for the reduced Vainberg-Tonti Lagrangian with the E_σ defined by Theorem 6.7.2. To simplify the expression for L, Lemma 6.7.3 can be applied.

Another (less straightforward but more convenient way) to prove Corollary 6.7.4 consists in a direct computation of the Euler-Lagrange expressions of the Lagrangian $L = T - V$, (6.100)–(6.102). □

The functions T and V of Corollary 6.7.4 will be called *kinetic energy* and *potential energy*, respectively.

Theorem 6.7.2 and Corollary 6.7.4 have interesting applications—some of them will be mentioned in Chapter 10.

(2) Another (more general) level of the inverse problem is suggested by the concept of equivalence of dynamical forms introduced above in this section. Namely, given a dynamical form E, one can ask *if there is a locally variational form E', equivalent with E*. In other words, given a *semispray connection* w (or, equivalently, a *semispray distribution* Δ_w) one can ask if there exists a locally variational form E related with w. If there is a locally variational form related with w we say that the semispray connection w (semispray distribution Δ_w) is *variational*. The integrating factor G such that $E' = GE$ is locally variational is then called a *variational integrating factor*, or *variational multiplier* for E. Hence, by Proposition 6.7.1, the setting of the inverse problem of the calculus of variations as a problem on searching for a variational integrating factor, geometrically means the *problem to decide if a semispray connection/semispray distribution is variational*. Therefore, this level of the inverse problem will be also referred to as the *inverse variational problem for distributions*.

Several points should be emphasized at this place.

First of all, a complete solution of the inverse problem in this setting is yet not known. There have been achieved some particular results, concerning mainly second order ODE. The most significant are results on algebraic and geometric characterization of conditions for the existence of variational multipliers and a solution of the inverse variational problem for second-order semispray connections (J. Douglas [1], P. Havas [1], W. Sarlet [1,2], M. Henneaux [2], O. Krupková [2,8], I. Anderson and G. Thompson [1], E. Martínez, J. F. Cariñena, W Sarlet [1], M. Crampin, W. Sarlet, E. Martínez, G. B. Byrnes and G. E. Prince [1], and others), the solution of the inverse variational problem for sprays, and of the problem on variationality of a linear connection (J. Klein [1,2], D. Krupka and A. Sattarov [1], I. Anderson and G. Thompson [1]), the solution of the problem on variationality of a Finsler connection (D. Krupka and A. Sattarov [1]), or a complete classification of variational semispray distributions on $R \times TM$ (O. Krupková [8]—the classification theorem and some of its consequences will be discussed later in Chapter 10, where Lagrangean systems on fibered manifolds $\pi : R \times M \to R$ will be studied).

Next, it is known that a variational integrating factor for a regular dynamical form E always exists if $m = 1$ (for more detail see Chapter 1), but need not exist if $m \geqslant 2$—this interesting fact has been first observed by J. Douglas [1]. Moreover, if for E a variational integrating factor exists, it need not be unique. Therefore there is a problem to find *all* locally variational forms equivalent with a given dynamical form. Consequently, one can study the following related question:

Given a locally variational form E, find all locally variational forms, equivalent with E.

Accordingly, two Lagrangians λ, λ' are called *alternative*, or *s-equivalent* (= solution-equivalent) if their Euler-Lagrange forms E_λ and $E_{\lambda'}$ are equivalent.

The latter problem geometrically means to pick up all locally variational forms from the equivalence class of dynamical forms related to a variational distribution. In the language of Lepagean two-forms this means for a distribution to find all Lepagean two-forms such that the given distribution is their characteristic distribution.

Again, this remains to be an open problem. For some particular results and examples we refer e.g. to N. Ya. Sonin [1], G. Darboux [1], M. Henneaux and L. C. Shepley [1], M. Henneaux [1], W. Sarlet [6], M. F. Rañada [2], J. F. Cariñena and M. F. Rañada [3], and references therein.

6.7.5. Example. On the fibered manifold $\pi : R \times R^3 \to R$ consider the dynamical form

$$E = E_i \, dx^i \wedge dt, \quad E_i = \delta_{ij}(\ddot{x}^j - k\dot{x}^j), \tag{6.103}$$

$i = 1, 2, 3$, where k is a constant. E represents the equivalence class $\{B_{ij}(\ddot{x}^j - k\dot{x}^j)\}$ (where B is an everywhere regular first-order matrix), corresponding to the semispray connection

$$w^i = k\dot{x}^i. \tag{6.104}$$

First of all, notice that the form E does not satisfy the Helmholtz conditions, i.e., it is not variational. Let us ask if the semispray distribution Δ_w is variational. This means we have to find out if there exists a variational integrating factor for E. Writing down the Helmholtz conditions for $E'_i = B_{ij}(\ddot{x}^j - k\dot{x}^j)$ we obtain the following equations for the B_{ij}:

$$B_{ij} = B_{ji}, \quad \frac{\partial B_{ij}}{\partial \dot{x}^l} = \frac{\partial B_{il}}{\partial \dot{x}^j}, \quad \frac{\partial B_{ij}}{\partial x^l} = \frac{\partial B_{il}}{\partial x^j},$$
$$\frac{\partial B_{ij}}{\partial t} + \frac{\partial B_{ij}}{\partial x^l}\dot{x}^l + \frac{\partial B_{ij}}{\partial \dot{x}^l}k\dot{x}^l + kB_{ij} = 0.$$

These equations are solvable with respect to the B_{ij}; one possible solution is

$$B_{ij} = \delta_{ij}e^{-kt}.$$

This means that the dynamical form E (6.103) is equivalent to the variational form

$$E'_i = \delta_{ij}e^{-kt}(\ddot{x}^j - k\dot{x}^j), \tag{6.105}$$

and the corresponding semispray connection Δ_w (6.104) is variational.

6.7.6. Example. Consider the dynamical form

$$E = (\ddot{x} + \dot{y})\, dx \wedge dt + (\ddot{y} + y)\, dy \wedge dt. \tag{6.106}$$

It can be shown that there exists no variational multiplier for E (see S. Hojman and L. F. Urrutia [1], or J. Douglas [1], for more details).

6.7.7. Example. On $\pi : R \times R \to R$ consider the dynamical form E, defined by

$$(m\ddot{x} + k\dot{x} + cx)\, dx \wedge dt, \tag{6.107}$$

where $m, k, c \neq 0$ are constants. E describes the motion of a damped harmonic oscillator of mass m, and obviously is not variational. Since $m = 1$, the corresponding semispray distribution Δ_w is variational. Let us find a variational multiplier for E. Denote $E' = fE$. Applying the Helmholtz conditions to E' we get for f the equation

$$\frac{\partial f}{\partial \dot{x}}(m\ddot{x} + k\dot{x} + cx) + kf - m\frac{df}{dt} = 0.$$

This equation has e.g. the following solutions:

$$f_1 = \frac{1}{m\dot{x}^2 + kx\dot{x} + cx^2}, \qquad f_2 = e^{kt/m}.$$

The function f_1 is not a variational multiplier for E, since there are points where $f_1 = 0$ (this means that the equations

$$m\ddot{x} + k\dot{x} + cx = 0$$

and

$$\frac{m\ddot{x} + k\dot{x} + cx}{m\dot{x}^2 + kx\dot{x} + cx^2} = 0$$

do not possess the same set of solutions). The function f_2 is a variational multiplier for E, i.e., we have E' of the form

$$e^{kt/m}(m\ddot{x} + k\dot{x} + cx)\, dx \wedge dt. \tag{6.108}$$

The dynamical forms (6.107) and (6.108) are equivalent possessing the same set of solutions, however, they describe *different* physical systems: While E describes a damped harmonic oscillator of mass m, E' refers to an oscillator with the nonconstant mass $m(t)$, accreting the mass according to the rule $me^{kt/m}$. Indeed, a Lagrangian for E' is

$$L = \tfrac{1}{2}e^{kt/m}(cx^2 - m\dot{x}^2),$$

which leads to the energy

$$\mathcal{E} = -H = L - \frac{\partial L}{\partial \dot{x}}\dot{x} = \tfrac{1}{2}e^{kt/m}(cx^2 + m\dot{x}^2),$$

differing from that one of the damped harmonic oscillator. (A discussion of the properties of the mechanical systems (6.107) and (6.108) from the point of view of quantum mechanics can be found in R. Ray [1]).

6.7.8. Example. A particle moving in a uniform magnetic field is described by the dynamical form

$$E = \sum_i (\ddot{q}^i - \omega\varepsilon_{ij}\dot{q}^j)\, dq^i \wedge dt, \quad i, j = 1, 2, \tag{6.109}$$

where $\varepsilon_{ij} = -\varepsilon_{ji}$ and ω is a constant. Let us discuss physical properties of some alternative Lagrangians for (6.109) (V. D. Skarzhinsky [1]). Restricting ourselves to Lagrangians quadratic in velocities we find with the help of the Helmholtz conditions the following family of alternative (= s-equivalent) Lagrangians, parametrized by 3 constants c_0, c_1, c_2:

$$L(t, q, \dot{q}) = c_0 L_0(q, \dot{q}) + c_1 L_1(t, \dot{q}) + c_2 L_2(t, \dot{q}), \tag{6.110}$$

where

$$\begin{aligned} L_0(q, \dot{q}) &= \tfrac{1}{2}\big((\dot{q}^1)^2 + (\dot{q}^2)^2\big) + \tfrac{1}{2}\omega(q^1\dot{q}^2 - \dot{q}^1 q^2), \\ L_1(t, \dot{q}) &= \tfrac{1}{2}\big((\dot{q}^1)^2 - (\dot{q}^2)^2\big)\sin\omega t + \dot{q}^1\dot{q}^2\cos\omega t, \\ L_2(t, \dot{q}) &= \tfrac{1}{2}\big((\dot{q}^1)^2 - (\dot{q}^2)^2\big)\cos\omega t + \dot{q}^1\dot{q}^2\sin\omega t. \end{aligned} \tag{6.111}$$

Notice that L_0 is a Lagrangian for the dynamical form E (6.109) (which is variational), whereas L_1, L_2 are Lagrangians for dynamical forms equivalent with E. (There exist also non-quadratic time-independent Lagrangians s-equivalent with (6.110)).

The canonical quantization procedure based on Lagrangians L_0, L_1 and L_2 leads to different quantum systems. The physical difference becomes obvious when measurability properties of physical observables, or structure of the energy spectrum is considered. For example, for L_0, the velocity operators $\dot{q}^1$ and $\dot{q}^2$ do not commute and cannot be measured simultaneously, while for L_1 and L_2 such measurements are quite possible. For L_0 the energy operator has a discrete spectrum $E_n = \hbar\omega(n+1/2)$ in contrast with the continuous energy spectra for L_1 and L_2 (V. V. Dodonov, V. I. Manko and V. D. Skarzhinsky [1], V. D. Skarzhinsky [1]).

6.7.9. Remark on the definition of Lagrangean system. We have seen that equivalent dynamical forms possess the same dynamics, having the same set of integral sections, the same set of first integrals, etc. Therefore it is legitimate to think about a more general understanding of a regular mechanical system as the equivalence class of dynamical forms related with a semispray connection/semispray distribution, and, accordingly, of a regular *Lagrangean system* as *the equivalence class of dynamical forms related with a variational connection/distribution*. There are, however, some arguments against such a wider understanding of Lagrangean systems the origin of which lies in the physical understanding of this concept. Indeed, examples show that *a semispray distribution describes the motion of a mechanical system, but does not represent the mechanical system itself.* As an illustration we may consider the equivalent dynamical forms (6.107) and (6.108) in Example 6.7.7: the motion of both is decribed by the same semispray connection

$$\ddot{x} \circ w = -k\dot{x} - cx,$$

however, from the physical point of view (6.107) and (6.108) are essentially different physical systems, possessing different energies, and having different quantum mechanical properties. An analogical discussion applies to the dynamical forms (6.103) and (6.105) in Example 6.7.5, or to different equivalent dynamical forms in Example 6.7.8. Therefore it semms that the integrating factor, being related directly to the kinetic energy of the system (cf. Corollary 6.7.4), carries an important physical information. Therefore, to avoid confusion, we prefer the definition of a Lagrangean system as given by 4.3.12.

Certainly, working with the equivalence class of dynamical forms rather than with a particular form, would be reasonable if one would be interested purely in the dynamics, without reference to the physical content of the equations in question—this concerns, in particular, the situations when first integrals of the given equations are to be found, or when some concrete integration methods have to be applied.

(3) Another generalization of the (local) inverse variational problem is obtained according to the following definition (D. Krupka [4]):

Let E be a dynamical form of order s on a fibered manifold π. E is called *weakly variational* if there exists an integer r and a locally variational form E' of order r on π such that the set of integral sections of E coincides with the set of extremals of E'.

In terms of the preceding paragraph where equivalence of dynamical forms of the same order has been studied, the weak variationality means that the corresponding distributions may be different (living on different jet prolongations of π).

In the most general setting of the inverse variational problem, in the defintion of the weak variationality, one can even drop the assumption that E' should live on the same fibered manifold as E, and the problem can be posed as follows:

Let E be a dynamical form of order s on a fibered manifold π. Decide if there exists a locally variational form E' (possibly on a different fibered manifold π') such that the integral sections of E are in *one-to-one correspondence* with the extremals of E'.

Surprizingly, such a general version of the inverse variatonal problem can be, for a large class of dynamical forms, solved very easily, and the solution is quite trivial. This concerns namely *regular* dynamical forms of *even orders*, and *regular* dynamical forms (of any order) on fibered manifolds $\pi : Y \to X$, $\dim X = 1$, where the dimension m of the fibers is *even*. More concretely, we have

6.7.10. Theorem. *Let $\pi : Y \to X$ be a fibered manifold,* $\dim X = 1$.

(1) *To every regular dynamical form E of order $2r$ on π there locally exists a regular variational form E' such that the integral sections of E are in one-to-one correspondence with the extremals of E'.*
(2) *If $m = \dim Y - 1$ is even then to every regular dynamical form E of order $s \geqslant 1$ on π there locally exists a regular variational form E' such that the integral sections of E are in one-to-one correspondence with the extremals of E'.*

Proof. We shall show that E' can be constructed as a (local) *first-order* variational form *on the fibered manifold* π_{s-1}, where s is the order of E; in other words, that *every regular mechanical system* (i.e., represented by a regular dynamical form) satisfying the assumptions of Theorem 6.7.10 can be viewed as a *Lagrangean system of order zero*.

Consider a regular dynamical form of order s for a fibered manifold π, let w be the related semispray connection (as we know, w is defined by $w^*E = 0$). Since the corresponding semispray distribution Δ_w is of rank one, it is completely integrable. We have $\operatorname{corank} \Delta_w = \dim J^{s-1}Y - 1 = sm$. Hence, by Frobenius Theorem, in a neighborhood of every point in $J^{s-1}Y$ there exists an adapted chart (U, χ), $\chi = (t, f^1, \dots, f^{sm})$ to the foliation defined by Δ_w, such that the functions $f^1, \dots, f^{sm}$ are first integrals of the semispray distribution Δ_w.

Suppose that s or m is even. Then sm is even; denote $sm = 2p$ and put

$$\alpha = df^1 \wedge df^{p+1} + \dots + df^p \wedge df^{2p}.$$

Then α is a closed two-form on $U \subset J^{s-1}Y$, $\operatorname{corank} \alpha = 1$, and Δ_w is the characteristic distribution of α. This means that α is a *regular zero-order Lagrangean system* on the fibered manifold π_{s-1}. The extremals of this Lagrangean system are sections δ of the fibered manifold π_{s-1} which are integral sections of the distribution Δ_w. However, Δ_w is a semispray distribution, so that every integral section of Δ_w is holonomic, i.e., of the form $\delta = J^{s-1}\gamma$ for a section γ of π. Consequently, there is a one-to-one correspondence between the integral sections of E and the extremals of the Lagrangean system α. Thus putting

$$E' = \tilde{p}_1 \alpha,$$

where $\tilde{p}_1$ is the 1-contactization operator with respect to the projection π_{s-1}, we get a locally variational form which is regular and of order one on $\pi_{s-1} : J^{s-1}Y \to X$. □

Notice that minimal-order Lagrangians for E' are of the form

$$L = \sum_{i=1}^{p} f^i \frac{\tilde{d} f^{p+i}}{dt} + \frac{\tilde{d} F}{dt},$$

where f^j, $1 \leqslant j \leqslant 2p$, are first integrals of Δ_w, F is an arbitrary function on $U \subset J^{s-1}Y$, and $\tilde{d}/dt$ is the total derivative with respect to the projection π_{s-1}, and that these Lagrangians are defined on open subsets of $J^1(J^{s-1}Y)$.

Theorem 6.7.10 is a higher-order generalization of a result obtained by Havas [2], namely that every system of second-order ODE, explicitly solved with respect to the second derivatives, can be viewed as a system of equations for extremals of a first-order Lagrangian if the coordinates and velocities are considered as independent variables. An explicit form of the corresponding Lagrangians has then been obtained by S. Hojman and L. F. Urrutia [1], and R. Hojman, S. Hojman and J. Sheinbaum [1].

Chapter 7.

SINGULAR LAGRANGEAN SYSTEMS

7.1. Introduction

In the preceding chapters we have assigned to Lagrangean systems certain distributions reflecting their dynamical properties (the Euler-Lagrange and the characteristic distribution). We have studied the most simple Lagrangean systems which are represented by Euler-Lagrange distributions of rank *one* (locally spanned by one semispray), and which we called *regular*. In this chapter we shall turn to study the other Lagrangean systems which we call *singular* or *degenerate*.

Recall that in the general situation:

– The Euler-Lagrange distribution need not have a constant rank, and consequently, need not be spanned by a system of continuous vector fields (which means that the dynamics of singular Lagrangean systems generally cannot be described by a vector field, or by a system of vector fields).

– The Euler-Lagrange distribution need not be horizontal (which means that there can exist points in the phase space where no motion is allowed).

– The characteristic distribution and the Euler-Lagrange distribution may be different.

– Maximal integral manifolds of the Euler-Lagrange distribution need not coincide with the solutions of given variational ODE (which means that the integration problem cannot be reduced to the standard procedure of finding maximal integral manifolds of a distribution).

– Integral sections of the Euler-Lagrange distribution (= Hamilton extremals) need not coincide with prolonged extremals; however, prolonged extremals form a subset in the set of integral sections. This means that generally one has to distinguish two levels of the integration problem: one for Hamilton extremals, describing the extended dynamics, and the other for prolonged extremals, describing the proper dynamics (the higher-order semispray problem).

Notice that dealing with singular Lagrangean systems we are faced with a new non-trivial question—to clarify *the structure of solutions* of the Euler-Lagrange distributions, i.e., the "dynamical pictures". For regular Lagrangean systems this question is solved trivially—the dynamics is represented by a one-dimensional foliation of the phase space. To answer this question in general it is reasonable first to make a *geometric classification of Lagrangean systems*, based on the properties of the Euler-Lagrange and characteristic distributions. In this way one could distinguish between Lagrangean systems with relatively simple Euler-Lagrange distributions (e.g. horizontal, of a constant rank, etc.), and the other Lagrangean systems the dynamics of which cannot be studied easily with help of Frobenius theory. In this chapter such a classification is presented. Classifying Lagrangean systems according to their dynamics, the concepts of *regularity, semiregularity, weak regularity*, etc., are naturally introduced to be properties of the *class of equivalent Lagrangians* rather than of a particular Lagrangian.

Based on this classification, a *constraint algorithm* which enables one to describe geometrically the constraints on the motion, and to find the extended and proper dynamics explicitly, is proposed. Consequently, one can study the structure of solutions of the Hamilton equations and their relations to extremals (the so called higher-order semispray problem). We shall also see that, behind the regular ones, there are also some other Lagrangean systems with quite nice dynamical properties, namely, the *semiregular* and *weakly regular* Lagrangean systems. We shall also introduce *generalized Legendre transformations* which can simplify the study of dynamics of certain Lagrangean systems.

Following the spirit of the standard terminology, we shall also speak of singular systems as of *constrained systems*. However, as we shall see below, the terminology "constrained" is not very suitable, since there may be singular systems the motion of which does not suffer from any constraints (the so-called *weakly regular Lagrangean systems*). These "singular unconstrained systems" will be studied in Sec. 7.6.

Before turning to detailed study of singular systems we remind the reader that the concept of regularity used in this book is wider than the standard one (cf. Definition 6.2.1 and the surrounding material, contrary to (6.1)). Consequently, many Lagrangians which traditionally are considered singular, and therefore are studied within the range of the Dirac theory of constrained systems, are regular in our understanding, and they can be treated in a much simpler way. Typical examples of such regular Lagrangean systems are those defined by higher-order Lagrangians equivalent with Lagrangians satisfying (6.1), or Lagrangean systems described by odd-order Euler-Lagrange equations explicitly solved with respect to the highest derivatives, or many of the Lagrangians affine in the highest derivatives (cf. examples in Section 6.6). Explicit examples of Lagrangean systems which are constrained in the standard sense, but regular in our geometric sense can be found in Chapters 4, 6 and 9. From similar reasons, the standard concept of a singular Lagrangian is not identical with ours. (The reader is encouraged to re-examine at least some of the examples of singular Lagrangians studied throughout the literature—a few of such references are listed in the Bibliography).

The concepts in this chapter are due to O. Krupková [9,2].

7.2. Classification of singular Lagrangean systems

Classification criteria for singular Lagrangean systems are directly suggested by properties of their Euler-Lagrange distribution. In this section we provide formal definitions, concrete examples of different singular Lagrangean systems will be studied later (see Sections 7.4–7.6, and the following chapters).

The first classification point should reflect the difference between Lagrangean systems which "admit a global Hamiltonian dynamics", i.e., such that the set of admissible initial conditions for the Hamilton equations is the whole phase space, and those the extended motion of which is "constrained" to a subspace of the phase space.

Let $s \geqslant 1$, let α be a Lagrangean system on $J^{s-1}Y$. Consider *the primary constraint set* $\tilde{\mathcal{P}}$ of α. Recall that $\tilde{\mathcal{P}}$ is the subset of the phase space where the Euler-Lagrange distribution is weakly horizontal (i.e., coincides with the characteristic distribution of α). By Theorem 5.4.1 (keeping the notations from Chapters 4 and 5),

$$\tilde{\mathcal{P}} = \{x \in J^{s-1}Y \mid \operatorname{rank} \mathrm{F} = \operatorname{rank}(\mathrm{F} \mid \mathrm{A})\}. \tag{7.1}$$

Note that $\tilde{\mathcal{P}}$ need not be a submanifold of the phase space.

7.2.1. Definition. A Lagrangean system for which the set $J^{s-1}Y - \tilde{\mathcal{P}}$ is nonempty will be called a *Lagrangean system with primary dynamical constraints.*

We can also say that *Lagrangean systems with primary dynamical constraints do not admit a "global" extended dynamics* (in the sense that there are *a priori* restrictions on initial points of the extended motion).

7.2.2. Definition. We say that a Lagrangean system α on $J^{s-1}Y$ possesses *no primary dynamical constraints* if its Euler-Lagrange distribution Δ is weakly horizontal, i.e., if $J^{s-1}Y - \tilde{\mathcal{P}} = \emptyset$.

Let us summarize the properties of Lagrangean systems with no primary dynamical constraints:

(1) At each point of the phase space there exists a nonvertical vector belonging to the Euler-Lagrange distribution Δ.
(2) At each point of the phase space $\operatorname{rank} \mathrm{F} = \operatorname{rank}(\mathrm{F} \mid \mathrm{A}) \geqslant 1$.
(3) The characteristic distribution coincides with the Euler-Lagrange distribution.
(4) $\operatorname{corank} \Delta = \operatorname{rank} \alpha$.

A Lagrangean system with no primary dynamical constraints is described by a weakly horizontal Euler-Lagrange distribution, which however need not be of a constant rank. Consequently (since it is spanned by smooth Pfaffian forms), it generally cannot be spanned by a system of continuous vector fields. This means that the dynamics cannot be found by standard techniques of integration of distributions, and one has to apply another procedure, the *constraint algorithm*, which will be subject of the next section.

Among Lagrangean systems with no primary dynamical constraints, an important role is played by those of a constant rank (resp. of a locally constant rank, i.e., of a constant rank on each connected component).

7.2.3. Definition. A Lagrangean system is called *semiregular* if its Euler-Lagrange distribution is weakly horizontal and has a locally constant rank. A Lagrangian λ is called *semiregular* if it defines a semiregular Lagrangean system (i.e., if α_{E_λ} is semiregular).

Recall that we denote by $\mathcal{P}$ the primary semispray-constraint set, i.e.,

$$\mathcal{P} = \{x \in J^{s-1}Y \mid \operatorname{rank} \mathrm{B} = \operatorname{rank}(\mathrm{B} \mid \mathrm{A})\}. \tag{7.2}$$

$\mathcal{P}$ is a subset of $\tilde{\mathcal{P}}$, generally different from $\tilde{\mathcal{P}}$. This suggests us another classification point.

7.2.4. Definition. A Lagrangean system for which the set $J^{s-1}Y - \mathcal{P}$ is nonempty will be called a *Lagrangean system with primary semispray constraints.*

We can also say that *Lagrangean systems with primary semispray constraints do not admit a "global" proper dynamics* (in the sense that there are *a priori* restrictions on initial points of the proper motion).

7.2.5. Definition. We say that a Lagrangean system α on $J^{s-1}Y$ possesses *no primary semispray constraints* if $J^{s-1}Y = \mathcal{P}$.

Recall that this means that $\operatorname{rank} \mathrm{B}(x) = \operatorname{rank}(\mathrm{B} \mid \mathrm{A})(x)$ at each point x of the phase space.

7.2.6. Definition. A Lagrangean system is called *weakly regular* if it carries no primary semispray constraints and rank B is locally constant on the phase space. A Lagrangian λ is called *weakly regular* if it defines a weakly regular Lagrangean system (i.e., if α_{E_λ} is weakly regular).

A semiregular Lagrangean system need not be weakly regular, and a weakly regular Lagrangean system need not be semiregular.

7.2.7. Definition. A Lagrangean system is called *strongly semiregular* if it is weakly regular and semiregular.

Similarly as in the case of extended dynamics, in most of the cases the proper dynamics cannot be found by standard techniques, and one must apply the "constraint algorithm" described below.

7.3. The constraint algorithm for the Euler-Lagrange distribution

In this section we shall introduce a general procedure which enables one to solve the Euler-Lagrange distribution explicitly. We shall call this procedure the *constraint algorithm for the Euler-Lagrange distribution.*

We have seen that in general it is not possible to characterize the dynamics of a Lagrangean system by a system of continuous vector fields. On the other hand, every Hamilton extremal defines a one-dimensional (immersed) submanifold M of the phase space and a *vector field* ξ *along* M such that for every $y \in M$, $\xi(y)$ belongs to the Euler-Lagrange distribution Δ at y. Clearly, such submanifolds can have nonempty intersections. Hence, the problem to be solved is to find *at each point* x of the phase space the bunch of submanifolds where the motion "is allowed to develop", i.e., such that every section of every of these submanifolds is a Hamilton extremal, and conversely, every Hamilton extremal passing through x is locally embedded in a manifold of this bunch. Since the extended dynamics and proper dynamics in general do not coincide, we have to distinguish two levels of the integration problem: (1) to find the *extended dynamics*, i.e., *all* integral sections (Hamilton extremals), and (2) to find the *proper dynamics*, i.e., *holonomic* integral sections (prolongations of extremals); the latter problem is called *(higher-order) semispray problem.*

7.3.1. Extended dynamics. First, we shall describe an algorithm for finding explicitly the dynamics of the Hamiltonian system, associated with a given $(s-1)$-th order Lagrangean system.

Step 1. Find the primary constraint set set $\tilde{\mathcal{P}}$ (recall that $\tilde{\mathcal{P}}$ need not be a submanifold of the phase space). If $\tilde{\mathcal{P}} = \emptyset$, there is no extended dynamics, hence no dynamics at all. If $\tilde{\mathcal{P}} \neq \emptyset$, choose a point $x \in \tilde{\mathcal{P}}$, and proceed to the next step.

Step 2. Denote $S_{(1)} = \tilde{\mathcal{P}}$, and $M_{(1)} = \cup_\iota M_{(1)\iota}$ the union of all connected submanifolds $M_{(1)\iota}$ of maximal dimensions, lying in $S_{(1)}$ and in a neighborhood U of x, and passing through x (here "maximal dimension" means that if N is a connected submanifold passing through x, lying in $S_{(1)} \cap U$, and $M_{(1)\iota} \subset N$ then $M_{(1)\iota} = N$).

Suppose that $M_{(1)} \neq \{x\}$. For each ι consider the restriction of the Euler-Lagrange distribution Δ to $M_{(1)\iota}$, i.e., the distribution $\Delta_{(1)\iota} = \Delta \cap TM_{(1)\iota}$ (called the *constrained to* $M_{(1)\iota}$ Euler-Lagrange distribution). If $\Delta_{(1)\iota}$ is weakly horizontal at each point of $M_{(1)\iota}$, we call the manifold $M_{(1)\iota}$ *a final constraint submanifold at* x; the problem now is reduced to the problem of integration of the distribution $\Delta_{(1)\iota}$. Otherwise, one of the following two possibilities occurs: (i) $\Delta_{(1)\iota}$ is trivial at x, or is not weakly horizontal at x; then exclude the manifold $M_{(1)\iota}$ from the bunch $M_{(1)}$. (ii) $\Delta_{(1)\iota}$ is weakly horizontal at x but is not weakly horizontal on $M_{(1)\iota}$; then proceed to the next step.

Step 3. Exclude the points where $\Delta_{(1)\iota}$ is not weakly horizontal from $M_{(1)\iota}$, and denote the resulting set by $S_{(2)\iota}$; clearly, $x \in S_{(2)\iota}$. Repeat the procedure described in Step 2 with $S_{(2)\iota}$ instead of $S_{(1)\iota}$.

After sufficiently many steps we obtain either a bunch of final constraint submanifolds at x, or we find that there is no final constraint submanifold passing through x. If there exists a point such that there is no final constraint submanifold passing through x, we say that the Lagrangean system possesses *secondary dynamical constraints*.

Collecting the points where there exists a bunch of final constraint submanifolds we obtain a *subset* of the phase space where the extended motion proceeds, we call it the set of *admissible initial conditions for the Hamilton equations*. The structure of this set can be complicated; in particular, it need not be a submanifold of the phase space. Considering then the collection of final constraint submanifolds together with to them constrained Euler-Lagrange distributions, we get the dynamical picture corresponding to the solutions of the Hamilton equations. Typically, this dynamical picture will be rather complicated, measuring the "rate of singularity" of a given Lagrangean system.

7.3.2. Proper dynamics: the higher-order semispray problem. Now, we shall be interested in an explicit description of the proper dynamics of the Lagrangean system. The procedure is as follows:

Find the set $\mathcal{P}$ (recall that $\mathcal{P}$ need not be a submanifold of the phase space). If $\mathcal{P} = \emptyset$, there is no motion. If $\mathcal{P} \neq \emptyset$, choose a point $x \in \mathcal{P}$. Denote $R_{(1)} = \mathcal{P}$. Proceed in the same way as described in 7.3.1, replacing $S_{(1)}$ by $R_{(1)}$, etc.

As a result of the procedure one gets at each point $x \in R_{(1)}$ (i) the bunch M_x of final constraint submanifolds, and (ii) a system $\mathcal{V}_x$ of vector fields along the manifolds of the bunch M_x which belong to the Euler-Lagrange distribution (at the points of M_x); by construction, all these vector fields are nonvertical.

Let $N \in M_x$, and let ξ be the vector field in $\mathcal{V}_x$ tangent to N. We shall say that ξ can be *identified with a semispray* along a submanifold $N_0 \subset N$ if there exists a submanifold

N_0 of N, a neighborhood U of N_0 and a semispray ζ on U such that at each point $y \in N_0$, $\xi(y) = \zeta(y)$. Let $x \in \mathcal{P}$ be a point. We shall call the point x *admissible* if there exists a neighborhood U_x of x and a vector field $\xi \in \mathcal{V}_x$ such that ξ can be identified with a semispray along a submanifold of an element of the bunch M_x. If every point in $\mathcal{P}$ is admissible we say that the Lagrangean system possesses *no secondary semispray constraints*. Otherwise, we shall say that the Lagrangean system possesses *secondary semispray constraints*.

Collecting the admissible points of $\mathcal{P}$ we obtain a *subset* of the phase space where the proper motion proceeds; we call this set the *set of admissible initial conditions for the Euler-Lagrange equations*. As expected, this set need not be a submanifold of the phase space. Now, consider the family $\{\mathcal{V}_x\}$ where x runs over the set of admissible points. To get the dynamical picture corresponding to the solutions of the Euler-Lagrange equations, it is sufficient to pick up those elements of $\{\mathcal{V}_x\}$ which (in the above sense) can be identified with semisprays along at least one-dimensional submanifolds of elements of the $\{M_x\}$.

7.3.3. Remarks. The constraint algorithm described above has been proposed in O. Krupková [9]. From the mathematical point of view it represents a general algorithm to find integral sections of a distribution of a non-constant rank. It can be viewed as a generalization and re-interpretation of the constraint algorithm developed within the range of the Dirac theory of constrained systems by M. J. Gotay and J. M. Nester [1], and M. J. Gotay, J. M. Nester and G. Hinds [1], and generalized to time-dependent and higher-order Lagrangians by D. Chinea, M. de León and J. C. Marrero [1], M. de León and J. C. Marrero [1] and M. de León, J. Marín-Solano and J. C. Marrero [1].

The problem of finding those solutions of the Hamilton equations which correspond to extremals is in the presymplectic geometry usually called the *SODE problem* (= second-order differential equation problem). It has been studied by different authors using different methods (cf. e.g. J. F. Cariñena and L. A. Ibort [1], J. F. Cariñena, C. López and N. Román-Roy [1,2], M. J. Gotay and J. M. Nester [1], and M. de León and J. C. Marrero [1], M. de León, J. Marín-Solano and J. C. Marrero [1]). Within our approach a new understanding to this problem is given, namely, as a problem of finding holonomic solutions of a distribution of a non-constant rank which cannot be spanned by smooth vector fields. Naturally, its generalization to higher-order Lagrangean systems is then called the *higher-order semispray problem*).

7.3.4. Remark. The concepts of primary and secondary dynamical constraints, resp. of primary and secondary semispray constraints are based on the geometric understanding of the dynamics by means of the Euler-Lagrange distribution, and are not in a direct correspondence with the more-or-less heuristic concepts of "primary" and "secondary constraints" of the Dirac theory of constrained systems (sse P. A. M. Dirac [1]).

7.4. Applications of the constraint algorithm

In this section we denote $\omega^i = dq^i - \dot{q}^i\, dt$.

7.4.1. Consider the Lagrangian

$$L = \tfrac{1}{2}\big((\dot{q}^1)^2 + (q^1)^2 q^2\big) \tag{7.3}$$

on $R \times R^2 \times R^2$ (J. F. Cariñena [1]). This is a first-order Lagrangian which does not satisfy any of the regularity conditions (6.10), (6.11), i.e., it defines a *singular first-order Lagrangean system* α. Since

$$p_1 = \dot{q}^1, \quad p_2 = 0, \quad H = \tfrac{1}{2}\big((\dot{q}^1)^2 - (q^1)^2 q^2\big),$$

we get

$$\alpha = -dH \wedge dt + dp_1 \wedge dq^1 = -(\dot{q}^1\, d\dot{q}^1 - \tfrac{1}{2}(q^1)^2\, dq^2 - q^1 q^2\, dq^1) \wedge dt + d\dot{q}^1 \wedge dq^1.$$

The Euler-Lagrange and the characteristic distributions are

$$\begin{aligned}
\Delta &\approx \operatorname{span}\{\tfrac{1}{2}(q^1)^2\, dt,\ \omega^1,\ d\dot{q}^1 - q^1 q^2\, dt\},\\
\mathcal{D} &\approx \operatorname{span}\{\tfrac{1}{2}(q^1)^2\, dt,\ \omega^1,\ d\dot{q}^1 - q^1 q^2\, dt,\ \dot{q}^1\, d\dot{q}^1 - \tfrac{1}{2}(q^1)^2\, dq^2 - q^1 q^2\, dq^1\}.
\end{aligned}$$

(A) Extended dynamics. Obviously, Δ is weakly horizontal at the points $q^1 = 0$, i.e., the primary constraint set is

$$\tilde{\mathcal{P}} = \{x \in J^1 Y \mid q^1(x) = 0\}.$$

In other words, this Lagrangean system possesses *primary dynamical constraints*.

The set $\tilde{\mathcal{P}}$ is a submanifold of the phase space. At each point of $\tilde{\mathcal{P}}$,

$$\Delta = \mathcal{D} = \operatorname{span}\left\{\frac{\partial}{\partial t} + \dot{q}^1 \frac{\partial}{\partial q^1},\ \frac{\partial}{\partial q^2},\ \frac{\partial}{\partial \dot{q}^2}\right\} \approx \operatorname{span}\{\omega^1,\ d\dot{q}^1\}.$$

Let us compute the constrained to $\tilde{\mathcal{P}}$ Euler-Lagrange distribution. At the points of $\tilde{\mathcal{P}}$ where $\dot{q}^1 \neq 0$ we have

$$\Delta|_{\tilde{\mathcal{P}}} = \operatorname{span}\left\{\frac{\partial}{\partial q^2},\ \frac{\partial}{\partial \dot{q}^2}\right\},$$

which is a vertical distribution, and at the points of $\tilde{\mathcal{P}}$ where $\dot{q}^1 = 0$ we get a weakly horizontal distribution

$$\Delta|_{\tilde{\mathcal{P}}} = \operatorname{span}\left\{\frac{\partial}{\partial t},\ \frac{\partial}{\partial q^2},\ \frac{\partial}{\partial \dot{q}^2}\right\}.$$

This means that the Lagrangean system possesses *secondary dynamical constraints*. The (final) constraint set is the manifold $M = \{x \in J^1 Y \mid q^1(x) = 0,\ \dot{q}^1(x) = 0\} \subset J^1 Y$.

Summarizing, we get the following dynamical picture for the Hamiltonian system associated with our Lagrangean system α: the extended dynamics is *constrained to a submanifold* $M = \{x \in J^1 Y \mid q^1(x) = \dot{q}^1(x) = 0\}$ of the phase space. On this submanifold, the extended motion is indeterministic (being not uniquely determined by the initial conditions), and is given by a weakly horizontal distribution of rank 3, spanned by the vector fields

$$\frac{\partial}{\partial t} + f \frac{\partial}{\partial q^2} + g \frac{\partial}{\partial \dot{q}^2},$$

where f, g are arbitrary functions on M.

(B) Proper dynamics. Computing the semispray-constraint set we get

$$\mathcal{P} = \{x \in J^1 Y \mid q^1(x) = 0\},$$

i.e., $\mathcal{P} = \tilde{\mathcal{P}}$. This means that the Lagrangean system possesses *primary semispray constraints*, which identify with the primary dynamical constraints. Similarly as above we get

$$\Delta|_{\mathcal{P}} = \operatorname{span}\left\{\frac{\partial}{\partial q^2}, \frac{\partial}{\partial \dot{q}^2}\right\}$$

at the points of $\mathcal{P}$ where $\dot{q}^1 \neq 0$, and

$$\Delta|_{\mathcal{P}} = \operatorname{span}\left\{\frac{\partial}{\partial t}, \frac{\partial}{\partial q^2}, \frac{\partial}{\partial \dot{q}^2}\right\}$$

at the points of $\mathcal{P}$ where $\dot{q}^1 = 0$, i.e., the Lagrangean system possesses *secondary semispray constraints*. As above, denote $M = \{x \in J^1Y \mid q^1(x) = \dot{q}^1(x) = 0\}$. The preceding weakly horizontal distribution along M which is of rank 3 has a subdistribution of rank 2, spanned by the following semisprays along M

$$\frac{\partial}{\partial t} + \dot{q}^2 \frac{\partial}{\partial q^2} + g \frac{\partial}{\partial \dot{q}^2}, \tag{7.4}$$

where g is a function on M. Hence, *the proper dynamics is constrained to* M and is *indeterministic* there: prolongations of extremals coincide with the integral sections of the distribution spanned by the vector fields (7.4).

The above dynamical picture gives us also information about properties of *extremals* on the configuration space. Namely, we can see that *all* solutions of the Euler-Lagrange equations are embedded in the submanifold $Q = \{x \in Y, \mid q^1(x) = 0\}$, and that *every* section lying in this submanifold is an extremal.

7.4.2. As a second example consider the Cawley's Lagrangian

$$L = \dot{q}^1\dot{q}^3 + \tfrac{1}{2}(q^2)^2 q^3 \tag{7.5}$$

on $R \times R^3 \times R^3$ (R. Cawley [1], J. F. Cariñena, C. López and N. Román-Roy [2]). This Lagrangian defines a first order Lagrangean system

$$\alpha = (q^2 q^3 \omega^2 + \tfrac{1}{2}(q^2)^2 \omega^3) \wedge dt - \omega^1 \wedge d\dot{q}^3 - \omega^3 \wedge d\dot{q}^1.$$

Computing the distributions $\mathcal{D}$ and Δ we get

$$\mathcal{D} \approx \operatorname{span}\{q^2 q^3\, dt,\ \omega^1,\ q^2 q^3 \omega^2,\ \omega^3,\ d\dot{q}^1 - \tfrac{1}{2}(q^2)^2\, dt,\ d\dot{q}^3\},$$
$$\Delta \approx \operatorname{span}\{q^2 q^3\, dt,\ \omega^1,\ \omega^3,\ d\dot{q}^1 - \tfrac{1}{2}(q^2)^2\, dt,\ d\dot{q}^3\}.$$

We can see that $\mathcal{D} \subset \Delta$ and $\mathcal{D} \neq \Delta$. The function $f = \dot{q}^3$ is a *first integral* of the distributions $\mathcal{D}$ and Δ.

(A) Extended dynamics. It holds rank F = rank(F | A) if and only if $q^2 q^3 = 0$, i.e., the Lagrangean system possesses *primary dynamical constraints*, and

$$\tilde{\mathcal{P}} = \{x \in J^1Y \mid q^2 q^3 = 0\},$$

which is not a submanifold of the phase space.

(i) Let $x \notin \tilde{\mathcal{P}}$. The Euler-Lagrange distribution is at x spanned by vertical vectors, hence, there is no dynamics at this point.

(ii) Let $x \in \tilde{\mathcal{P}}$. Then we have

$$\mathcal{D} = \Delta = \operatorname{span}\left\{\frac{\partial}{\partial t} + \dot{q}^1\frac{\partial}{\partial q^1} + \dot{q}^3\frac{\partial}{\partial q^3} + \tfrac{1}{2}(q^2)^2\frac{\partial}{\partial \dot{q}^1},\ \frac{\partial}{\partial q^2},\ \frac{\partial}{\partial \dot{q}^2}\right\}$$
$$\approx \operatorname{span}\left\{\omega^1,\ \omega^3,\ d\dot{q}^1 - \tfrac{1}{2}(q^2)^2\,dt,\ d\dot{q}^3\right\},$$

i.e., $\operatorname{rank}\mathcal{D} = \operatorname{rank}\Delta = 3$ at x.

For $x \in \tilde{\mathcal{P}}$, $q^2 \neq 0$ we get that the bunch of submanifolds of $\tilde{\mathcal{P}}$ passing through x consists from a single submanifold $M_x = \{q^3 = 0\}$. The constrained to M_x Euler-Lagrange distribution is

$$\Delta|_{M_x} = \operatorname{span}\left\{\frac{\partial}{\partial q^2},\ \frac{\partial}{\partial \dot{q}^2}\right\},$$

at the points of M_x where $\dot{q}^3 \neq 0$, which is nowhere weakly horizontal, and

$$\Delta|_{M_x} = \operatorname{span}\left\{\frac{\partial}{\partial t} + \dot{q}^1\frac{\partial}{\partial q^1} + \tfrac{1}{2}(q^2)^2\frac{\partial}{\partial \dot{q}^1},\ \frac{\partial}{\partial q^2},\ \frac{\partial}{\partial \dot{q}^2}\right\},$$

at the points of M_x where $\dot{q}^3 = 0$, which is weakly horizontal. This means that the Lagrangian system possesses *secondary dynamical constraints*, and we must exclude from $\tilde{\mathcal{P}}$ the points such that $q^2 \neq 0$, $q^3 = 0$, $\dot{q}^3 \neq 0$. At each of the remaining points we have a (unique) final constraint submanifold $M_1 = \{y \in M_x \mid \dot{q}^3 = 0\}$. Along M_1, the Euler-Lagrange distribution is reduced to the *completely integrable* distribution of rank 3, spanned by

$$\frac{\partial}{\partial t} + \dot{q}^1\frac{\partial}{\partial q^1} + \tfrac{1}{2}(q^2)^2\frac{\partial}{\partial \dot{q}^1},\quad \frac{\partial}{\partial q^2},\quad \frac{\partial}{\partial \dot{q}^2}.$$

Suppose that $x \in \tilde{\mathcal{P}}$, $q^3 \neq 0$. We get $M_x = \{q^2 = 0\}$, and the constrained to M_x Euler-Lagrange distribution is

$$\Delta|_{M_x} = \operatorname{span}\left\{\frac{\partial}{\partial t} + \dot{q}^1\frac{\partial}{\partial q^1} + \dot{q}^3\frac{\partial}{\partial q^3},\ \frac{\partial}{\partial \dot{q}^2}\right\};$$

it is weakly horizontal. M_x is a (unique) *final constraint submanifold at the point* x and $\Delta|_{M_x}$ is a *completely integrable* distribution of rank 2.

The only remaining points in the phase space to be considered are $x \in \tilde{\mathcal{P}}$, $q^2 = 0$, $q^3 = 0$. The bunch of submanifolds at x now consists of two manifolds, $M_x^1 = \{q^3 = 0\}$ and $M_x^2 = \{q^2 = 0\}$. The Euler-Lagrange distribution Δ is weakly horizontal in the points of M_x^1 where $\dot{q}^3 = 0$. Along M_x^2, Δ is weakly horizontal and of rank 2, hence M_x^2 is a final constraint submanifold at x.

Summarizing the results we get the following picture of the extended dynamics: extended motion is *constrained to the subset*

$$\{q^2 = 0\} \cup \{q^3 = \dot{q}^3 = 0\} \tag{7.6}$$

of the phase space, which is a union of two intersecting closed submanifolds. Along the submanifold $\{q^2 = 0\}$, Hamilton extremals are integral sections of the completely integrable distribution of rank 2, spanned by the vector fields

$$\frac{\partial}{\partial t} + \dot{q}^1\frac{\partial}{\partial q^1} + \dot{q}^3\frac{\partial}{\partial q^3},\quad \frac{\partial}{\partial \dot{q}^2}, \tag{7.7}$$

i.e., the extended motion there is *semiregular* and proceeds within *two-dimensional leaves*. Along the submanifold $\{q^3 = \dot{q}^3 = 0\}$ we get Hamilton extremals as integral sections of the completely integrable distribution

$$\frac{\partial}{\partial t} + \dot{q}^1 \frac{\partial}{\partial q^1} + \tfrac{1}{2}(q^2)^2 \frac{\partial}{\partial \dot{q}^1}, \quad \frac{\partial}{\partial q^2}, \quad \frac{\partial}{\partial \dot{q}^2} \tag{7.8}$$

of rank 3, i.e., the extended motion there is *semiregular* and proceeds within *three-dimensional leaves* of the corresponding foliation.

(B) Proper dynamics. The system possesses *primary semispray constraints* which coincide with the primary dynamical constraints, since

$$\operatorname{rank} \mathrm{B} = \operatorname{rank}(\mathrm{B} \mid \mathrm{A}) \quad \text{if and only if} \quad q^2 q^3 = 0\,.$$

This means that we are lead to consider the semispray problem for the subset (7.6) and the distributions (7.7) and (7.8). We can see that the system possesses *secondary semispray constraints*: to a point $x \in \{q^2 = 0\}$ there is a semispray ζ such that $\zeta(x)$ belongs to the distribution (7.7) at x if and only if $\dot{q}^2(x) = 0$. The Euler-Lagrange distribution constrained to the closed submanifold $\{q^2 = \dot{q}^2 = 0\}$ is of rank 1, and it is spanned by the vector field

$$\frac{\partial}{\partial t} + \dot{q}^1 \frac{\partial}{\partial q^1} + \dot{q}^3 \frac{\partial}{\partial q^3} \tag{7.9}$$

(which is a semispray along $\{q^2 = \dot{q}^2 = 0\}$). The distribution (7.8) on $\{q^3 = \dot{q}^3 = 0\}$ has a semispray subdistribution of rank 2, spanned by the vector fields

$$\frac{\partial}{\partial t} + \dot{q}^1 \frac{\partial}{\partial q^1} + \dot{q}^2 \frac{\partial}{\partial q^2} + \tfrac{1}{2}(q^2)^2 \frac{\partial}{\partial \dot{q}^1} + g \frac{\partial}{\partial \dot{q}^2}, \tag{7.10}$$

where g is a function on $\{q^3 = \dot{q}^3 = 0\}$.

In other words, the proper motion is *constrained to the subset*

$$\{q^2 = \dot{q}^2 = 0\} \cup \{q^3 = \dot{q}^3 = 0\} \tag{7.11}$$

of the phase space. On the submanifold $\{q^2 = \dot{q}^2 = 0\}$ it is *regular*, described by the vector field (7.9) (through each point there passes exactly one maximal integral section = 1-jet prolongation of an extremal). Consequently, on this subset the motion is uniquelly determined by the initial conditions. On the submanifold $\{q^3 = \dot{q}^3 = 0\}$ it is *weakly regular*, described by the vector fields (7.10). Since the distribution (7.10) *is not completely integrable*, we do not have a foliation defined by this distribution. However, since (7.10) is a subdistribution of the semiregular completely integrable distribution (7.8), the Lagrangean system is in fact *strongly semiregular* on $\{q^3 = \dot{q}^3 = 0\}$, which means that the prolonged extremals are embedded in the 3-dimensional leaves of the foliation of (7.8).

Notice that *on the configuration space*, the solutions of the Euler-Lagrange equations (*extremals*) are constrained to the subset $\{q^2 = 0\} \cup \{q^3 = 0\}$.

7.4.3. Finally, let us study the dynamics of the following Lagrangian:

$$L = \dot{q}^1 \dot{q}^2 + \tfrac{1}{2} q^2 (\dot{q}^3)^2 \tag{7.12}$$

(R. Cawley [1], see also J. F. Cariñena, C. López and N. Román-Roy [2]). This Lagrangian defines the first order Lagrangean system

$$\alpha = (\tfrac{1}{2}(\dot{q}^3)^2\,\omega^2 - \dot{q}^2\dot{q}^3\,\omega^3) \wedge dt + \dot{q}^3\,\omega^2 \wedge \omega^3 - \omega^1 \wedge d\dot{q}^2 - \omega^2 \wedge d\dot{q}^1 - q^2\,\omega^3 \wedge d\dot{q}^3.$$

The corresponding distributions $\mathcal{D}$ and Δ are generated by means of the one-forms

$$\omega^1,\ \omega^2,\ q^2\omega^3,\ d\dot{q}^2,\ d\dot{q}^1 - \dot{q}^3\,dq^3 + \tfrac{1}{2}(\dot{q}^3)^2\,dt,\ \dot{q}^2\,d\dot{q}^1 - \tfrac{1}{2}(\dot{q}^3)^2\,dq^2,\ q^2\,d\dot{q}^3 + \dot{q}^3\,dq^2,$$

and

$$\omega^1,\ \omega^2,\ q^2\omega^3,\ d\dot{q}^2,\ d\dot{q}^1 - \dot{q}^3\,dq^3 + \tfrac{1}{2}(\dot{q}^3)^2\,dt,\ q^2\,d\dot{q}^3 + \dot{q}^3\,dq^2,$$

respectively. Notice that the function $f = \dot{q}^2$ is a *first integral* of the distributions $\mathcal{D}$ and Δ.

(A) Extended dynamics. For the primary constraint set we get

$$\tilde{\mathcal{P}} = \{q^2 \neq 0\} \cup \{q^2 = 0,\ \dot{q}^2\dot{q}^3 = 0\}, \tag{7.13}$$

i.e., the Lagrangean system possesses *primary dynamical constraints*.

(i) For $x \notin \tilde{\mathcal{P}}$ there is no dynamics.

(ii) Consider the set $U = \{x \in \tilde{\mathcal{P}},\ q^2 \neq 0\}$ which is an open submanifold of the phase space. On this submanifold we have

$$\begin{aligned}\mathcal{D} = \Delta &= \operatorname{span}\left\{\frac{\partial}{\partial t} + \dot{q}^1\frac{\partial}{\partial q^1} + \dot{q}^2\frac{\partial}{\partial q^2} + \dot{q}^3\frac{\partial}{\partial q^3} + \tfrac{1}{2}(\dot{q}^3)^2\frac{\partial}{\partial \dot{q}^1} - \frac{\dot{q}^2\dot{q}^3}{q^2}\frac{\partial}{\partial \dot{q}^3}\right\}\\ &\approx \operatorname{span}\{\omega^1,\ \omega^2,\ \omega^3,\ d\dot{q}^2,\ d\dot{q}^1 - \dot{q}^3\,dq^3 + \tfrac{1}{2}(\dot{q}^3)^2\,dt,\ q^2\,d\dot{q}^3 + \dot{q}^3\,dq^2\}.\end{aligned} \tag{7.14}$$

We can see that the extended motion on U coincides with the proper motion, and it is *regular* (through each point of U there passes exactly one maximal prolonged extremal), thus uniquely determined by the initial conditions.

(iii) Let $x \in \tilde{\mathcal{P}},\ q^2 = 0,\ \dot{q}^3 \neq 0$. We get at x

$$\begin{aligned}\Delta &= \operatorname{span}\left\{\frac{\partial}{\partial t} + \dot{q}^1\frac{\partial}{\partial q^1} - \tfrac{1}{2}(\dot{q}^3)^2\frac{\partial}{\partial \dot{q}^1},\ \frac{\partial}{\partial \dot{q}^3},\ \frac{\partial}{\partial q^3} + \dot{q}^3\frac{\partial}{\partial \dot{q}^1}\right\}\\ &\approx \operatorname{span}\{\omega^1,\ dq^2,\ d\dot{q}^2,\ d\dot{q}^1 - \dot{q}^3\,dq^3 + \tfrac{1}{2}(\dot{q}^3)^2\,dt\}.\end{aligned} \tag{7.15}$$

Obviously, $\Delta(x) \in T_xM_x$, where $M_x = \{q^2 = 0,\ \dot{q}^2 = 0\}$, i.e., at the points of M_x, $\Delta = \Delta|_{M_x}$. This means that the bunch of submanifolds at x is built from one manifold M_x. Since the distribution $\Delta|_{M_x}$ is weakly horizontal, we get that M_x is a final constraint submanifold at x. Note that $\Delta|_{M_x}$ is of rank 3, and it *is not completely integrable*.

(iv) Let $x \in \tilde{\mathcal{P}},\ q^2 = 0,\ \dot{q}^2 \neq 0$. Consider the manifold $M_x^1 = \{q^2 = 0,\ \dot{q}^3 = 0\}$ passing through x. At the points of M_x^1,

$$\Delta = \operatorname{span}\left\{\frac{\partial}{\partial t} + \dot{q}^1\frac{\partial}{\partial q^1} + \dot{q}^2\frac{\partial}{\partial q^2},\ \frac{\partial}{\partial q^3},\ \frac{\partial}{\partial \dot{q}^3}\right\} \approx \operatorname{span}\{\omega^1,\ \omega^2,\ d\dot{q}^1,\ d\dot{q}^2\}.$$

The constrained to M_x^1 Euler-Lagrange distribution is a distribution of rank 1, spanned by the vector field $\partial/\partial q^3$. Since it is not weakly horizontal, M_x^1 is not a final constraint submanifold at x. However, by (ii), in a neighborhood of x the Euler-Lagrange distribution is of rank one at each point not belonging to M_x^1; moreover,

$$\Delta(x) = \operatorname{span}\left\{\frac{\partial}{\partial t} + \dot{q}^1\frac{\partial}{\partial q^1} + \dot{q}^2\frac{\partial}{\partial q^2},\ \frac{\partial}{\partial q^3},\ \frac{\partial}{\partial \dot{q}^3}\right\},$$

i.e., the generator ζ of (7.14) can be extended to the point x by putting

$$\zeta(x) = \frac{\partial}{\partial t} + \dot{q}^1 \frac{\partial}{\partial q^1} + \dot{q}^2 \frac{\partial}{\partial q^2},$$

which belongs to $\Delta(x)$. This means that the bunch of submanifolds passing through x contains also a one-dimensional manifold M_x^2—the integral manifold of ζ, and this is a final constraint submanifold at x.

(v) Finally consider the points $x \in \tilde{\mathcal{P}}$ such that $q^2 = 0$, $\dot{q}^2 = 0$, $\dot{q}^3 = 0$. The bunch of submanifolds passing through x consists from the manifolds $M_x = \{q^2 = \dot{q}^2 = 0\}$ which is a final constraint submanifold since Δ_{M_x} is of the form (7.15), $M_x^1 = \{q^2 = 0,\ \dot{q}^3 = 0\}$ which is not a final constraint submanifold at x, and (analogously as in (iv)) M_x^2 which is a final constraint submanifold at x. Omitting the points where Δ is not weakly horizontal from M_x^1 we get a final constraint submanifold $M_x' = \{q^2 = \dot{q}^2 = \dot{q}^3 = 0\} \subset M_x$. Notice that the vector field ζ tangent to M_x^2 is at x of the form

$$\zeta(x) = \frac{\partial}{\partial t} + \dot{q}^1 \frac{\partial}{\partial q^1},$$

i.e., it belongs to the distribution $\Delta|_{M_x}(x)$.

Notice that the considered Lagrangean system possesses *no secondary dynamical constraints*.

Let us summarize the results: The extended motion is *constrained to the subset* $\tilde{\mathcal{P}}$ (7.13), which is not a submanifold of the phase space. At the points of $\{q^2 \neq 0\} \cup \{q^2 = \dot{q}^3 = 0,\ \dot{q}^2 \neq 0\}$ the extended motion is *regular* (deterministic) and coincides with the proper motion; it is described by the vector field ζ where

$$\zeta(x) = \frac{\partial}{\partial t} + \dot{q}^1 \frac{\partial}{\partial q^1} + \dot{q}^2 \frac{\partial}{\partial q^2} + \dot{q}^3 \frac{\partial}{\partial q^3} + \tfrac{1}{2}(\dot{q}^3)^2 \frac{\partial}{\partial \dot{q}^1} - \frac{\dot{q}^2 \dot{q}^3}{q^2} \frac{\partial}{\partial \dot{q}^3}$$

for $x \in \{q^2 \neq 0\}$, and

$$\zeta(x) = \frac{\partial}{\partial t} + \dot{q}^1 \frac{\partial}{\partial q^1} + \dot{q}^2 \frac{\partial}{\partial q^2}$$

for $x \in \{q^2 = \dot{q}^3 = 0,\ \dot{q}^2 \neq 0\}$. At the points of the closed submanifold $N = \{q^2 = \dot{q}^2 = 0\} \subset \tilde{\mathcal{P}}$ the motion is *indeterministic*, described by the distribution of rank 3 spanned by the vector fields (7.15), which is *not completely integrable*.

(B) Proper dynamics. The system possesses *primary semispray constraints* which coincide with the primary dynamical constraints. From the above dynamical picture we can see that it remains to consider the motion on the submanifold $N = \{q^2 = \dot{q}^2 = 0\}$. The distribution $\Delta|_N$ has a subdistribution of rank 2 which can be spanned by the following *semisprays* (along N)

$$\frac{\partial}{\partial t} + \dot{q}^1 \frac{\partial}{\partial q^1} + \dot{q}^3 \frac{\partial}{\partial q^3} + \tfrac{1}{2}(\dot{q}^3)^2 \frac{\partial}{\partial \dot{q}^1}, \quad \frac{\partial}{\partial \dot{q}^3}.$$

This distribution is *not completely integrable*. Summarizing we get that there are *no secondary semispray constraints*, the set of admissible initial conditions for the Euler-Lagrange equations is $\mathcal{P} = \tilde{\mathcal{P}}$, and the proper motion proceeds (constrained to $\mathcal{P}$) as follows: at the points of N it is *weakly regular* (hence indeterministic), and it does not develop within leaves; in all the other points of $\mathcal{P}$ it is *regular* (develops as rays).

As a consequence we get that on the configuration space the motion is not constrained, i.e., *through every point of Y there passes an extremal.*

7.5. Semiregular Lagrangean systems

By definition, a semiregular Lagrangean system is characterized by a weakly horizontal Euler-Lagrange distribution of a constant rank. Taking into account the results of the previous sections and of Chapter 5, we get that semiregular Lagrangean systems have the following properties:

(1) The characteristic distribution coincides with the Euler-Lagrange distribution.
(2) rank F = rank(F | A) = const.
(3) The system carries no dynamical constraints.
(4) The set of admissible initial conditions for the Hamilton equations is the whole phase space, i.e., there is a "global" extended dynamics.
(5) corank Δ = rank α = const.
(6) The Euler-Lagrange distribution is locally defined either by means of $k = \operatorname{rank} \alpha$ smooth Pfaffian forms or by $\dim J^{s-1}Y - k$ *smooth* vector fields.
(7) *The Euler-Lagrange distribution is completely integrable* (since it is a characteristic distribution of a closed two-form of a constant rank).
(8) The extended dynamics is described by a flat generalized connection.
(9) The Euler-Lagrange distribution defines a regular foliation $\mathcal{F}$ of the phase space with ($\dim J^{s-1}Y$ − rank α)-dimensional leaves; every Hamilton extremal is an embedding of an open subset of the base X into a leaf of $\mathcal{F}$, and, conversely, every section of π_{s-1} into a leaf of $\mathcal{F}$ is a Hamilton extremal.
(10) Every point in the phase space is an initial point for a non-uniquely determined extended motion which develops within a leaf of $\mathcal{F}$ passing through the initial point.

Not every Hamilton extremal must correspond to an extremal; this means that solving the Euler-Lagrange distribution of a semiregular Lagrangean system (i.e., finding the corresponding foliation) one gets only a rough characterization of the proper motion. To get the proper dynamics explicitly, i.e., to solve the (*higher-order*) *semispray problem*, one has to apply the constraint algorithm.

7.5.1. Remark. The concept of *semiregularity* is a generalization of the concept of regularity. Note that in the paper L. Klapka [1] a different concept of semiregularity was introduced, namely, a semiregular Lagrangian was defined be a Lagrangian such that its Cartan form θ_λ satisfies the condition rank $d\theta_\lambda$ = const. Suggested by the geometric properties of the Euler-Lagrange distribution, we prefer the "softer" definition, which from the point of view of classifying the dynamics, is a more natural generalization of the concept of regularity.

7.5.2. Remark. Semiregularity *along a submanifold* of the phase space does *not* imply complete integrability. More precisely, if the Euler-Lagrange distribution Δ gives rise to a distribution $\Delta|_M$ *along a submanifold M* of the phase space which is weakly horizontal and of a constant rank than $\Delta|_M$ *need not be completely integrable* (cf. Example 7.4.3)).

7.5.3. Example. Consider the Lagrangian

$$L = \tfrac{1}{2}(\dot{q}^1)^2 - \dot{q}^2 q^3 \tag{7.16}$$

(V. V. Nesterenko and A. M. Chervyakov [1], see also J. F. Cariñena, C. López and N. Román-Roy [2]). This Lagrangian defines a first order Lagrangean system

$$\alpha = (\dot{q}^3\,\omega^2 - \dot{q}^2\,\omega^3) \wedge dt - \omega^1 \wedge d\dot{q}^1 + \omega^2 \wedge \omega^3,$$

where $\omega^i = dq^i - \dot{q}^i\,dt$.

(A) Extended dynamics. Since $\operatorname{rank} \mathrm{F} = \operatorname{rank}(\mathrm{F} \mid \mathrm{A})$ on the phase space, there are *no primary dynamical constraints* and the distributions $\mathcal{D}$ and Δ coincide; we have

$$\mathcal{D} = \Delta = \operatorname{span}\left\{\frac{\partial}{\partial t} + \dot{q}^1\frac{\partial}{\partial q^1},\ \frac{\partial}{\partial \dot{q}^2},\ \frac{\partial}{\partial \dot{q}^3}\right\} \approx \operatorname{span}\{\omega^1,\ dq^2,\ dq^3,\ d\dot{q}^1\}.$$

The Euler-Lagrange distribution is weakly horizontal and of rank 3 on the phase space (hence completely integrable), i.e., the Lagrangian system is *semiregular*. The set $\tilde{\mathcal{P}}$ coincides with the set of admissible initial conditions for the Hamilton equations, and equals J^1Y. Since Δ is completely integrable, the extended dynamics is described by the 3-dimensional foliation of the phase space defined by the distribution Δ.

(B) Proper dynamics. It holds $\operatorname{rank} \mathrm{B} = \operatorname{rank}(\mathrm{B} \mid \mathrm{A})$ if and only if $\dot{q}^2 = \dot{q}^3 = 0$, which means that the Lagrangean system possesses *primary semispray constraints*. The semispray-constraint set is

$$\mathcal{P} = \{\dot{q}^2 = \dot{q}^3 = 0\}.$$

$\mathcal{P}$ is a closed submanifold of the phase space. The constrained to $\mathcal{P}$ Euler-Lagrange distribution is a horizontal distribution of rank 1, spanned by the vector field

$$\frac{\partial}{\partial t} + \dot{q}^1\frac{\partial}{\partial q^1}\,.$$

Obviously, to each point x of $\mathcal{P}$ there exists a semispray ζ such that $\zeta(x)$ belongs to $\Delta|_{\mathcal{P}}$ at x, i.e., there are *no secondary semispray constraints*. Hence, the proper motion is *constrained* to the submanifold $\mathcal{P}$ of the phase space, and is there *regular* (through each point of $\mathcal{P}$ there passes exactly one maximal prolonged extremal).

We have seen that in general, the dynamics of Lagrangean systems can be found with the help of the constraint algorithm. This essentially simplifies if the Lagrangean system is *semiregular* (or *regular*). In this case, the integration problem is reduced to that one of finding the foliation defined by the Euler-Lagrange distribution, i.e., of finding at each point of the phase space a system of adapted coordinates to the foliation, built from independent first integrals of the distribution. In Chapter 9 we shall discuss several methods for constructing such adapted coordinates. These methods will be a generalization of the well-known integration methods from classical mechanics (Liouville theorem, Hamilton-Jacobi equation, method of canonical transformations).

7.6. Weakly regular Lagrangean systems

Taking into account Definition 7.2.6, and the results of Chapter 5 we can see that weakly regular Lagrangean systems have the following properties:

(1) The Euler-Lagrange and the characteristic distributions are weakly horizontal and coincide; however, they need not have a constant rank.
(2) At each point x of the phase space, $\operatorname{rank} \mathrm{F}(x) = \operatorname{rank}(\mathrm{F} \mid \mathrm{A})(x)$ (but the rank need not be a constant).
(3) $\operatorname{rank} \mathrm{B} = \operatorname{rank}(\mathrm{B} \mid \mathrm{A}) = \text{const.}$
(4) The Euler-Lagrange distribution has a subdistribution of rank $r = m+1-\operatorname{rank} \mathrm{B}$, which can be locally spanned by r *semisprays*.
(5) The system carries no dynamical constraints.
(6) The system carries no semispray constraints.
(7) The set of admissible initial conditions for the Hamilton equations is the whole phase space, i.e., there is a "global" extended dynamics.
(8) The set of admissible initial conditions for the Euler-Lagrange equations is the whole phase space, i.e., there is a "global" proper dynamics.
(9) Every point in the phase space is an initial point for a non-uniquely determined proper motion.
(10) Through each point of the configuration space there passes an extremal.

7.6.1. Proposition. *If* $\operatorname{rank} \mathrm{B} < m$ *then the semispray subdistribution of the Euler-Lagrange distribution is not completely integrable.*

Proof. If $\operatorname{rank} \mathrm{B} < m$ then the subdistribution is of rank $r > 1$ and it can be spanned by the vector fields

$$\frac{\partial}{\partial t} + \sum_{i=0}^{s-2} q_{i+1}^{\sigma} \frac{\partial}{\partial q_i^{\sigma}}, \qquad \frac{\partial}{\partial q_{s-1}^{\sigma_j}}, \qquad 1 \leqslant j \leqslant r-1.$$

Computing the Lie brackets we get that this system of vector fields is not involutive. □

Summarizing, we get that the dynamical picture for weakly regular Lagrangean systems is the following: The proper motion is not restricted to a submanifold of the phase space, and is indeterministic. The Euler-Lagrange distribution need not be of a constant rank, but has a subdistribution of a constant rank locally spanned by $m+1-\operatorname{rank} \mathrm{B}$ semisprays, which is not completely integrable (unless $\operatorname{rank} \mathrm{B} \neq m$). Prolonged extremals coincide with integral sections of this subdistribution. Since the semispray subdistribution does not give rise to a foliation of the phase space, the proper motion from a fixed initial point cannot be represented as proceeding within a leaf.

A particular case of a weakly regular Lagrangean system is a *strongly semiregular* Lagrangean system. By definition, every strongly semiregular Lagrangean system is described by a weakly horizontal Euler-Lagrange distribution of a constant rank, hence completely integrable, which has a semispray subdistribution of a constant rank (not completely integrable). This means that

(1) there are no restrictions on the initial conditions,
(2) both the extended and the proper motion is indeterministic,

(3) the extended motion is described by the leaves of the foliation $\mathcal{F}$ defined by the Euler-Lagrange distribution,

(4) there is no subfoliation of $\mathcal{F}$ corresponding to the semispray subdistribution, i.e., the proper motion cannot be characterized "developing as a leaf"; however, choosing initial conditions, it proceeds in a leaf of $\mathcal{F}$ passing through the initial point (cannot leave the leaf).

7.6.2. Example. Consider a (2 + 1)-dimensional configuration manifold and the Lagrangian

$$L = \tfrac{1}{2}(\dot{q}^1 + \dot{q}^2)^2 \tag{7.17}$$

(O. Krupková [9]). This is a first order Lagrangean system

$$\alpha = d\dot{q}^1 \wedge \omega^1 + d\dot{q}^2 \wedge \omega^2 + d\dot{q}^1 \wedge \omega^2 + d\dot{q}^2 \wedge \omega^1,$$

where $\omega^i = dq^i - \dot{q}^i\, dt$. We get

$$\begin{aligned}\mathcal{D} = \Delta &= \operatorname{span}\left\{\frac{\partial}{\partial t} + (\dot{q}^1 + \dot{q}^2)\frac{\partial}{\partial q^1},\ \frac{\partial}{\partial q^1} - \frac{\partial}{\partial q^2},\ \frac{\partial}{\partial \dot{q}^1} - \frac{\partial}{\partial \dot{q}^2}\right\}\\ &\approx \operatorname{span}\{d\dot{q}^1 + d\dot{q}^2,\ \omega^1 + \omega^2\},\end{aligned}$$

which is a weakly horizontal, completely integrable distribution of rank 3 on the phase space. This means that the Lagrangean system possesses *no primary dynamical constraints*, and it is *semiregular* (the extended motion proceeds within the 3-dimensional leaves of the corresponding foliation).

Since $\operatorname{rank} B = \operatorname{rank}(B \mid A) = 1$, there are *no primary semispray constraints*, and the Lagrangean system is *weakly regular*, hence *strongly semiregular*. The semispray subdistribution of the Euler-Lagrange distribution is of rank 2, and is spanned by the vector fields

$$\frac{\partial}{\partial t} + \dot{q}^1\frac{\partial}{\partial q^1} + \dot{q}^2\frac{\partial}{\partial q^2},\quad \frac{\partial}{\partial \dot{q}^1} - \frac{\partial}{\partial \dot{q}^2},$$

or equivalently, by

$$\frac{\partial}{\partial t} + \dot{q}^1\frac{\partial}{\partial q^1} + \dot{q}^2\frac{\partial}{\partial q^2} + g\left(\frac{\partial}{\partial \dot{q}^1} - \frac{\partial}{\partial \dot{q}^2}\right),$$

where g is a function. This subdistribution is not completely integrable, i.e., the proper motion of a point in the phase space does not proceed within a 2-dimensional leaf (however, it cannot leave the 3-dimensional leaf of Δ). The motion is not constrained to a (proper) submanifold, but the set of admissible initial conditions for the Euler-Lagrange equations is the whole phase space.

7.7. Generalized Legendre transformations

In Section 6.4. we have shown that regular Lagrangean systems admit distinguished coordinate transformations on the phase space, called Legendre transformations. Now, we shall see that also *some of the singular Lagrangean systems* admit certain coordinate transformations on the phase space which utilize (at least) a part of momenta as new coordinates.

Let α be a Lagrangean system of order $s-1$, let p_ν^k, $1 \leqslant \nu \leqslant m$, $0 \leqslant k \leqslant s-c-1$, be its momenta. Consider the square matrix

$$\Pi = \left(\frac{\partial p_\nu^k}{\partial q_i^\sigma} \right), \tag{7.18}$$

where ν, k label rows and σ, i label columns, $1 \leqslant \nu \leqslant m$, $0 \leqslant k \leqslant s-c-1$ and $1 \leqslant \sigma \leqslant m$, $c \leqslant i \leqslant s-1$ (cf. (6.24).

Suppose that

$$\operatorname{rank} \Pi = \text{const} = K \tag{7.19}$$

on a connected component of $J^{s-1}Y$. This condition means that K of the momenta are independent; without the loss of generality we can suppose that $p_1, \dots, p_K$ are the independent momenta. Now, (in a neighborhood of every point) we can introduce new coordinates as follows: if (V, ψ), $\psi = (t, q^\sigma)$ is a fiber chart at a point $y = \gamma(x)$, put $(\bar{V}, \bar{\psi})$,

$$\bar{\psi} = (t, q^\sigma, \dots q_{c-1}^\sigma, p_1, \dots, p_K, g^{K+1}, \dots, g^{m(s-c)}), \tag{7.20}$$

where $\bar{V} \subset V_{s-1}$ is an appropriate neighborhood of $J_x^{s-1}\gamma$, $p_1, \dots, p_K$ are the independent momenta of α defined on $\bar{V}$, and $g^{K+1}, \dots, g^{m(s-c)}$ are arbitrary functions on $\bar{V}$, such that $\bar{\psi} \circ \psi_{s-1}^{-1}$ is a coordinate transformation. Notice that, for simplicity, one can take in place of the g's, $m(s-c) - K$ appropriate *(higher) velocities* q_j's, where $j \geqslant c$. If this is the case, we denote these velocities by $v^{K+1}, \dots, v^{m(s-c)}$.

7.7.1. Definition. Every chart on the phase space which is of the form (7.20) will be called a *generalized Legendre chart* for the Lagrangean system α. The corresponding coordinates $(t, q_i^\sigma, p_j, g^l)$, where $1 \leqslant \nu \leqslant m$, $0 \leqslant i \leqslant c-1$, $1 \leqslant j \leqslant K$, and $K+1 \leqslant l \leqslant m(s-c)$, will be called *generalized Legendre coordinates* for α. The mapping $\mathfrak{Leg} = \bar{\psi} \circ \psi_{s-1}^{-1}$ will be then referred to as a *generalized Legendre transformation.*

7.7.2. Remark. Lagrangean systems which admit generalized Legendre transformations coincide neither with semiregular nor with weakly regular Lagrangean systems.

More precisely, a semiregular Lagrangean system admits generalized Legendre transformations, but a Lagrangean system which admits generalized Legendre transformations need not be semiregular. A weakly regular *first-order* Lagrangean system admits generalized Legendre transformations (since in this case, $\Pi = \mathrm{B}$), but a Lagrangean system which admits generalized Legendre transformations need not be weakly regular. For higher-order Lagrangean systems, the situation is even more complicated, since in this case, weakly regular Lagrangean systems need not admit generalized Legendre transformations ($\operatorname{rank} \mathrm{B} = \text{const}$. does not imply that $\operatorname{rank} \Pi = \text{const}$).

In generalized Legendre coordinates the Hamilton equations take a form which is a generalization of the *canonical form* of Hamilton equations for regular Lagrangean systems. As an illustration, let us write down a generalized canonical form for a *first-order* Lagrangean system (i.e., for $s = 2$); formulas for higher-order Lagrangean systems are obtained analogously. We encourage the reader to make explicit computations for the case $s = 2r$, $\operatorname{rank} \mathrm{B} = \text{const} = k$, $\operatorname{rank} \Pi = \text{const} = rk$.

Let α be a first-order Lagrangean system, $\lambda = L\,dt$ its (local) first-order Lagrangian, and H and p_σ, $1 \leqslant \sigma \leqslant m$, the corresponding Hamiltonian and momenta. Suppose that

$$\operatorname{rank}(B_{\sigma\nu}) = \operatorname{rank}\left(\frac{\partial^2 L}{\partial \dot q^\sigma \partial \dot q^\nu}\right) = \operatorname{rank}\left(\frac{\partial p_\sigma}{\partial \dot q^\nu}\right) = K \geqslant 1. \tag{7.21}$$

Consider a generalized Legendre transformation in the form

$$(t, q^1, \ldots, q^m, \dot q^1, \ldots, \dot q^m) \to (t, q^1, \ldots, q^m, p_1 \ldots, p_K, v^{K+1}, \ldots v^m),$$

and denote the remaining momenta $p_{K+1}, \ldots, p_m$ by $\phi_{K+1}, \ldots, \phi_m$. With the help of (7.21), we get

$$\frac{\partial \phi_j}{\partial v^l} = 0, \quad \frac{\partial H}{\partial v^l} = 0, \quad K+1 \leqslant j, l \leqslant m.$$

Expressing the Lagrangean system α in the generalized Legendre coordinates we get

$$\alpha = -\left(\sum_{\sigma=1}^{m} \frac{\partial H}{\partial q^\sigma} dq^\sigma + \sum_{i=1}^{K} \frac{\partial H}{\partial p_i} dp_i\right) \wedge dt$$
$$+ \sum_{i=1}^{K} dp_i \wedge dq^i + \sum_{j=K+1}^{m} \left(\frac{\partial \phi_j}{\partial t} dt + \sum_{\sigma=1}^{m} \frac{\partial \phi_j}{\partial q^\sigma} dq^\sigma + \sum_{i=1}^{K} \frac{\partial \phi_j}{\partial p_i} dp_i\right) \wedge dq^j.$$

Hence, the *Euler-Lagrange distribution* Δ is generated by means of the one-forms

$$-\frac{\partial H}{\partial q^i} dt - dp_i + \sum_{l=K+1}^{m} \frac{\partial \phi_l}{\partial q^i} dq^l, \quad -\frac{\partial H}{\partial p_i} dt + dq^i + \sum_{l=K+1}^{m} \frac{\partial \phi_l}{\partial p_i} dq^l,$$

$$-\left(\frac{\partial H}{\partial q^j} + \frac{\partial \phi_j}{\partial t}\right) dt - \sum_{i=1}^{K} \frac{\partial \phi_j}{\partial q^i} dq^i - \sum_{l=K+1}^{m} \left(\frac{\partial \phi_j}{\partial q^l} - \frac{\partial \phi_l}{\partial q^j}\right) dq^l - \sum_{i=1}^{K} \frac{\partial \phi_j}{\partial p_i} dp_i, \tag{7.22}$$

$1 \leqslant i \leqslant K, \quad K+1 \leqslant j \leqslant m,$

and the *characteristic distribution* $\mathcal{D}$ is generated by means of (7.22), and the one-form

$$\sum_{i=1}^{K} \frac{\partial H}{\partial q^i} dq^i + \sum_{i=1}^{K} \frac{\partial H}{\partial p_i} dp_i + \sum_{j=K+1}^{m} \left(\frac{\partial H}{\partial q^j} + \frac{\partial \phi_j}{\partial t}\right) dq^j. \tag{7.23}$$

Now, the Hamilton equations read

$$-\frac{\partial H}{\partial q^i} - \frac{dp_i}{dt} + \sum_{l=K+1}^{m} \frac{\partial \phi_l}{\partial q^i} \frac{dq^l}{dt} = 0, \quad -\frac{\partial H}{\partial p_i} + \frac{dq^i}{dt} + \sum_{l=K+1}^{m} \frac{\partial \phi_l}{\partial p_i} \frac{dq^l}{dt} = 0,$$

$$-\left(\frac{\partial H}{\partial q^j} + \frac{\partial \phi_j}{\partial t}\right) - \sum_{i=1}^{K} \frac{\partial \phi_j}{\partial q^i} \frac{dq^i}{dt} - \sum_{l=K+1}^{m} \left(\frac{\partial \phi_j}{\partial q^l} - \frac{\partial \phi_l}{\partial q^j}\right) \frac{dq^l}{dt} - \sum_{i=1}^{K} \frac{\partial \phi_j}{\partial p_i} \frac{dp_i}{dt} = 0,$$

$1 \leqslant i \leqslant K, \quad K+1 \leqslant j \leqslant m,$

(7.24)

along δ.

The distribution Δ need not be weakly horizontal, and need not have a constant rank. We shall compute the *weak horizontality conditions* from the requirement $\Delta = \mathcal{D}$ (Theorem 5.4.1). Considering the matrix of the generators of $\mathcal{D}$, and requiring that the form (7.23) be a linear combination of (7.22), we get the following result:

7.7.3. Proposition. *The distribution Δ is weakly horizontal at a point x if and only if, at this point, the rank of the matrix*

$$\mathrm{P} = \left(\sum_{i=1}^{K} \left(\frac{\partial \phi_l}{\partial p_i} \frac{\partial \phi_j}{\partial q^i} - \frac{\partial \phi_l}{\partial q^i} \frac{\partial \phi_j}{\partial p_i} \right) + \frac{\partial \phi_l}{\partial q_j} - \frac{\partial \phi_j}{\partial q_l} \right)$$

equals to the rank of the matrix

$$\mathrm{P}' = \left(\; \mathrm{P} \; \middle| \; \sum_{i=1}^{K} \left(\frac{\partial H}{\partial p_i} \frac{\partial \phi_j}{\partial q^i} - \frac{\partial H}{\partial q^i} \frac{\partial \phi_j}{\partial p_i} \right) + \frac{\partial H}{\partial q_j} + \frac{\partial \phi_j}{\partial t} \right),$$

where j labels rows and l labels columns.

Notice that

$$\operatorname{corank} \Delta = \operatorname{rank} \mathrm{F} = 2K + \operatorname{rank} \mathrm{P}.$$

Using the one-forms (7.22), we can write down equations for the components of vector-field generators of Δ. If we denote

$$\zeta = \zeta^0 \frac{\partial}{\partial t} + \sum_{\sigma=1}^{m} \zeta^\sigma \frac{\partial}{\partial q^\sigma} + \sum_{i=1}^{K} \zeta_i \frac{\partial}{\partial p_i} + \sum_{j=K+1}^{m} \hat{\zeta}^j \frac{\partial}{\partial v^j},$$

we get

$$\zeta_i = -\frac{\partial H}{\partial q^i} \zeta^0 + \sum_{l=K+1}^{m} \frac{\partial \phi_l}{\partial q^i} \zeta^l, \quad \zeta^i = \frac{\partial H}{\partial p_i} \zeta^0 - \sum_{l=K+1}^{m} \frac{\partial \phi_l}{\partial p_i} \zeta^l, \quad 1 \leqslant i \leqslant K,$$

and

$$\begin{aligned} &\sum_{l=K+1}^{m} \left(\sum_{i=1}^{K} \left(\frac{\partial \phi_l}{\partial p_i} \frac{\partial \phi_j}{\partial q^i} - \frac{\partial \phi_l}{\partial q^i} \frac{\partial \phi_j}{\partial p_i} \right) + \frac{\partial \phi_l}{\partial q^j} - \frac{\partial \phi_j}{\partial q^l} \right) \zeta^l \\ &= \left(\sum_{i=1}^{K} \left(\frac{\partial H}{\partial p_i} \frac{\partial \phi_j}{\partial q^i} - \frac{\partial H}{\partial q^i} \frac{\partial \phi_j}{\partial p_i} \right) + \frac{\partial H}{\partial q^j} + \frac{\partial \phi_j}{\partial t} \right) \zeta^0, \quad K+1 \leqslant j \leqslant m. \end{aligned} \tag{7.25}$$

First, suppose that $\operatorname{rank} \mathrm{P} = \operatorname{rank} \mathrm{P}' = 0$. Then Δ is semiregular, and it is spanned by the following $2(m-K)+1$ independent vector fields:

$$\begin{aligned} &\frac{\partial}{\partial t} + \sum_{i=1}^{K} \left(\frac{\partial H}{\partial p_i} \frac{\partial}{\partial q^i} - \frac{\partial H}{\partial q^i} \frac{\partial}{\partial p_i} \right), \\ &\sum_{i=1}^{K} \left(-\frac{\partial \phi_l}{\partial p_i} \frac{\partial}{\partial q^i} + \frac{\partial \phi_l}{\partial q^i} \frac{\partial}{\partial p_i} \right) + \frac{\partial}{\partial q^l}, \quad \frac{\partial}{\partial v^l}, \quad K+1 \leqslant l \leqslant m. \end{aligned} \tag{7.26}$$

Let $\operatorname{rank} \mathrm{P} \neq 0$. If α is *semiregular* (i.e., weakly horizontal and of a constant rank) then (in a neighborhood of every point) equations (7.25) are solvable with respect to the ζ^l's providing us with differentiable solutions depending on $m - K - \operatorname{rank} \mathrm{P}$ parameters. We have

$$\zeta^l = \zeta_0^l + \sum_i a_j \bar{\zeta}_{(j)}^l,$$

where ζ_0 is a (fixed) solution of equations (7.25) and $\bar{\zeta}_{(j)}$, $1 \leqslant j \leqslant m - K - \operatorname{rank} \mathrm{P}$, are independent solutions of the homogeneous equations $\mathrm{P}\,\zeta = 0$. In this way, we get the Euler-Lagrange distribution spanned by the vector fields

$$\frac{\partial}{\partial t} + \sum_{i=1}^{K}\Big(\frac{\partial H}{\partial p_i}\frac{\partial}{\partial q^i} - \frac{\partial H}{\partial q^i}\frac{\partial}{\partial p_i}\Big) + \sum_{l=K+1}^{m} \zeta_0^l\Big(\sum_{i=1}^{K}\Big(-\frac{\partial \phi_l}{\partial p_i}\frac{\partial}{\partial q^i} + \frac{\partial \phi_l}{\partial q^i}\frac{\partial}{\partial p_i}\Big) + \frac{\partial}{\partial q^l}\Big),$$

$$\sum_{l=K+1}^{m} \bar{\zeta}_{(j)}^l\Big(\sum_{i=1}^{K}\Big(-\frac{\partial \phi_l}{\partial p_i}\frac{\partial}{\partial q^i} + \frac{\partial \phi_l}{\partial q^i}\frac{\partial}{\partial p_i}\Big) + \frac{\partial}{\partial q^l}\Big), \quad 1 \leqslant j \leqslant m - K - \operatorname{rank} \mathrm{P},$$

$$\frac{\partial}{\partial v^l}, \quad K+1 \leqslant l \leqslant m. \tag{7.27}$$

The vector fields (7.26) (resp. (7.27)) describe the *extended dynamics* of the given semiregular Lagrangean system.

7.7.4. Remark. We emphasize the following practical meaning of *Legendre transformations*, or *generalized Legendre transformations*: if one is searching for an explicit form of integral sections of an Euler-Lagrange distribution, the use of appropriate Legendre coordinates can in certain cases simplify the integration, namely, if some of the momenta are first integrals. Then often some of the exact integration methods can be applied to solve the integration problem (cf. Chapter 9).

Chapter 8.

SYMMETRIES OF LAGRANGEAN SYSTEMS

8.1. Introduction

Since Emmy Noether's famous theorem has been published in 1918, stating that to every symmetry of a Lagrangian there corresponds a conservation law, there has been a continuous non-decreasing interest in studying symmetries and applying them for solving many problems in the theory of differential equations, the calculus of variations, and theoretical physics. Significantly, Lagrangean systems with prescribed symmetries have been studied, and exact integration methods for the motion equations using symmetries have been investigated. According to the well-known property of the Euler-Lagrange equations of classical mechanics, a sufficient number of appropriate conservation laws are *equivalent* to the original equations of motion, and transfer the motion equations to the form $dF/dt = 0$, i.e., $F = \text{const}$ (thus, a conservation law is a *first integral* of the original equations). It turns out, however, that for non-classical (e.g., singular) Lagrangean systems integration methods based on symmetries are not so straightforward (cf. Chapter 9).

In this chapter we shall study symmetries related with Lagrangean systems, and relations between these symmetries. Also we shall be interested in conservation laws associated with symmetries. Applications to exact integration of the Euler-Lagrange equations will then be subject of the next chapter.

We remind the reader that Lagrangean systems throughout this book are understood in accordance with Definition 4.3.12; roughly speaking, a Lagrangean system is the *class of equivalent Lagrangians* (differing by total derivative of a function). Having in mind various applications of the theory of symmetries, such an approach gives one the possibility to utilize for investigations of a Lagrangian system symmetries of different equivalent Lagrangians, of the corresponding Lepagean one-forms, of the Euler-Lagrange form, of the Lepagean two-form itself, or even symmetries of the associated distributions (the Euler-Lagrange and the characteristic distribution). Thus, in principle, more symmetries and conservation laws are at hands for concrete applications in comparison with the situation when one understands a Lagrangean system to be identified with a particular Lagrangian. Indeed, symmetries of a Lagrangian need not identify with symmetries of another equivalent one, moreover, there can be Lagrangians with symmetries "better adapted" to a concrete problem.

We have seen that the dynamics of Lagrangean systems can be described by the characteristic distribution on the phase space. This means that the integration problem for variational equations could be simplified if one would know *first integrals* of this distribution. We remind the reader that, however, if the Lagrangean system is *not regular*, then its extremals do not coincide with the maximal integral manifolds of the characteristic distribution. Therefore, the concept of a *constant of the motion* (i.e., a function which is conserved along extremals) is not identical with that of a *first integral* of this distribution. Consequently, contrary to regular Lagrangean systems, for singular Lagrangean systems one has to distiguish between the constants of the motion and first integrals. We shall see, however, that with some modifications, the theory of symmetries is useful for simplifying the integration problem in the non-regular case, too.

The geometric theory of symmetries of Lagrangean systems, presented in the sequel, is a part of a *geometric theory of invariant variational problems*, whose foundations were laid by A. Trautman [1,2] in the 60'*s*. His methods have been developed within the range of the calculus of variations on fibered manifolds by D. Krupka [1,2]. Our approach and exposition follows the papers by D. Krupka [2], and O. Krupková [4,10].

There is a plenty of papers dealing with various questions concerning symmetries and first integrals of Lagrangean systems in mechanics, some of them are listed in the Bibliography to this book. There are also interesting results relating the theory of symmetries with the inverse problem of the calculus of variations (see e.g. W. Sarlet [3–6], Z. Oziewicz [1], I. Anderson and G. Thompson [1]). Symmetries are also widely utilized in symplectic geometry in such constructions as e.g. the symplectic reduction, or momentum map. The interested reader can consult e.g. the books by R. Abraham and J. E. Marsden [1], or P. Libermann and Ch.-M. Marle [1].

Throughout the chapter, we consider a fixed fibered manifold $\pi : Y \to X, \dim X = 1$.

8.2. Classification of symmetries, conserved functions

Let $s \geqslant 1$, let α be a Lagrangean system of order $s-1$ on π. By Definition 4.3.12, α is a Lepagean two-form on $J^{s-1}Y$, and we have seen that it is a unique closed counterpart of a locally variational form E on J^sY. Thus, one has many (equivalent) possibilities to represent a Lagrangean system. In what follows, we shall utilize the representations in the form of

– the closed two-form α,

– the locally variational form $E = p_1\alpha$,

– the equivalence class of (local) minimal-order Lagrangians for E (recall that the minimal order $r_0 = c$ if $s = 2c$, and $r_0 = c + 1$ if $s = 2c + 1$); the equivalence relation is given by

$$L \sim L' \quad \text{iff} \quad L' = L + \frac{df}{dt}, \qquad \text{where } f \text{ is a function of order } r_0 - 1,$$

– the equivalence class of (local) Lagrangians for E (this concerns Lagrangians of all finite orders $\geqslant$ the minimal order); the equivalence relation is given by

$$L \sim L' \quad \text{iff} \quad L' = L + \frac{df}{dt},$$

where f is a function,

– the equivalence class of the corresponding (local) Lepagean one-forms (= Cartan forms); the equivalence relation is given by

$$\theta_\lambda \sim \theta_{\lambda'} \quad \text{iff} \quad \theta_{\lambda'} = \theta_\lambda + df,$$

where f is a function, and it concerns forms of all finite orders $\geqslant s-1$.

Thus to get information on symmetries related with a Lagrangean system one can study symmetries of the closed two-form α, of the dynamical form E, of local Lagrangians, or of local Lepagean one-forms. Often also symmetries of the corresponding distributions (the Euler-Lagrange distribution Δ and the characteristic distribution $\mathcal{D}$ of α) can be useful. Such symmetries are vector fields defined on open subsets of different jet prolongations of the fibered manifold π. Namely, symmetries of α, $\mathcal{D}$ and Δ are local vector fields on the phase space $J^{s-1}Y$. Symmetries of E are local vector fields defined on J^sY, symmetries of a (local) Lagrangian of order r are defined on an open subset of J^rY, and symmetries of a (local) Lepagean one-form (Cartan form) θ of order k are defined on an open subset of J^kY (recall that $k \geqslant s-1$). From the point of view of applications, the most interesting symmetries are those "living" on the *phase space* where the dynamics proceeds. Among such symmetries, there is an important subfamily of those vector fields which are $(s-1)$th prolongations of some vector fields defined on the *configuration space* Y. To avoid confusion, we shall use the following terminology:

8.2.1. Definition. Let α be a Lagrangean system of order $s-1$. A π-projectable vector field ξ on an open subset of Y will be called *a point symmetry associated with the Lagrangean system* α if at least one of the following conditions is satisfied:

(1) $J^{s-1}\xi$ is a symmetry of the two-form α,
(2) $J^s\xi$ is a symmetry of the locally variational form $E = p_1\alpha$,
(3) there exists an integer r and a (local) Lagrangian of order r for α such that $J^r\xi$ is a symmetry of λ,
(4) there exists an integer r and a (local) Lepagean one-form θ of order r for α such that $J^r\xi$ is a symmetry of θ,
(5) $J^{s-1}\xi$ is a symmetry of the characteristic distribution $\mathcal{D}$ of α,
(6) $J^{s-1}\xi$ is a symmetry of the Euler-Lagrange distribution Δ of α.

8.2.2. Definition. Let α be a Lagrangean system of order $s-1$. A vector field ξ on an open subset of $J^{s-1}Y$ will be called *a dynamical symmetry associated with the Lagrangean system* α, if at least one of the following conditions is satisfied:

(1) ξ is a symmetry of the two-form α, i.e., $\partial_\xi\alpha = 0$,
(2) there exists a (local) Lagrangian of order $r \leqslant s-1$ for α such that ξ is a symmetry of $\pi^*_{s-1,r}\lambda$,
(3) there exists a (local) Lepagean one-form θ of order $s-1$ for α such that ξ is a symmetry of θ,
(4) ξ is a symmetry of the characteristic distribution $\mathcal{D}$ of α,
(5) ξ is a symmetry of the Euler-Lagrange distribution Δ of α.

Roughly speaking, by a *point symmetry* associated with a Lagrangean system we shall mean a symmetry defined on the *configuration space* Y, and by a *dynamical symmetry* we shall mean a symmetry defined on the *phase space*.

If we have a differential form defined on the phase space $J^{s-1}Y$ and a (local) π-projectable vector field ξ on Y is its point symmetry, then $J^{s-1}\xi$ is its dynamical symmetry; however, clearly not every dynamical symmetry has to correspond to a point symmetry.

Now, let us turn to the concepts of a constant of the motion and of a first integral of a Lagrangean system.

8.2.3. Definition. Let α be a Lagrangean system of order $s-1$, $s \geqslant 1$. Let f be a function defined on an open subset U of the phase space $J^{s-1}Y$.

f is called a *first integral of the Lagrangean system* α if f is a first integral of the characteristic distribution $\mathcal{D}$ of α, i.e., if

$$df \in \mathcal{D},$$

or, equivalently, if for every integral mapping φ of $\mathcal{D}$,

$$\varphi^* df = d(f \circ \varphi) = 0\,.$$

f is called a *constant of the motion of the Lagrangean system* α if for every extremal γ

$$f \circ J^{s-1}\gamma = \text{const}\,.$$

It is worthwhile to add a few comments to the above terminology:

First, notice that we have defined constants of the motion to be functions defined on an open subset of the *phase space*. The reason for this is purely practical: If f is a function of order $r > s-1$ such that

$$f \circ J^r\gamma = \text{const},$$

then the above conservation law represents a differential equation of order *greater* than or *equal* to the order of the original variational equations. Hence, such a conservation law does not lead to lowering the order of the original integration problem.

Next, notice that generally there is a difference between first integrals and constants of the motion. More precisely, we have the following:

8.2.4. Proposition. *Every first integral of the characteristic distribution is a constant of the motion.*

If the Lagrangean system is regular then the set of first integrals coincides with the set of constants of the motion.

Proof. Every first integral is, by definition, constant along Hamilton extremals. Since prolonged extremals form a subset in the set of Hamilton extremals, every first integral is a constant of the motion.

Suppose that α is regular, and let f be a constant of the motion. Then the characteristic distribution $\mathcal{D}$ is locally spanned by one nowhere zero vector field ζ such that the prolonged extremals coincide with the integral sections of ζ. Hence, $i_\zeta df = 0$, proving that $f \in \mathcal{D}$. □

Since generally the set of integral mappings of the characteristic distribution does not coincide with the set of prolonged extremals, it is not surprizing that there are (singular)

Lagrangean systems such that not every constant of the motion is a first integral of the characteristic distribution. An easy example is provided by the following Lagrangean system:

8.2.5. Example (J. Hrivňák [1]). Consider a Lagrangian L on the fibered manifold $R \times R^3 \to R$,

$$L = \tfrac{1}{2}(\dot{x}^2 + \dot{y}^2) - U(x, y, z),$$

where (x, y, z) are the canonical coordinates on R^3. The Euler-Lagrange form of L is

$$E_\lambda = -\left(\frac{\partial U}{\partial x} + \ddot{x}\right) dx \wedge dt - \left(\frac{\partial U}{\partial y} + \ddot{y}\right) dy \wedge dt - \frac{\partial U}{\partial z} dz \wedge dt.$$

Since along the extremals $E_\lambda = 0$, we conclude that extremals are constrained to the submanifold M of Y defined by

$$\frac{\partial U}{\partial z} = 0.$$

Computing the generators of the Euler-Lagrange and the characteristic distribution we get

$$\Delta \approx \operatorname{span}\left\{-d\dot{x} - \frac{\partial U}{\partial x} dt,\ -d\dot{y} - \frac{\partial U}{\partial y} dt,\ dx - \dot{x}\, dt,\ dy - \dot{y}\, dt,\ -\frac{\partial U}{\partial z} dt\right\},$$

$$\mathcal{D} \approx \operatorname{span}\left\{-d\dot{x} - \frac{\partial U}{\partial x} dt,\ -d\dot{y} - \frac{\partial U}{\partial y} dt,\ dx - \dot{x}\, dt,\ dy - \dot{y}\, dt,\ -\frac{\partial U}{\partial z} dt,\ -\frac{\partial U}{\partial z} dz\right\}.$$

We shall find a constant of the motion which is not a first integral of the characteristic distribution. For simplicity, let us consider the potential U in the form

$$U = \frac{1}{r^2},$$

where $r = \sqrt{x^2 + y^2 + z^2}$, i.e., we consider motion in the central gravitational field. Since now

$$\frac{\partial U}{\partial z} = -\frac{2z}{r^4},$$

the motion is constrained to the submanifold M, given by the equation

$$z = 0.$$

Thus, any function $f(z)$ is a constant of the motion. Suppose that f is a non-trivial constant of the motion, i.e., that $df \neq 0$. Since

$$df = \frac{\partial f}{\partial z} dz,$$

we can see that the 1-form df cannot be expressed as a linear combination of 1-forms belonging to Δ, i.e., f is not a first integral of Δ. Similar arguments lead to a conclusion that f is a first integral of the distribution $\mathcal{D}$ only in the case that

$$\frac{\partial f}{\partial z} = 0 \quad \text{at the points } z = 0$$

(indeed, otherwise f could not be expressible in the form $df = gz\, dz$ for a nowhere zero function g). Consequently, every function $f(z)$ which does not satisfy the above condition is a constant of the motion but is not a first integral of $\mathcal{D}$.

Finally, let us say a few words to the motivation for the definition of a first integral of a Lagrangean system, as a first integral of its *characteristic* distribution $\mathcal{D}$. Since $\mathcal{D}$ is a subdistribution of the Euler-Lagrange distribution Δ, we can see that the condition $df \in \Delta$ implies $df \in \mathcal{D}$, i.e., for every Lagrangean system, the set of first integrals of Δ is a *subset* of the set of first integrals of $\mathcal{D}$. We conclude from Theorem 5.4.1 that for Lagrangean systems with *no dynamical constraints* these sets coincide, in general, however, they are different. An argument for the choice of the distribution $\mathcal{D}$ to define the concept of a first integral of a Lagrangean system is provided by the following considerations:

8.2.6. Example. Consider a fibered manifold $\pi : R \times M \to R$, let α be a Lagrangean system of order $s-1$, where $s \geqslant 1$, on π. We say that the Lagrangean system α is *time-independent*, or *autonomous* if (in a neighborhood of every point) there exists a Lagrangian λ for α such that

$$\frac{\partial L}{\partial t} = 0.$$

Consequently, time-independent Lagrangean systems possess time-independent Hamiltonians and time-independent momenta.

Using the expression of α in the canonical form (Theorem 4.5.2),

$$\alpha = -dH \wedge dt + \sum_{k=0}^{s-c-1} dp_{\nu}^{k} \wedge dq_{k}^{\nu}$$

we get that

$$i_{\partial/\partial t}\,\alpha = -dH,$$

i.e., that the time-independent Hamiltonians are *first integrals* of the characteristic distribution $\mathcal{D}$ of α. On the other hand, taking into account the definition of the Euler-Lagrange distribution, we can see that $dH \in \Delta$ if and only if $\Delta = \mathcal{D}$, i.e., iff the Lagrangean system carries no dynamical constraints.

This means, however, that there are time-independent Lagrangean systems for which H is *not* a first integral of the Euler-Lagrange distribution. An example of such a Lagrangean system is represented by the Cawley's Lagrangian

$$L = \dot{q}^1\dot{q}^3 + \tfrac{1}{2}(q^2)^2 q^3$$

(the dynamics of this Lagrangean system has been investigated in Example 7.4.2). Indeed, for this Lagrangian

$$\mathcal{D} \approx \operatorname{span}\{q^2q^3\,dt,\ \omega^1,\ q^2q^3\,\omega^2,\ \omega^3,\ d\dot{q}^1 - \tfrac{1}{2}(q^2)^2\,dt,\ d\dot{q}^3\},$$
$$\Delta \approx \operatorname{span}\{q^2q^3\,dt,\ \omega^1,\ \omega^3,\ d\dot{q}^1 - \tfrac{1}{2}(q^2)^2\,dt,\ d\dot{q}^3\},$$

showing that $\mathcal{D} \neq \Delta$. Computing the Hamiltonian we get

$$H = \dot{q}^1\dot{q}^3 - \tfrac{1}{2}(q^2)^2 q^3.$$

Now,

$$\begin{aligned} dH &= \dot{q}^1\,d\dot{q}^3 + \dot{q}^3\,d\dot{q}^1 - q^2q^3\,dq^2 - \tfrac{1}{2}(q^2)^2\,dq^3 \\ &= \dot{q}^1\,d\dot{q}^3 + \dot{q}^3(d\dot{q}^1 - \tfrac{1}{2}(q^2)^2\,dt) - q^2q^3\,\omega^2 - \dot{q}^2q^2q^3\,dt - \tfrac{1}{2}(q^2)^2\,\omega^3, \end{aligned}$$

and we can see that $dH \in \mathcal{D}$, but $dH \notin \Delta$.

In what follows, we shall be interested in relations between symmetries and first integrals of Lagrangean systems.

8.3. Point symmetries associated with Lagrangean systems

Let α be a Lagrangean system of order $s-1$, $s \geqslant 1$. Consider its (local) Lagrangian λ of order r, the Lepagean equivalent θ_λ, and the Euler-Lagrange form E_λ of λ (recall that, within obvious conventions, $E_\lambda = p_1\alpha$ and $d\theta_\lambda = \alpha$).

8.3.1. Definition. Let ξ be a π-projectable vector field on Y. The condition that ξ is a point symmetry of λ, i.e.,

$$\partial_{J^r\xi}\lambda = 0,$$

is called *Noether equation*. The condition that ξ is a point symmetry of E_λ, i.e.,

$$\partial_{J^{2r}\xi}E_\lambda = 0,$$

is called *Noether–Bessel-Hagen equation*.

In fibered coordinates, the Noether equation reads

$$\frac{\partial L}{\partial t}\xi^0 + \sum_{k=0}^{r}\frac{\partial L}{\partial q_k^\nu}\xi_k^\nu + L\frac{d\xi^0}{dt} = 0, \tag{8.1}$$

and for the Noether–Bessel-Hagen equation one gets

$$\frac{\partial E_\sigma}{\partial t}\xi^0 + \sum_{k=0}^{2r}\frac{\partial E_\sigma}{\partial q_k^\nu}\xi_k^\nu + E_\nu\frac{\partial \xi^\nu}{\partial q^\sigma} + E_\sigma\frac{d\xi^0}{dt} = 0. \tag{8.2}$$

In these formulas, L (resp. E_σ's) are the fiber-chart components of λ (resp. E_λ), and $(\xi^0, \xi_k^\nu))$ are the components of the prolongation of ξ. Recall that

$$\xi_i^\sigma = \frac{d\xi_{i-1}^\sigma}{dt} - q_i^\sigma\frac{d\xi^0}{dt}, \quad i \geqslant 1. \tag{8.3}$$

Let us stop for a moment at the definition of symmetry of a Lagrangian (resp. an Euler-Lagrange form). It is important to note that to get a transparent definition one has to consider Lagrangian as a *differential form*, not as a *function*. Working with functions (very frequent in the existing literature) is ambiguous in this context: in particular, (8.1) is then often used as a *definition* of symmetry of a Lagrange function, which makes this concept mysterious from the geometrical point of view. From similar reasons, a transparent definition of symmetry of the Euler-Lagrange expressions has to be stated for the Euler-Lagrange *differential two-form*, and not for the Euler-Lagrange expressions themselves.

Let us study relations between various point symmetries of a Lagrangean system. We shall start with a proposition which provides us with important conditions equivalent with the Noether–Bessel-Hagen equation.

8.3.2. Proposition. *Let ξ be a π-projectable vector field on Y, let λ be a (local) Lagrangian of order r, E_λ its Euler-Lagrange form. The vector field ξ is a point symmetry of E_λ if and only if one of the following equivalent conditions holds:*

(1) $E_{\partial_{J^r\xi}\lambda} = 0$,

(2) *there exists a unique closed one-form ρ of order $r-1$ such that*

$$\partial_{J^r\xi}\lambda = h\rho .$$

Proof. The first part of Proposition 8.3.2. trivially follows from Theorem 4.7.5. The equivalence of (1) and (2) is a consequence of Corollary 4.3.8. □

Notice that, in the notations of the above proposition, ρ is the Lepagean equivalent of the Lagrangian $\partial_{J^r\xi}\lambda$. This means that, by the first variation formula,

$$h\rho = \partial_{J^r\xi}\lambda = \partial_{J^{2r}\xi}\, h\theta_\lambda = h(\partial_{J^{2r-1}\xi}\theta_\lambda)\,,$$

i.e., the form $\partial_{J^{2r-1}\xi}\theta_\lambda$ is projectable onto an open subset of $J^{r-1}Y$ and

$$\rho = \partial_{J^{2r-1}\xi}\theta_\lambda\,. \tag{8.4}$$

Now, we can prove a theorem on relations between different point symmetries associated with a Lagrangean system α:

8.3.3. Theorem. *Let ξ be a (local) π-projectable vector field on Y.*

(1) *ξ is a point symmetry of θ_λ if and only if it is a point symmetry of λ.*

(2) *ξ is a point symmetry of α if and only if it is a point symmetry of E_λ.*

(3) *If ξ is a point symmetry of θ_λ then it is a point symmetry of α.*

(4) *If ξ is a point symmetry of α then it is a point symmetry of its characteristic distribution $\mathcal{D}$.*

(5) *If ξ is a point symmetry of λ then it is a point symmetry of E_λ.*

Proof. (1) The implication $\Rightarrow$ follows from the first variation formula, which for a Lepagean one-form θ of order r and its corresponding Lagrangian $h\theta$ reads

$$\partial_{J^{r+1}\xi}h\theta = h(\partial_{J^r\xi}\theta)\,.$$

Conversely, if ξ is a point symmetry of λ then, by Corollary 4.7.6, $\partial_{J^{2r-1}\xi}\theta_\lambda = \theta_{\lambda_0}$, where $\lambda_0 = 0$. However, $\theta_{\lambda_0} = 0$, i.e., ξ is a point symmetry of θ_λ.

(2) Consider a one-form θ (defined on an appropriate open subset of $J^{s-1}Y$) such that $\alpha = d\theta$. By assumption, the form $i_{J^{s-1}\xi}d\theta$ is closed. Hence the one-form $\rho = \partial_{J^{s-1}\xi}\theta = i_{J^{s-1}\xi}d\theta + di_{J^{s-1}\xi}\theta$ is closed, and for the Lagrangian $\lambda = h\theta$ we get

$$\partial_{J^s\xi}\lambda = h(\partial_{J^{s-1}\xi}\theta) = h\rho\,.$$

By Proposition 8.3.2., ξ is a point symmetry of E_λ. Conversely, if ξ is a point symmetry of E_λ then, by Corollary 4.7.6, $\partial_{J^{s-1}\xi}\alpha$ is the Lepagean equivalent of the identically zero locally variational form, hence, $\partial_{J^{s-1}\xi}\alpha = 0$.

The assertions (3) and (4) have been proved in Chapter 2.

(5) follows directly from Theorem 4.7.5. □

Summarizing briefly the results on point symmetries associated with Lagrangean systems, we get the following diagram:

$$\begin{array}{ccc} \xi \text{ is a point symmetry of } \theta_\lambda & \Longleftrightarrow & \xi \text{ is a point symmetry of } \lambda \\ \Downarrow & & \Downarrow \\ \xi \text{ is a point symmetry of } \alpha_{E_\lambda} & \Longleftrightarrow & \xi \text{ is a point symmetry of } E_\lambda \\ \Downarrow & & \\ \xi \text{ is a point symmetry of } \mathcal{D}\,. & & \end{array}$$

In other words, the set of point symmetries of a Lepagean two-form/Lagrangean system α coincides with the set of point symmetries of the corresponding locally variational form $E = p_1\alpha$, and contains the set of all point symmetries of any Lagrangian (resp. Cartan form) of E.

8.4. Point symmetries and first integrals

Now, we shall study relations between point symmetries and first integrals.

Let us turn to the fundamental *First Theorem of Emmy Noether* which in the geometric setting turns out to be an easy consequence of the first variation formula and the Noether equation.

8.4.1. Noether Theorem. *Let λ be (local) a Lagrangian of order r (defined on an open subset $W \subset J^rY$), let θ_λ be its Lepagean equivalent. Let a π-projectable vector field ξ on Y be a point symmetry of the Lagrangian λ. Let γ be an extremal of λ defined on $\pi_r(W) \subset X$. Then*

$$i_{J^{2r-1}\xi}\theta_\lambda \circ J^{2r-1}\gamma = \text{const}\,. \tag{8.5}$$

Proof. By the integral first variation formula (3.19), over any piece $\Omega \subset X$ it holds

$$\int_\Omega J^r\gamma^*\partial_{J^r\xi}\lambda = \int_\Omega J^{2r-1}\gamma^* i_{J^{2r-1}\xi}d\theta_\lambda + \int_\Omega dJ^{2r-1}\gamma^* i_{J^{2r-1}\xi}\theta_\lambda.$$

Since γ is an extremal, the first term on the right-hand side of the above equation vanishes by Theorem 3.3.2. Since ξ is a point symmetry of λ, the left-hand side equals to 0, and we get

$$\int_\Omega J^{2r-1}\gamma^* di_{J^{2r-1}\xi}\theta_\lambda = 0$$

over any piece Ω; hence, we get

$$J^{2r-1}\gamma^* di_{J^{2r-1}\xi}\theta_\lambda = 0,$$

and finally (8.5). □

Notice that the function

$$f_{(L,\xi)} \equiv i_{J^{2r-1}\xi}\theta_\lambda \tag{8.6}$$

is of order $2r - 1$ where r is the order of the Lagrangian. The equation

$$f_{(L,\xi)} \circ J^{2r-1}\gamma = \text{const} \tag{8.7}$$

is called a *conservation law*, and it is an ODE of order $2r - 1$ for γ.

Using Corollary 4.5.7 we get that if $r \leqslant s$ then $f_{(L,\xi)}$ is defined *on the phase space*, i.e., $f_{(L,\xi)}$ is a *constant of the motion*. In this case, the conservation law (8.7) is an ODE of order $s-1$, i.e., *lower* than the order of the original variational equations. Thus, practically, *knowing a symmetry of a Lagrangian of order* $\leqslant$ *the order of the variational equations, lowers the order of the original integration problem*—this is a key idea in the theory of symmetries. On the contrary, knowing a symmetry of a Lagrangian of order greater than or equal to the order of the given variational equations ususally results in a higher-order conservation law. The practical meaning of such conservation laws, however, is not clear. Indeed, in such a case, (8.7) represents a differential equation of order equal to or even greater than the order of the original variational equations which hardly could simplify the original integration problem.

We have seen so far that for regular Lagrangean systems constants of the motion coincide with first integrals of the characteristic distribution $\mathcal{D}$, and hence with first integrals of the Euler-Lagrange distribution Δ (since in this case both the distributions coincide). This means, however, that for *regular* Lagrangean systems Noether Theorem provides us with *first integrals* of $\mathcal{D}$, each corresponding to a symmetry of a Lagrangian (of a suitably low order).

Since

$$\int_\Omega J^r\gamma^*\partial_{J^r\xi}\lambda = \int_\Omega J^{2r-1}\gamma^*\partial_{J^{2r-1}\xi}\theta_\lambda,$$

Noether Theorem remains true if instead of a symmetry of the Lagrangian λ one takes a symmetry of its Lepagean equivalent θ_λ. In other words, Noether Theorem implies that if ξ is a point symmetry of θ_λ then $i_{J^{2r-1}\xi}\theta_\lambda$ is conserved along prolonged extremals, i.e., for θ_λ of order $s-1$, it is a *constant of the motion*. However, taking into account the results of Sec. 2.3, we can see that a *stronger* assertion holds:

8.4.2. Proposition. *Let α be a Lagrangean system of order $s-1$, $s \geqslant 1$. Let ξ be a point symmetry of the Lepagean equivalent θ_λ of a (local) Lagrangian λ of α. If θ_λ is of order $s-1$ then $i_{J^{s-1}\xi}\theta_\lambda$ is a first integral of the characteristic distribution of the Lagrangean system α.*

Using the above proposition, Theorem 8.3.3 and Corollary 4.5.7 we immediately get the *Noether Theorem* in the following stronger formulation:

8.4.3. Theorem. *Let α be a Lagrangean system on $J^{s-1}Y$, let λ be a (local) Lagrangian of order $r \leqslant s$. If a π-projectable vector field ξ on Y is a point symmetry of λ then $i_{J^{2r-1}\xi}\theta_\lambda$ is a first integral of the characteristic distribution of the Lagrangean system α.*

8.4.4. Remark. Expressing the function $f_{(L,\xi)}$ in fibered coordinates, we get

$$f_{(L,\xi)} = L\xi^0 + \sum_{i=0}^{r-1} f_\sigma^{i+1}(\xi_i^\sigma - q_{i+1}^\sigma\xi^0), \tag{8.8}$$

where the f_σ^j's, resp. the ξ_i^σ's are defined by (3.12), resp. (8.3). Hence, for a *minimal-order*

Lagrangian $\lambda_{\min}$, (8.8) takes the form

$$f_{(L_{\min},\xi)} = -H\xi^0 + \sum_{i=0}^{r-1} p^i_\sigma \xi^\sigma_i, \tag{8.9}$$

where the H and p^σ_i's are the Hamiltonian and momenta of the given Lagrangean system, corresponding to $\lambda_{\min}$.

8.4.5. Example. The *Euclidean group* (sometimes called also *Aristotle group*) is defined to be the following 7-parameter group of transformations of $R \times R^3$:

$$\bar{t} = t + c \quad \text{time translation,}$$

$$\bar{x}^i = x^i + a^i, \quad i = 1, 2, 3 \quad \text{space translations,}$$

$$\bar{x}^i = x^i + \varepsilon_{ijk} b^j x^k, \quad i = 1, 2, 3 \quad \text{space rotations.}$$

The Euclidean group is a subgroup of the *Galilei group* on $R \times R^3$ which is defined to be the 10-parameter group of transformations, containing time and space translations, space rotations and 3 *Galilei transformations*

$$\bar{x}^i = x^i + v^i t.$$

Computing the vector field generators of the above transformations we get

$$\frac{\partial}{\partial t}, \quad \frac{\partial}{\partial x^i}, \quad i = 1, 2, 3 \tag{8.10}$$

for the time and space translations,

$$x^3 \frac{\partial}{\partial x^2} - x^2 \frac{\partial}{\partial x^3}, \quad -x^3 \frac{\partial}{\partial x^1} + x^1 \frac{\partial}{\partial x^3}, \quad x^2 \frac{\partial}{\partial x^1} - x^1 \frac{\partial}{\partial x^2}, \tag{8.11}$$

for the space rotations, and

$$t \frac{\partial}{\partial x^i}, \quad i = 1, 2, 3 \tag{8.12}$$

for the Galilei transformations.

Let us find the corresponding first integrals. For a first-order Lagrangian L invariant with respect to the time-translation we get from (8.6) the first integral

$$i_{\partial/\partial t}\, \theta_\lambda = L - \frac{\partial L}{\partial \dot{x}^i} \dot{x}^i = -H,$$

i.e., the Hamiltonian is conserved along extremals (the *energy conservation law*). For L invariant with respect to a space translation along the x^i-axis we obtain

$$i_{\partial/\partial x^i}\, \theta_\lambda = \frac{\partial L}{\partial \dot{x}^i} = p_i,$$

i.e., the momentum p_i is conserved. In case that p_i is conserved for all i, we get the *momentum conservation law*. If L is invariant under space rotations, the first integral is the vector

$$\left(x^3 \frac{\partial L}{\partial \dot{x}^2} - x^2 \frac{\partial L}{\partial \dot{x}^3}, \; -x^3 \frac{\partial L}{\partial \dot{x}^1} + x^1 \frac{\partial L}{\partial \dot{x}^3}, \; x^2 \frac{\partial L}{\partial \dot{x}^1} - x^1 \frac{\partial L}{\partial \dot{x}^2}\right) = -r \times p.$$

$M = r \times p$ is called *angular momentum*, and the corresponding conservation law is referred to as the *angular momentum conservation law*. If L is invariant with respect to Galilei transformations then the vector

$$\left(t\frac{\partial L}{\partial \dot{x}^1}, t\frac{\partial L}{\partial \dot{x}^2}, t\frac{\partial L}{\partial \dot{x}^3}\right) = tp$$

is conserved.

Let us turn to the meaning of point symmetries of a locally variational form. Using Proposition 8.3.2 and the first variation formula, in a similar way as in the proof of the Noether Theorem we get

8.4.6. Generalized Noether Theorem. *Let E be a locally variational form of order s, let a π-projectable vector field ξ on Y be a point symmetry of E. If λ is a (local) Lagrangian of order r for E on an open set W, and ρ is the unique closed one-form of order $r-1$ such that $\partial_{J^r\xi}\lambda = h\rho$, and if γ is an extremal of E defined on $\pi_r(W) \subset X$, then*

$$J^{2r-1}\gamma^*(di_{J^{2r-1}\xi}\theta_\lambda - \rho) = 0. \tag{8.13}$$

This means that a point symmetry of a locally variational form E gives rise to *conservation laws*

$$(i_{J^{2r-1}\xi}\theta_\lambda - g) \circ J^{2r-1}\gamma = \text{const}, \tag{8.14}$$

where g is a function (defined on an appropriate open set) such that $dg = \rho$. Notice that since ρ is given by (8.4), we get

$$di_{J^{2r-1}\xi}\theta_\lambda - \rho = -i_{J^{2r-1}\xi}\, d\theta_\lambda,$$

which is a closed one-form, projectable onto an open subset of $J^{s-1}Y$. Consequently, knowing a point symmetry of E, the Generalized Noether Theorem provides us, for *every* Lagrangian, with *a constant of the motion*.

According to Chapter 2 we have

8.4.7. Proposition. *Let α be a Lagrangean system of order $s-1$, $s \geqslant 1$. Let ξ be a π-projectable vector field on Y.*

(1) *If ξ is a point symmetry of α then in a neighborhood of every point in $J^{s-1}Y$ one has $i_{J^{s-1}\xi}\alpha = df$, where f is a first integral of the characteristic distribution $\mathcal{D}$ of α.*

(2) *Suppose that $\mathcal{D} \neq \Delta$. If ξ is π-vertical and a point symmetry of α then in a neighborhood of every point of $J^{s-1}Y$ one has $i_\xi\alpha = df$, where f is a first integral of the Euler-Lagrange distribution Δ.*

Now, using Theorem 8.3.3, the above proposition, and the fact that for every Lagrangian λ defined on an open set $W \subset J^rY$ it holds $d\theta_\lambda = \alpha|_W$, we get the *Generalized Noether Theorem* in the form

8.4.8. Theorem. *Let α be a Lagrangean system on $J^{s-1}Y$. Let $E = p_1\alpha$, and let a π-projectable vector field ξ on Y be a point symmetry of E. Then for every Lagrangian λ of E, the function $i_{J^{2r-1}\xi}\theta_\lambda - g$, where r is the order of λ and g is given by $dg = \partial_{J^{2r-1}\xi}\theta_\lambda$, is* *defined on an open subset of $J^{s-1}Y$, and it is a first integral of the chracteristic distribution of the Lagrangean system α.*

In the next chapter we shall investigate exact integration methods for regular and semiregular Lagrangean systems, which can be applied if one knows *a suitable family of first integrals of the characteristic distribution* (the generalized Liouville, and Hamilton-Jacobi integration methods). Notice that in this context, Theorems 8.4.3 and 8.4.8 have a valuable practical meaning: since (even for singular Lagrangean systems) the conserved functions obtained by means of the Noether Theorem are first integrals of the characteristic distribution, they can be used for exact integration of the Euler-Lagrange equations by means of the Liouville theorem, or the Hamilton-Jacobi method, as explained in the next chapter.

8.5. Applications of Noether equation and of Noether–Bessel-Hagen equation

Noether equation is of fundamental importance for theoretical physics. It can be used as an effective tool to solve the following problems:

(1) *Given a Lagrangian, find all its point symmetries and the corresponding first integrals.* In this case one has to solve the Noether equation with respect to the vector field ξ. First integrals are then found on the basis of the Noether Theorem (formula (8.6)).

(2) *Given a group of transformations of Y, find all corresponding invariant Lagrangians.* In this case, the symmetries are prescribed, and one has to solve the Noether equation with respect to L.

The second of the above procedures is often used, namely in field theory, to discover Lagrangians for concrete physical systems (different fields and their interaction). In this case the symmetry group for the theory is known and the problem is to find all Lagrangians possessing the prescribed symmetries. (For an interested reader we note that in field theory one often deals with a symmetry group acting on the space-time, which is the *base* X of the corresponding fibered manifold. In this case, however, one must work with the vector field generators of the invariant transformations *lifted* to the total space Y).

Similarly, the *Noether–Bessel-Hagen equation* can be used to solve the following problems:

(1) *Find (all) infinitesimal transformations of Y which leave given Euler-Lagrange expressions (a given Euler-Lagrange form) invariant.* In this case, (8.2) has to be considered as a system of PDE's for *symmetries* ξ of a given Euler-Lagrange form. (Notice that by Theorem 8.3.3 (2), solving this problem one gets all point symmetries of the given *Lagrangean system*).

(2) *Find (all) dynamical forms possessing prescribed point symmetries.* In this case, the ξ's are known, and (8.2) becomes a system of PDE's for the components E_σ of *dynamical forms* E.

(3) *Find (all) Euler-Lagrange expressions (locally variational forms) possessing prescribed point symmetries.* In this case, the ξ's are known, and one is looking for dynamical forms which satisfy the equation (8.2) *and* the variationality conditions (4.9). In other words, one has to solve a system of PDE's, consisting of (8.2) and (4.9), with respect to the E_σ's. (Alternatively, one can combine (8.2) with Theorem 6.7.2). This procedure has been proposed and applied in M. Štefanová and O. Štěpánková [1].

Notice that, in mechanics, by Theorem 8.3.3 (2) the latter problem is *equivalent* with that of searching for all *Lagrangean systems* possessing prescribed point symmetries.

8.5.1. Example. Let us compute all point symmetries for the classical free-particle Lagrangian

$$L = \tfrac{1}{2}mv^2 \equiv \tfrac{1}{2}mg_{ij}\dot{x}^i\dot{x}^j$$

on $J^1(R \times R^3)$.

Noether equation for L is of the form

$$mg_{ij}\dot{x}^j(\dot{\xi}^i - \dot{x}^i\dot{\xi}^0) + \tfrac{1}{2}mg_{ij}\dot{x}^i\dot{x}^j\dot{\xi}^0 = 0,$$

i.e.,

$$g_{ij}\dot{x}^i\dot{\xi}^j - \tfrac{1}{2}g_{ij}\dot{x}^i\dot{x}^j\dot{\xi}^0 = 0.$$

Using that ξ^0 is a function of t only, and

$$\dot{\xi}^j = \frac{\partial \xi^j}{\partial t} + \frac{\partial \xi^j}{\partial x^k}\dot{x}^k,$$

we get that the Noether equation is equivalent to the conditions

$$\frac{\partial \xi^j}{\partial t} = 0, \quad g_{ij}\dot{x}^i\frac{\partial \xi^j}{\partial x^k}\dot{x}^k - \tfrac{1}{2}g_{ij}\dot{x}^i\dot{x}^j\dot{\xi}^0 = 0,$$

i.e., to

$$\frac{\partial \xi^j}{\partial t} = 0, \quad \dot{\xi}^0 = 0, \quad \dot{x}^1\dot{x}^k\frac{\partial \xi^1}{\partial x^k} + \dot{x}^2\dot{x}^k\frac{\partial \xi^2}{\partial x^k} + \dot{x}^3\dot{x}^k\frac{\partial \xi^3}{\partial x^k} = 0.$$

Hence, $\xi^0 = a = \text{const}$, and

$$\frac{\partial \xi^1}{\partial x^1} = \frac{\partial \xi^2}{\partial x^2} = \frac{\partial \xi^3}{\partial x^3} = 0, \qquad \frac{\partial \xi^i}{\partial x^j} + \frac{\partial \xi^j}{\partial x^i} = 0, \; i \neq j.$$

Solving these equations one gets that symmetries of the given Lagrangian are generated by the vector fields

$$\frac{\partial}{\partial t}, \quad \frac{\partial}{\partial x^i}, \quad -\varepsilon_{ijk}x^j\frac{\partial}{\partial x^k}, \qquad i = 1, 2, 3,$$

i.e., the invariance group is the Euclidean group on $R \times R^3$.

8.5.2. Example. Let us compute all *first-order* Lagrangians for the fibered manifold $R \times R^3 \to R$, invariant with respect to the Euclidean and Galilei group.

Substituting the generators of the Euclidean group into the Noether equation we get the following system of 7 PDE for $L(t, x^j, \dot{x}^j)$:

$$\frac{\partial L}{\partial t} = 0, \qquad \frac{\partial L}{\partial x^i} = 0,$$

$$\varepsilon_{ijk}x^j\frac{\partial L}{\partial x^k} + \varepsilon_{ijk}\dot{x}^j\frac{\partial L}{\partial \dot{x}^k} = 0.$$

Thus Lagrangians invariant with respect to the Euclidean group depend only upon velocities as follows

$$\varepsilon_{ijk}\dot{x}^j\frac{\partial L}{\partial \dot{x}^k} = 0.$$

The latter system of PDE is satisfied by any function

$$L = L(v^2), \quad v^2 = (\dot{x}^1)^2 + (\dot{x}^2)^2 + (\dot{x}^3)^2.$$

Similarly, substituting the generators of the Galilei group into the Noether equation we get the following system of 10 PDE for $L(t, x^j, \dot{x}^j)$:

$$\frac{\partial L}{\partial t} = 0, \quad \frac{\partial L}{\partial x^i} = 0,$$

$$\varepsilon_{ijk} x^j \frac{\partial L}{\partial x^k} + \varepsilon_{ijk} \dot{x}^j \frac{\partial L}{\partial \dot{x}^k} = 0,$$

$$t \frac{\partial L}{\partial x^i} + \frac{\partial L}{\partial \dot{x}^i} = 0,$$

which obviously has only trivial solutions $L = \text{const}$. Hence, there is no non-trivial first-order Lagrangian invariant with respect to the Galilei group on $R \times R^3$.

Similar arguments as above lead to a conclusion that there exist non-trivial *second-order* Lagrangians invariant with respect to the Galilei group,

$$L = L((\ddot{x}^1)^2 + (\ddot{x}^2)^2 + (\ddot{x}^3)^2).$$

Such Lagrangians, however, give rise to Lagrangean systems of order 4.

8.5.3. Example. The kinetic energy Lagrangian,

$$T = \tfrac{1}{2} v^2,$$

describing the motion of the classical free particle of unit mass, is apparently not invariant with respect to the Galilei transformations. Now, we shall be interested in finding an *equivalent* first-order Lagrangian L such that the vector fields

$$\xi_{(i)} = t \frac{\partial}{\partial x^i}, \quad i = 1, 2, 3,$$

generating the Galilei transformations in $R \times R^3$, are symmetries of L. Hence, we require that

$$L = \tfrac{1}{2} v^2 + \frac{df}{dt}$$

for a function $f(t, x^i)$, and that L satisfies the Noether equation

$$\frac{\partial L}{\partial x^i} t + \frac{\partial L}{\partial \dot{x}^i} = 0.$$

In this way we get the following equations for f:

$$\frac{\partial^2 f}{\partial t\, \partial x^i} t + \frac{\partial^2 f}{\partial x^i \partial x^j} \dot{x}^j t + \dot{x}^i + \frac{\partial f}{\partial x^i} = 0.$$

The above equations are equivalent to

$$\frac{\partial^2 f}{\partial t\, \partial x^i} t + \frac{\partial f}{\partial x^i} = 0, \quad \frac{\partial^2 f}{\partial x^i \partial x^j} t + \delta^i_j = 0,$$

and the latter set of these equations leads to

$$\frac{\partial^2 f}{\partial x^i \partial x^j} = 0, \ i \neq j, \quad \frac{\partial^3 f}{(\partial x^i)^3} = 0, \ i = 1, 2, 3.$$

To satisfy the equations for f we can put e.g.

$$f = -\frac{r^2}{2t}.$$

Then we obtain the Lagrangian

$$L = \tfrac{1}{2}v^2 - \frac{rv}{t} + \frac{r^2}{2t^2}, \tag{8.15}$$

describing the classical free particle of unit mass, and invariant with respect to the Galilei transformations, as required. Writing down explicitly the corresponding first integrals of the motion equations $\ddot{x}^i = 0$ we get

$$i_{\xi_{(i)}}\theta_\lambda = t\dot{x}^i - x^i, \quad i = 1, 2, 3.$$

Notice that the Lagrangian (8.15) is invariant with respect to the space rotations, satisfying the Noether equations

$$\varepsilon_{ijk}x^j \frac{\partial L}{\partial x^k} + \varepsilon_{ijk}\dot{x}^j \frac{\partial L}{\partial \dot{x}^k} = 0,$$

but it is not invariant with respect to the time and space translations.

8.5.4. Example. Consider the Cawley's Lagrangian

$$L = \dot{q}^1\dot{q}^3 + \tfrac{1}{2}(q^2)^2 q^3$$

(recall that we have discussed the dynamical properties of this Lagrangian in Example 7.4.2).

We can see immediately that the vector fields $\partial/\partial t$ and $\partial/\partial q^1$ are symmetries of L, and the corresponding first integrals are $p_1 = \dot{q}^3$ and

$$H = \dot{q}^1\dot{q}^3 - \tfrac{1}{2}(q^2)^2 q^3.$$

Let us however find *all* point symmetries of L, and the corresponding first integrals (i.e., all Noetherian first integrals).

The Noether equation for L reads

$$q^2q^3\xi^2 + \tfrac{1}{2}(q^2)^2\xi^3 + \dot{q}^3(\dot{\xi}^1 - \dot{q}^1\dot{\xi}^0) + \dot{q}^1(\dot{\xi}^3 - \dot{q}^3\dot{\xi}^0) + (\dot{q}^1\dot{q}^3 + \tfrac{1}{2}(q^2)^2q^3)\dot{\xi}^0 = 0.$$

Taking into account that the functions $\xi^0, \xi^i, i = 1, 2, 3$ do not depend on $\dot{q}^i, i = 1, 2, 3$, and that ξ^0 is a function of t only, we can see that the above equation is equivalent to

$$\begin{gathered}
q^2q^3\xi^2 + \tfrac{1}{2}(q^2)^2\xi^3 + \tfrac{1}{2}(q^2)^2q^3\dot{\xi}^0 = 0, \\
\frac{\partial \xi^1}{\partial t} = 0, \quad \frac{\partial \xi^1}{\partial q^2} = 0, \quad \frac{\partial \xi^1}{\partial q^3} = 0, \\
\frac{\partial \xi^3}{\partial t} = 0, \quad \frac{\partial \xi^3}{\partial q^1} = 0, \quad \frac{\partial \xi^3}{\partial q^2} = 0, \\
\frac{\partial \xi^1}{\partial q^1} + \frac{\partial \xi^3}{\partial q^3} - \frac{d\xi^0}{dt} = 0.
\end{gathered}$$

Solving these equations we obtain that the latter one is equivalent to

$$\frac{\partial \xi^1}{\partial q^1} = 0, \quad \frac{\partial \xi^3}{\partial q^3} = 0, \quad \frac{d\xi^0}{dt} = 0.$$

Hence

$$\xi^1 = a, \quad \xi^3 = b, \quad \xi_0 = c,$$

where a, b, c are arbitrary constants, and

$$q^2 q^3 \xi^2 + \tfrac{1}{2}(q^2)^2 b = 0.$$

Thus we have obtained that all the point symmetries of L are of the form

$$c\frac{\partial}{\partial t} + a\frac{\partial}{\partial q^1} - b\Big(\frac{q^2}{2q^3}\frac{\partial}{\partial q^2} - \frac{\partial}{\partial q^3}\Big)$$

on the open submanifold $q^3 \neq 0$ of the configuration space-time, and

$$c\frac{\partial}{\partial t} + a\frac{\partial}{\partial q^1}$$

on $q^3 = 0$. The corresponding first integrals are then H for the time translations, $p_1 = \dot{q}^3$ for the translations along the q^1-axis, and

$$-\frac{q^2}{2q^3}\frac{\partial L}{\partial \dot{q}^2} + \frac{\partial L}{\partial \dot{q}^3} = \dot{q}^1$$

for the transformations generated by the vector field

$$-\frac{q^2}{2q^3}\frac{\partial}{\partial q^2} + \frac{\partial}{\partial q^3}.$$

8.5.5. Example. Time-independent Lagrangean systems in classical mechanis are described by Lagrangians of the form

$$L = T - U,$$

where T is the kinetic energy and $U = U(q^\sigma)$ is the potential energy of the system. Using that

$$\frac{\partial T}{\partial \dot{q}^\sigma}\dot{q}^\sigma = 2T, \quad p_\sigma = \frac{\partial T}{\partial \dot{q}^\sigma}, \quad H = -L + p_\sigma \dot{q}^\sigma = -L + \frac{\partial T}{\partial \dot{q}^\sigma}\dot{q}^\sigma = T + U,$$

we get the Noether equation in the form

$$-\frac{\partial U}{\partial q^\sigma}\xi^\sigma + p_\sigma \dot{\xi}^\sigma - H\dot{\xi}^0 = 0.$$

Now we shall investigate invariance with respect to the Euclidean group (Example 8.4.5). Obviously, all time-independent systems obey the energy conservation law. Considering the space translation in the direction of the q^σ-axis, the Noether equation gives

$$\frac{\partial U}{\partial q^\sigma} = 0,$$

i.e., the system obeys the momentum conservation law only in the direction where the potential energy is not changed. Finally, for simplicity, taking into account the rotation round the q^3-axis, the Noether equation gives

$$q^1\frac{\partial U}{\partial q^2} - q^2\frac{\partial U}{\partial q^1} = 0.$$

Hence, if the potential U possesses a rotational symmetry, the angular momentum is conseved in the direction of the axis of rotation.

Let us turn to illustrate applications of the Noether–Bessel-Hagen equation on easy examples from classical mechanics. It should be pointed out, however, that similar techniques can be used also for investigations in field theory.

8.5.6. Example. We shall find all dynamical forms on $J^2(R \times R^3)$ which are invariant with respect to the Galilei group, and variational.

In canonical ccordinates on $R \times R^3$ denote

$$E = E_i\, dx^i \wedge dt.$$

Since we suppose that E is variational, the E_i must depend linearly upon the accelerations, i.e.,

$$E_i = A_i + B_{ij}\ddot{x}^j,$$

where A_i and B_{ij} are functions of $t, x^k, \dot{x}^k$, satisfying the Helmholtz conditions (1.8).

Next, denote

$$\xi = \xi^0 \frac{\partial}{\partial t} + \xi^i \frac{\partial}{\partial x^i}.$$

Recall that for the 2nd prolongation of the vector field ξ we have

$$J^2\xi = \xi^0 \frac{\partial}{\partial t} + \xi^i \frac{\partial}{\partial x^i} + \Xi_1^i \frac{\partial}{\partial \dot{x}^i} + \Xi_2^i \frac{\partial}{\partial \ddot{x}^i},$$

where

$$\Xi_1^i = \dot{\xi}^i - \dot{x}^i \dot{\xi}^0, \quad \Xi_2^i = \dot{\Xi}_1^i - \ddot{x}^i \dot{\xi}^0 = \ddot{\xi}^i - 2\ddot{x}^i \dot{\xi}^0 - \dot{x}^i \ddot{\xi}^0.$$

The Noether–Bessel-Hagen equations for the components of E read

$$E_k \frac{\partial \xi^k}{\partial x^i} + E_i \dot{\xi}^0 + i_{J^2\xi}\, dE_i = 0,$$

i.e.,

$$E_k \frac{\partial \xi^k}{\partial x^i} + E_i \dot{\xi}^0 + \frac{\partial E_i}{\partial t} + \frac{\partial E_i}{\partial x^k}\xi^k + \frac{\partial E_i}{\partial \dot{x}^k}(\dot{\xi}^i - \dot{x}^i \dot{\xi}^0) + \frac{\partial E_i}{\partial \ddot{x}^k}(\ddot{\xi}^i - 2\ddot{x}^i \dot{\xi}^0 - \dot{x}^i \ddot{\xi}^0) = 0. \quad (8.16)$$

Let us substitute into the above equation the generators of the Galilei group (Example 8.4.5). The invariance of E with respect to the space translation gives the Noether–Bessel-Hagen equations (8.16) in the form

$$\frac{\partial E_i}{\partial t} = 0,$$

showing that the E_i do not depend upon t. Substituting the generators of the space translations into (8.16) we get

$$\frac{\partial E_i}{\partial x^k} = 0,$$

i.e., the E_i do not depend upon x^1, x^2, x^3. Similarly, from the invariance of $E_i(\dot{x}^k, \ddot{x}^k)$ with respect to the Galilei transformations we get

$$\frac{\partial E_i}{\partial \dot{x}^k} = 0.$$

Thus we have obtained that

$$E_i = A_i + B_{ij}\ddot{x}^j, \tag{8.17}$$

where A_i, B_{ij} are constants. Now, the Helmholtz conditions applied to (8.17) give us the symmetry of the matrix (B_{ij}),

$$B_{ij} = B_{ji}.$$

It remains to apply the requirement on symmetry of E with respect to the generators of infinitesimal space rotations. Substituting these generators into (8.16) we get the following conditions on the E_i's:

$$\begin{aligned} B_{k1}\ddot{x}^2 - B_{k2}\ddot{x}^1 &= E_2\delta^1_k - E_1\delta^2_k, \\ B_{k1}\ddot{x}^3 - B_{k3}\ddot{x}^1 &= E_3\delta^1_k - E_1\delta^3_k, \\ B_{k2}\ddot{x}^3 - B_{k3}\ddot{x}^2 &= E_3\delta^2_k - E_2\delta^3_k, \end{aligned}$$

which means that

$$A_2\delta^1_k - A_1\delta^2_k = 0, \quad A_3\delta^1_k - A_1\delta^3_k = 0, \quad A_3\delta^2_k - A_2\delta^3_k = 0,$$

and

$$\begin{aligned} -B_{k2} &= B_{21}\delta^1_k - B_{11}\delta^2_k, & B_{k1} &= B_{22}\delta^1_k - B_{12}\delta^2_k, & 0 &= B_{23}\delta^1_k - B_{13}\delta^2_k, \\ -B_{k3} &= B_{31}\delta^1_k - B_{11}\delta^3_k, & 0 &= B_{32}\delta^1_k - B_{12}\delta^3_k, & B_{k1} &= B_{33}\delta^1_k - B_{13}\delta^3_k, \\ 0 &= B_{31}\delta^2_k - B_{21}\delta^3_k, & -B_{k3} &= B_{32}\delta^2_k - B_{22}\delta^3_k, & B_{k2} &= B_{33}\delta^2_k - B_{23}\delta^3_k. \end{aligned}$$

Hence,

$$A_1 = A_2 = A_3 = 0, \quad B_{ij} = m\delta_{ij},$$

where m is a constant. Summarizing the results, we have obtained that *there is a unique (up to a constant) dynamical form E on $J^2(R \times R^3)$, invariant with respect to the Galilei group and variational.* Its components E_i are of the form

$$E_i = m\ddot{x}^i.$$

i.e., it describes the motion of the *free particle* of mass m in $R \times R^3$. The corresponding Lepagean two-form is then

$$\alpha = \sum_i m(dx^i - \dot{x}^i dt) \wedge d\dot{x}^i.$$

Notice that if we would be interested to find all dynamical forms invariant with respect to the Galilei group (and not necessarily variational) we would obtain

$$E_i = f(a^2)\ddot{x}^i, \tag{8.18}$$

where f is an arbitrary function of

$$a^2 = (\ddot{x}^1)^2 + (\ddot{x}^2)^2 + (\ddot{x}^3)^2.$$

8.5.7. Example. Let us find the invariance group for the dynamical form E,

$$E_i = m\ddot{x}^i, \tag{8.19}$$

describing the motion of the free particle of mass m in $R \times R^3$.

This means that we have to solve the Noether–Bessel-Hagen equation with respect to the generators ξ. Substituting (8.19) into the Noether–Bessel-Hagen equation (8.2) we get

$$\ddot{x}^k \frac{\partial \xi^k}{\partial x^i} + \ddot{x}^i \dot{\xi}^0 + \frac{\partial^2 \xi^i}{\partial t^2} + 2\frac{\partial^2 \xi^i}{\partial t\, \partial x^j}\dot{x}^j + \frac{\partial^2 \xi^i}{\partial x^j \partial x^k}\dot{x}^j \dot{x}^k + \frac{\partial \xi^i}{\partial x^j}\ddot{x}^j - 2\ddot{x}^i \dot{\xi}^0 - \dot{x}^i \ddot{\xi}^0 = 0.$$

Taking into account that ξ^i is a function of t and x^j only, and $\xi^0 = \xi^0(t)$, we can see that the expressions on the left hand side of the above equation are linear in $\ddot{x}^j$, and bilinear in $\dot{x}^j$, with the coefficients depending on t, x^k. Thus, these coefficients must equal zero, and we get the following system of PDE for the vector field ξ:

$$\frac{\partial \xi^k}{\partial x^i} + \frac{\partial \xi^i}{\partial \xi^k} = \delta_{ik}\dot{\xi}^0, \quad \frac{\partial^2 \xi^i}{\partial x^j \partial x^k} = 0, \quad 2\frac{\partial^2 \xi^i}{\partial t\, \partial x^j} = \delta_{ij}\ddot{\xi}^0, \quad \frac{\partial^2 \xi^i}{\partial t^2} = 0.$$

This means that

$$\begin{aligned}
&\frac{\partial \xi^k}{\partial x^i} + \frac{\partial \xi^i}{\partial \xi^k} = 0, \quad i \neq k,\\
&\frac{\partial \xi^1}{\partial x^1} = \frac{\partial \xi^2}{\partial x^2} = \frac{\partial \xi^3}{\partial x^3} = \tfrac{1}{2}\dot{\xi}^0,\\
&\frac{\partial^2 \xi^i}{\partial t^2} = 0, \quad \frac{\partial^2 \xi^i}{\partial x^j \partial x^k} = 0, \quad \frac{\partial^2 \xi^i}{\partial t\, \partial x^j} = 0, \quad i \neq j,\\
&\frac{\partial^2 \xi^1}{\partial t\, \partial x^1} = \frac{\partial^2 \xi^2}{\partial t\, \partial x^2} = \frac{\partial^3 \xi^i}{\partial t\, \partial x^3} = \tfrac{1}{2}\ddot{\xi}^0.
\end{aligned}$$

Solving these equations we easily get

$$\begin{aligned}
\xi^0 &= at^2 + 2\alpha t + k,\\
\xi^1 &= ax^1 t + b^1 t + \alpha x^1 + Ax^2 + Bx^3 + \beta^1,\\
\xi^2 &= ax^2 t + b^2 t - Ax^1 + \alpha x^2 + Cx^3 + \beta^2,\\
\xi^3 &= ax^3 t + b^3 t - Bx^1 - Cx^2 + \alpha x^3 + \beta^3,
\end{aligned} \tag{8.20}$$

where $a, b^1, b^2, b^3, \alpha, A, B, C, \beta^1, \beta^2, \beta^3, k$ are arbitrary constants. (8.20) are the components of a vector field on $R \times R^3$ which is a general solution to our problem. Hence, the dynamical form E (8.19) is invariant with respect to a 12-parameter group G of transformations of $R \times R^3$, generated by the following vector fields:

$$\begin{aligned}
&\frac{\partial}{\partial t}, \quad \frac{\partial}{\partial x^i}, \quad i = 1, 2, 3,\\
&x^3\frac{\partial}{\partial x^2} - x^2\frac{\partial}{\partial x^3}, \quad -x^3\frac{\partial}{\partial x^1} + x^1\frac{\partial}{\partial x^3}, \quad x^2\frac{\partial}{\partial x^1} - x^1\frac{\partial}{\partial x^2},\\
&t\frac{\partial}{\partial x^i}, \quad i = 1, 2, 3,\\
&2t\frac{\partial}{\partial t} + x^1\frac{\partial}{\partial x^1} + x^2\frac{\partial}{\partial x^2} + x^3\frac{\partial}{\partial x^3},\\
&t^2\frac{\partial}{\partial t} + tx^1\frac{\partial}{\partial x^1} + tx^2\frac{\partial}{\partial x^2} + tx^3\frac{\partial}{\partial x^3}.
\end{aligned} \tag{8.21}$$

Notice, that the Euclidean group and the Galilei group are subgroups of G.

8.5.8. Example. Examples 8.5.6 and 8.5.7 above show that for a dynamical form on $J^2(R \times R^3)$ invariant with respect to the Galilei group there should be a relation between variationality and invariance with respect to the 2-parameter group generated by the last two vector fields of (8.21). Indeed, we shall prove the following assertion:

Let E be a dynamical form on $J^2(R \times R^3)$, invariant with respect to the Galilei group. E is variational if and only the vector fields

$$\begin{aligned} \Xi_1 &= 2t\frac{\partial}{\partial t} + x^1\frac{\partial}{\partial x^1} + x^2\frac{\partial}{\partial x^2} + x^3\frac{\partial}{\partial x^3}, \\ \Xi_2 &= t^2\frac{\partial}{\partial t} + tx^1\frac{\partial}{\partial x^1} + tx^2\frac{\partial}{\partial x^2} + tx^3\frac{\partial}{\partial x^3} \end{aligned} \tag{8.22}$$

are symmetries of E.

If E is variational and invariant with respect to the Galilei group then by Example 8.5.6,

$$E = m\delta_{ij}\ddot{x}^j \, dx^j \wedge dt.$$

Hence, by Example 8.5.7, the vector fields (8.22) are symmetries of E.

Conversely, suppose that E is invariant with respect to the group G (8.21). Then E is of the form (8.18) (since it is invariant with respect to the Galilei group), and (8.22) are symmetries of E, i.e.,

$$\partial_{J^2\Xi_1} E = 0, \quad \partial_{J^2\Xi_2} E = 0. \tag{8.23}$$

Explicitly, this means that

$$f\ddot{x}^k\delta^i_k + 2f\ddot{x}^i - 3\left(\frac{\partial f}{\partial \ddot{x}^k}\ddot{x}^i + f\delta^i_k\right)\ddot{x}^k = 0 \tag{8.24}$$

for Ξ_1, and

$$tf\ddot{x}^k\delta^i_k + 2tf\ddot{x}^i - 3t\left(\frac{\partial f}{\partial \ddot{x}^k}\ddot{x}^i + f\delta^i_k\right)\ddot{x}^k = 0$$

for Ξ_2. Consequently, the Noether–Bessel-Hagen equations (8.23) are not independent, since $\partial_{J^2\Xi_2} E = t\partial_{J^2\Xi_1} E$. Now, we have to investigate the equations (8.24), which, however, simplify to

$$\frac{\partial f}{\partial \ddot{x}^k}\ddot{x}^i\ddot{x}^k = 0.$$

Since $f = f(a^2)$, the above equation implies that

$$\frac{df}{da^2}a^2 = 0,$$

i.e., $f(a^2) = \text{const}$ (denote $f(a^2) = m$). Thus, every second-order dynamical form E invariant with respect to G is of the form $E_i = m\ddot{x}^i$, and apparently satisfies the Helmholtz conditions.

8.5.9. Example. Finally, let us find all locally variational forms on $J^2(R \times R^3)$, invariant with respect to the Euclidean group.

Denote

$$E = E_i(t, x^k, \dot{x}^k, \ddot{x}^k)\, dx^i \wedge dt$$

(summation over $i = 1, 2, 3$). Substituting the generators of the Euclidean group (8.10), (8.11) into the Noether–Bessel-Hagen equation (8.2) we get that the E_i do not depend on t and x^k, $k = 1, 2, 3$, and they satisfy

$$\varepsilon_{ijk}\dot{x}^j \frac{\partial E_l}{\partial \dot{x}^k} + \varepsilon_{ijk}\ddot{x}^j \frac{\partial E_l}{\partial \ddot{x}^k} + \sum_p \varepsilon_{ilp} E_p = 0. \tag{8.25}$$

Moreover, we require that the $E_i(\dot{x}^k, \ddot{x}^k)$ obey the Helmholtz conditions (1.8). Using Theorem 6.7.2 we immediately get that

$$E_i = A_i(\dot{x}^k) + B_{ij}(\dot{x}^k)\ddot{x}^j, \tag{8.26}$$

where

$$B_{ij} = B_{ji}, \quad \frac{\partial B_{ij}}{\partial \dot{x}^l} = \frac{\partial B_{il}}{\partial \dot{x}^j}, \tag{8.27}$$

and

$$A_i = \alpha_{ij}\dot{x}^j + \beta_i, \quad \alpha_{ij} = -\alpha_{ji}, \tag{8.28}$$

where α_{ij}, β_i are constants.

Substituting the E_i (8.26)–(8.28) into (8.25) we get the equations

$$\sum_p (\varepsilon_{ijp}\alpha_{lp} + \varepsilon_{ipl}\alpha_{pj}) = 0,$$

which lead to

$$\alpha_{ij} = 0, \quad i, j = 1, 2, 3,$$

and

$$\varepsilon_{ijk}\dot{x}^j \frac{\partial B_{lm}}{\partial \dot{x}^k} + \sum_p (\varepsilon_{imp} B_{lp} + \varepsilon_{ipl} B_{pm}) = 0.$$

Denote

$$\eta = v^2 = (\dot{x}^1)^2 + (\dot{x}^2)^2 + (\dot{x}^3)^2.$$

Then the latter equations together with the first set of (8.27) (symmetry of the B_{ij}) give us after straightforward computations

$$B_{ij} = h(\eta)\delta_{ik}\delta_{jl}\dot{x}^k\dot{x}^l + \delta_{ij} f(\eta),$$

where f, h are arbitrary functions of v^2. Finally, applying the second set of the identities (8.27) we get

$$h(\eta) = 2\frac{df}{d\eta}.$$

Summarizing the results, we have obtained that all the Euler-Lagrange expressions invariant with respect to the Euclidean group are of the form

$$E_i = \sum_j \Big(2\frac{df}{d\eta}\dot{x}_i\dot{x}_j + \delta_{ij} f(\eta)\Big)\ddot{x}^j. \tag{8.29}$$

Hence, all Lagrangean systems invariant with respect to the Euclidean group are by (4.7) the following

$$\alpha = \sum_{i,j} \Big(2\frac{df}{dv^2}\dot{x}_i\dot{x}_j + \delta_{ij} f(v^2)\Big)(dx^i - \dot{x}^i dt) \wedge d\dot{x}^j.$$

Notice that in the same way the invariace with respect to the *Poincaré group* can be treated. Namely, taking the 4-dimensional Minkowski space-time with the canonical coordinates (x^i), $i = 0, 1, 2, 3$, as the configuration manifold, we can consider the fibered manifold $\pi : R \times R^4 \to R$, where the base R has the meaning of the space of parameters for the curves in R^4. Denote the canonical coordinate on the base by τ, and normalize the speed of light c to 1. The *Poincaré group* on R^4 is the 10-parametric transformation group, generated by the vector fields

$$\frac{\partial}{\partial x^i}, \qquad i = 0, 1, 2, 3$$

for the space-time translations, and

$$\begin{gathered} x^3\frac{\partial}{\partial x^2} - x^2\frac{\partial}{\partial x^3}, \qquad -x^3\frac{\partial}{\partial x^1} + x^1\frac{\partial}{\partial x^3}, \qquad x^2\frac{\partial}{\partial x^1} - x^1\frac{\partial}{\partial x^2}, \\ x^3\frac{\partial}{\partial x^0} + x^0\frac{\partial}{\partial x^3}, \qquad x^0\frac{\partial}{\partial x^1} + x^1\frac{\partial}{\partial x^0}, \qquad x^2\frac{\partial}{\partial x^0} + x^0\frac{\partial}{\partial x^2}, \end{gathered}$$

for the space-time rotations. The problem now is to find all 2nd order variational forms invariant with respect to the Poincaré group, and independent on the parameter τ. Obviously, by the same arguments and computations as in the case of the Euclidean invariance we get the E_i, $i = 0, 1, 2, 3$ of the form (8.29), where, of course, $\eta = -(\dot{x}^0)^2 + (\dot{x}^1)^2 + (\dot{x}^2)^2 + (\dot{x}^3)^2$. Inside the light cone we then obtain for E_i, $i = 1, 2, 3$, the familiar expression with $f = m(1 - w)^{-1/2}$, where $w = (\dot{x}^1)^2 + (\dot{x}^2)^2 + (\dot{x}^3)^2$.

8.5.10. Remark. In theoretical physics, one often needs to know all *Cartan forms* θ_λ, resp. all *Lepagean two-forms* α) possessing prescribed point symmetries. Conversely, there is a need to find all point symmetries (infinitesimal invariant transformations) of a given Cartan form, resp. of a given Lepagean two-form α. These problems can be solved directly, using the equations

$$\partial_{J^r\xi}\theta_\lambda = 0, \qquad \text{resp.} \quad \partial_{J^r\xi}\alpha = 0, \tag{8.30}$$

which have to be considered as equations for the form θ_λ (resp. α) in the former case, and as equations for ξ in the latter case.

However, having in mind Theorem 8.3.3, item (1), resp. (2), saying that in the case of *mechanics* the point symmetreis of θ_λ are the same as those of the Lagrangian λ, resp. the point symmetries of α are the same as those of the Euler-Lagrange form $E = p_1\alpha$, one can equivalently make use of the Noether equation (8.1), resp. of the Noether–Bessel-Hagen equation (8.2) combined with the Helmholtz conditions, as described in Sec. 8.5.

Examples using directly the second equation of (8.30) can be found in D. R. Grigore [1,2]; we note that it is used there to obtain all (first-oder) Lagrangean systems invariant with respect to the Euclidean group—this procedure is equivalent to that one applied in Example 8.5.9.

8.6. Dynamical symmetries

Among general dynamical symmetries of a Lagrangean system α of order $s - 1$, a significant role is played by symmetries of the *Lepagean two-form* α, symmetries of *Lepagean one-forms of order* $s - 1$ (i.e., symmetries of Lepagean equivalents of

Lagrangians of order $\leqslant s$), and by symmetries of the *characteristic distribution* $\mathcal{D}$ of α. Since a Lagrangean system on $J^{s-1}Y$ is a particular case of a closed two-form, we have for these kinds of dynamical symmetries all the results on symmetries and first integrals of closed two-forms presented in Chapter 2. In particular, as concerns Proposition 2.3.6, notice that for every nontrivial Lagrangean system (i.e., such that $\alpha \neq 0$) possessing at least one semispray belonging to the characteristic distribution $\mathcal{D}$, the requirement $\rho \notin \mathcal{D}$ of this proposition is satisfied for every ρ such that $d\rho = \alpha$; this follows easily from the fact that ρ is the Lepagean equivalent of a Lagrangian of order $\leqslant s$.

For general dynamical symmetries we have not an analog of Theorem 8.3.3; we only have the following relations:

$$\xi \text{ is a dynamical symmetry of } \theta_\lambda$$
$$\Downarrow$$
$$\xi \text{ is a dynamical symmetry of } \alpha_{E_\lambda}$$
$$\Downarrow$$
$$\xi \text{ is a dynamical symmetry of } \mathcal{D}\,,$$

and all these symmetries provide us with first integrals of the characteristic distribution.

In the case that the characteristic distribution $\mathcal{D}$ and the Euler-Lagrange distribution Δ are different (Lagrangean systems with dynamical constraints), one might be interested in first integrals of Δ. From the definition of the Euler-Lagrange distribution we immediately get that every symmetry ξ of α which is π_{s-1}-*vertical* gives rise to a first integral f of Δ, defined by $i_\xi\alpha = df$. Similarly, every π_{s-1}-*vertical* symmetry ξ of a Lepagean one-form θ of order $s-1$ gives rise to a first integral $i_\xi\theta$ of Δ.

If the Lagrangean system α is *semiregular* we can apply Theorem 2.2.3 to study symmetries of the characteristic distribution $\mathcal{D}$ (recall that the Euler-Lagrange distribution Δ in this case coincides with $\mathcal{D}$). We clearly have the following assertion:

8.6.1. Proposition. *Let $\zeta_1, \dots, \zeta_k$ be generators of the characteristic distribution $\mathcal{D}$. A vector field ξ on $J^{s-1}Y$ is a symmetry of $\mathcal{D}$ if and only if for every $i = 1, \dots, k$,*

$$[\xi, \zeta_i] = f_i^j \zeta_j \tag{8.31}$$

for some functions f_i^j, $1 \leqslant i, j \leqslant k$.

In particular, if the Lagrangean system α is regular, $\mathcal{D} = \operatorname{span}\{\zeta\}$, we get that a vector field ξ on $J^{s-1}Y$ is a symmetry of $\mathcal{D}$ if and only if

$$[\xi, \zeta] = f\zeta. \tag{8.32}$$

for a function f.

8.6.2. Remark. In the theory of symmetries of Lagrangean systems different authors often use different terminology. For example, assertions on symmetries and first integrals involving the forms θ_λ and $d\theta_\lambda$ are called Noether theorems, or generalized Noether theorems ; a vector field on $J^{2r-1}Y$ which is a symmetry of the Poincaré-Cartan form θ_λ is called a *Cartan symmetry*, and a vector field ξ on the configuration space Y such that $J^{2r-1}\xi$ is a Cartan symmetry is called *Noether symmetry* ; dynamical symmetries of a Lagrangean system, which are symmetries of the two-form α (resp. $d\theta_\lambda$), are called *Noetherian symmetries*; authors studying symmetries of (in the standard sense) regular first order Lagrangean systems often consider symmetries ξ satisfying the condition (8.32) and call them *generalized dynamical symmetries of* ζ), etc.

The meaning of the theory of symmetries for differential equations consists in the possibility to search for solutions with the help of first integrals. For singular Lagrangean systems the theory of symmetries can be applied in a way similar to the case of regular Lagrangean systems. One only has to compute the dynamical picture using the constraint algorithm (Chapter 7) and then to restrict to a submanifold of the phase space along which the dynamics is described by a system of (at least C^1)-*vector fields*. Then constants of the motion can be used to find explicit solutions of the Euler-Lagrange equations. Another possibility (applicable for semiregular systems) is to consider symmetries of the *characteristic distribution*. In integration methods of this kind we shall be interested in the next chapter.

Chapter 9.

GEOMETRIC INTEGRATION METHODS

9.1. Introduction

Variational equations in higher-order mechanics can be interpreted as equations for integral sections of the Euler-Lagrange distribution, or of the characteristic distribution of the corresponding Lepagean 2-form α. If the locally variational form (the Lagrangean system) is *semiregular* then the Euler-Lagrange distribution coincides with the characteristic distribution, and it is of a constant rank and *completely integrable*. This means that by Frobenius Theorem there is ensured the existence of local adapted charts to the corresponding foliation, i.e., charts consisting from *complete sets of independent first integrals* of the Euler-Lagrange distribution. Clearly, to solve the integration problem explicitly, one has to find on the phase space an atlas consisting of adapted charts to the foliation defined by the given Lagrangean system. The aim of this chapter is to explain some *methods* how to find such adapted charts, i.e., how to search for complete sets of independent first integrals. These integration methods, often referred to as *exact integration methods* for variational equations, represent a *generalization to semiregular and higher-order Lagrangean systems* of the classical Liouville integration method based on the *Liouville Theorem*, of the Hamilton-Jacobi integration methods based either on the *Jacobi Theorem* or on the *Van Hove Theorem*, and of the method of *canonical transformations*. The exposition in this chapter follows O. Krupková [3,4,7].

A large part of this section is devoted to various generalizations of the classical Hamilton-Jacobi theory. At the level of classical mechanics, the Hamilton-Jacobi equation

$$\frac{\partial S}{\partial t} + H\left(t, q^\sigma, \frac{\partial S}{\partial q^\sigma}\right) = 0$$

can be understood as an equation for a *generating function of a canonical transformation*, or for a *complete integral* of the Hamilton equations, or for *fields of extremals*. While in case of (in the standard sense) regular Lagrangians the above points seem to be only more or less different aspects of a unique theory, when geometrized and generalized to non-classical Lagrangians, they turn to become essentially *different integration methods* for the Euler-Lagrange equations, differing even in the range of applications. This feature becomes even more apparent if the Hamilton-Jacobi integration method is generalized to systems which are *not variational* (to this point, the interested reader can consult e.g. the papers O. Krupková and A. Vondra [1], and A. Vondra [3], where a geometric Hamilton-Jacobi theory for semispray distributions on fibered manifolds $\pi : Y \to X$, $\dim X = n$, is proposed, or O. Krupková [6], where an extension of the Hamilton-Jacobi theory to (arbitrary) completely integrable distributions on (arbitrary) manifolds is discussed).

In this chapter we shall work both with general closed two-forms and with Lepagean two-forms (= Lagrangean systems). To make the exposition more transparent, we shall use the following notation:

α a general closed two-form,

α_E a Lepagean two-form corresponding to a locally variational form E.

Also, if $\mathcal{D}$ is a distribution, the corresponding codistribution (annihilator of $\mathcal{D}$) will be denoted by $\mathcal{D}^0$. Recall from Chapter 2 that if $\mathcal{D}^0 = \operatorname{span}\{\eta_\iota\}$, we write $\mathcal{D} \approx \operatorname{span}\{\eta_\iota\}$.

9.2. The Liouville integration method

This section brings a generalization of the Liouville theorem of the classical calculus of variations to an arbitrary closed two-form of a constant rank.

A generalization of the classical Liouville theorem to Lagrangians for which $d\theta_\lambda$ has a constant rank, based on the Darboux theorem, is due to L. Klapka [1]. Here we follow a geometrical approach proposed by O. Krupková [4], [6], and based on the understanding of complete integrals as distributions (cf. S. Benenti [2]), and of the dynamics as a characteristic distribution of a closed two-form. Such an approach enables one to generalize the Liouville theorem to the family of (all) closed two forms of a constant rank, and, in particular to semiregular Lepagean two-forms.

Consider a manifold M of dimension n, and a *closed two-form α of a constant rank* on M. Denote by $\mathcal{D}$ the characteristic distribution of α. The rank of α is even; we set

$$\operatorname{rank} \alpha = 2p, \quad \operatorname{rank} \mathcal{D} = r, \tag{9.1}$$

i.e., $2p + r = n = \dim M$. Recall that $\mathcal{D}$ is completely integrable. By Darboux theorem, at each point of M there exists a *Darboux chart* related to α, i.e., a chart (U, χ), $\chi = (a^K, b_K, y^I)$, $1 \leqslant K \leqslant p$, $1 \leqslant I \leqslant \operatorname{rank} \mathcal{D}$, such that

$$\alpha = da^K \wedge db_K \tag{9.2}$$

on U. A Darboux chart is clearly an adapted chart to the distribution $\mathcal{D}$, since $\mathcal{D} = \operatorname{span}\{\partial/\partial y^I\}$; in other words, the functions a^K, b_K, $1 \leqslant K \leqslant p$, are *independent first integrals* of $\mathcal{D}$.

Now, the problem of finding solutions of the characteristic distribution $\mathcal{D}$ can be considered as a problem of finding a covering of M by Darboux charts. The classical idea how to proceed is the following: use the symmetries of α to find exactly p independent first integrals, and then *compute* the remaining ones "*by quadratures*". However, not any family of p independent first integrals will be suitable for computing the other p first integrals; therefore we shall be interested not only in the procedure alone, but also in getting a geometric characterization of families of first integrals which are appropriate in this sense (cf. the classical concept of "first integrals in involution"). The considerations will be based on *relations between first integrals and symmetries* of closed two-forms.

We have seen in Chapter 2 that if ξ is a symmetry of α and ζ belongs to the characteristic distribution $\mathcal{D}$ then $[\xi, \zeta]$ belongs to $\mathcal{D}$. This means that knowing an appropriate set of symmetries $\xi_1, \xi_2, \ldots, \xi_k$ (where $k \geqslant 1$) of α, we can get a completely integrable distribution $\mathfrak{I} = \text{span}\{\xi_K, \zeta_L, 1 \leqslant K \leqslant k, 1 \leqslant L \leqslant r\}$, where $\zeta_1, \dots, \zeta_r$ are (local) vector fields spanning $\mathcal{D}$. Evidently, $\mathcal{D}$ is a subdistribution of $\mathfrak{I}$. On the other hand, the symmetries $\xi_1, \dots, \xi_k$ give rise to first integrals $a^1, \dots, a^k$ of $\mathcal{D}$. There is a very important case when the symmetries $\xi_1, \dots \xi_k$ are linearly independent, and $k = p$. We shall show that if moreover the ξ's are tangent to the leaves $a^i = c^i = \text{const}$, the distribution $\mathfrak{I}$ can be obtained directly from the two-form α (without searching for vector generators of $\mathcal{D}$), and the distribution $\mathcal{D}$ can be easily solved—the "remaining" p independent first integrals can be computed "by quadratures", i.e., by means of differentiation and integration from α.

9.2.1. Definition. Let α be a closed two-form of a constant rank $= 2p$ on M. Let $\mathfrak{I}$ be a distribution defined on an open subset U of M. $\mathfrak{I}$ will be called a *complete integral* of α on U if

(1) $\operatorname{corank} \mathfrak{I} = p$ on U and $\mathfrak{I}$ is completely integrable,
(2) α belongs to the differential ideal generated by $\mathfrak{I}$.

Note that the condition (2) in the definition of a complete integral means that $\alpha = 0$ on the maximal integral manifolds of $\mathfrak{I}$.

The *existence* of local complete integrals is ensured by the Darboux theorem. Namely, if (U, χ), $\chi = (a^K, b_K, y^I)$, $1 \leqslant K \leqslant p$, $1 \leqslant I \leqslant r$, is a Darboux chart of α, then putting

$$\mathfrak{I} \approx \text{span}\{da^K, 1 \leqslant K \leqslant p\} \tag{9.3}$$

we get a complete integral of α defined on U.

The geometrical meaning of complete integrals is described by the following

9.2.2. Proposition. *Let $\mathfrak{I}$ be a complete integral of α, defined on an open set U. Then on U, the characteristic distribution $\mathcal{D}$ is a subdistribution of $\mathfrak{I}$.*

Proof. Since $\mathfrak{I}$ is completely integrable, at each point of U there exists a chart (W, χ), $\chi = (x^J, a^K)$, $1 \leqslant J \leqslant n - p$, $1 \leqslant K \leqslant p$, adapted to $\mathfrak{I}$, i.e., such that $\mathfrak{I} \approx \text{span}\{da^K, 1 \leqslant K \leqslant p\}$. In this chart

$$\alpha = da^K \wedge (g_{KJ}\, dx^J + h_{KL}\, da^L), \tag{9.4}$$

where we can suppose $h_{KL} = -h_{LK}$. The distribution $\mathcal{D}$ is locally generated by means of the 1-forms $i_{\partial/\partial x^J} \alpha$, $i_{\partial/\partial a^K} \alpha$, i.e.,

$$g_{KJ}\, da^K, \quad g_{KJ}\, dx^J + 2h_{KL}\, da^L, \quad 1 \leqslant J \leqslant n - p, \quad 1 \leqslant K \leqslant p. \tag{9.5}$$

Since $\operatorname{rank} \alpha = 2p$, the rank of the matrix of the generators (9.5) is $2p$ which means that the matrix (g_{KJ}) has the maximal rank (equal to p). Hence,

$$\mathcal{D} \approx \text{span}\{da^K, g_{KJ}\, dx^J + 2h_{KL}\, da^L, 1 \leqslant K \leqslant p\}.$$

This completes the proof. □

9.2.3. Corollary. *Let $\mathfrak{I}$ be a complete integral of α. Then every leaf of the corresponding foliation is foliated by the leaves of the characteristic distribution $\mathcal{D}$.*

9.2.4. Corollary. *Let $\mathcal{J}$ be a complete integral of α. Let (W, χ), $\chi = (x^J, a^K)$, where $1 \leqslant J \leqslant n-p$, $1 \leqslant K \leqslant p$, be an adapted chart to $\mathcal{J}$, i.e., such that $\mathcal{J}^0 = \operatorname{span}\{da^K,\ 1 \leqslant K \leqslant p\}$. Then the functions a^K, $1 \leqslant K \leqslant p$, are independent first integrals of the characteristic distribution $\mathcal{D}$.*

We shall see below that complete integrals will be used to find explicit solutions of the characteristic distribution. Therefore we shall now be interested in characterizing those systems of first integrals of $\mathcal{D}$ which define complete integrals of α.

9.2.5. Definition. Let α be a closed two-form of constant rank $2p$ on M. Let a^K, $1 \leqslant K \leqslant p$, be independent first integrals of the characteristic distribution $\mathcal{D}$. We say that the integrals a^K are *in involution* if $\mathcal{J} \approx \operatorname{span}\{da^K,\ 1 \leqslant K \leqslant p\}$ is a complete integral of α.

The following theorem answers the question under what conditions a family of first integrals of $\mathcal{D}$ (resp. a family of symmetries of α) defines a complete integral.

9.2.6. Theorem. *Let a^K, $1 \leqslant K \leqslant p$, be independent first integrals of $\mathcal{D}$, and let ξ_K, $1 \leqslant K \leqslant p$, be some corresponding symmetries of α, i.e., $i_{\xi_K}\alpha = da^K$, for all K. (Alternately, let ξ_K, $1 \leqslant K \leqslant p$, be nontrivial symmetries of α, i.e., such that $i_{\xi_K}\alpha \neq 0$ for all K, linearly independent at each point of their domain of definition, and let a^K, $1 \leqslant K \leqslant p$, be some corresponding first integrals of $\mathcal{D}$.) Then the following conditions are equivalent:*

(1) *The distribution $\mathcal{J} = \operatorname{span}\{i_{\xi_K}\alpha,\ 1 \leqslant K \leqslant p\} \approx \operatorname{span}\{da^K,\ 1 \leqslant K \leqslant p\}$ is a complete integral of α.*
(2) $i_{\xi_K} i_{\xi_L}\alpha = i_{\xi_K}\, da^L = 0$ *for all* $1 \leqslant K,\ L \leqslant p$.
(3) *The codistribution $\operatorname{span}\{da^K,\ 1 \leqslant K \leqslant p\}$ is the annihilator of the distribution $\operatorname{span}\{\xi_K,\ \zeta_L,\ 1 \leqslant K \leqslant p,\ 1 \leqslant L \leqslant r\}$, where the vector fields $\zeta_1, \dots, \zeta_r$ span $\mathcal{D}$.*

Proof. Suppose (1). Let (x^J, a^K) be a chart adapted to $\mathcal{J}$. In this chart,

$$\alpha = da^K \wedge (g_{KJ}\, dx^J + h_{KL}\, da^L), \quad h_{KL} = -h_{LK}.$$

From the condition $i_{\xi_K}\alpha = da^K$, $1 \leqslant K \leqslant p$, and using that the matrix (g_{KJ}) has maximal rank p, we get that the vector fields ξ_K, $1 \leqslant K \leqslant p$, are of the form

$$\xi_K = \xi_K^J \frac{\partial}{\partial x^J}, \quad \xi_K^J g_{LJ} = \delta_{KL}.$$

Now, $i_{\xi_K} i_{\xi_L}\alpha = i_{\xi_K} da^L = 0$ for all $1 \leqslant K,\ L \leqslant p$, proving (2).

Suppose (2) and consider the codistribution $\mathcal{J}^0 = \operatorname{span}\{da^K,\ 1 \leqslant K \leqslant p\}$. Then (2) means that for all K, the ξ_K belong to $\mathcal{J}$, and by assumption $\xi_K \notin \mathcal{D}$. Since the integrals a^K are independent, and the condition (2) holds, the vector fields $\xi_1, \dots, \xi_p$ are linearly independent. For every ζ belonging to $\mathcal{D}$ we have $i_\zeta da^K = 0$ for all K, i.e., $\mathcal{D}$ is a subdistribution of $\mathcal{J}$. Hence, if $\mathcal{D} = \operatorname{span}\{\zeta_L,\ 1 \leqslant L \leqslant r\}$ we have $\mathcal{J} = \operatorname{span}\{\xi_K,\ \zeta_L,\ 1 \leqslant K \leqslant p,\ 1 \leqslant L \leqslant r\}$.

Suppose (3). We shall show that the distribution $\mathcal{J} \approx \operatorname{span}\{da^K,\ 1 \leqslant K \leqslant p\} = \operatorname{span}\{\xi_K,\ \zeta_L,\ 1 \leqslant K \leqslant p,\ 1 \leqslant L \leqslant r\}$ is a complete integral of α; clearly, it remains to show that $\alpha = 0$ on the leaves of $\mathcal{J}$. By assumption, $\mathcal{J}$ and $\mathcal{D}$ are completely integrable.

Let x be a point in the domain of definition of the functions a^K. We can consider a chart (x^J, a^K), $1 \leqslant J \leqslant p+r$, $1 \leqslant K \leqslant p$, at x, which is adapted to the distributions $\mathcal{J}$ and $\mathcal{D}$, i.e., such that

$$\mathcal{D} = \operatorname{span}\left\{\frac{\partial}{\partial x^{p+1}}, \ldots, \frac{\partial}{\partial x^{p+r}}\right\},$$

and

$$\xi_K = \xi_K^J \frac{\partial}{\partial x^J}, \quad 1 \leqslant K \leqslant p,$$

where the square matrix (ξ_K^L), $1 \leqslant K, L \leqslant p$, is regular. In this chart,

$$\alpha = f_{IJ}\, dx^I \wedge dx^J + g_{IK}\, dx^I \wedge da^K + h_{KM}\, da^K \wedge da^M$$

where the functions f_{IJ} and h_{KM} are supposed antisymmetric. From the condition $i_\zeta \alpha = 0$ for all $\zeta \in \mathcal{D}$ we get $f_{IJ} = 0$, $g_{IK} = 0$ for all $I \geqslant p+1$, and from $i_{\xi_K}\alpha = da^K$ for $1 \leqslant K \leqslant p$ we get

$$f_{IL}\,\xi_K^I = 0, \quad 1 \leqslant L \leqslant p, \quad g_{IL}\,\xi_K^I = \delta_{KL}.$$

Since (ξ_K^I) is regular, the equation $f_{IL}\,\xi_K^I = 0$ gives $f_{IL} = 0$, $1 \leqslant I, L \leqslant p$. Consequently,

$$\alpha = g_{KM}\, dx^K \wedge da^M + h_{KM}\, da^K \wedge da^M,$$

proving our assertion. □

Definition 9.2.5 (as well as the condition (3) of Theorem 9.2.6) expresses the geometric content of the concept of "a system of first integrals in involution", which in the classical and higher-order mechanics is defined by the condition (2) of the above theorem.

Notice that the condition (2) of Theorem 9.2.6 can be rewritten in the form

$$\{f_K, f_L\} = 0,$$

where f_K, f_L are first integrals corresponding to the symmetries ξ_K, ξ_L, and $\{.,.\}$ denotes the Poisson bracket (cf. Section 2.3).

The following theorem is a generalization of the *Liouville theorem* of classical mechanics to closed two-forms of a constant rank, and it is of fundamental importance for integration of distributions:

9.2.7. Theorem. *Let α be a closed two-form of constant rank $2p$ on M, let $\mathcal{J}$ be a complete integral of α. Then at each point of* dom $\mathcal{J}$ *there exists a chart (U, ϕ), $\phi = (a^K, b_K, y^S)$, $1 \leqslant K \leqslant p$, $1 \leqslant S \leqslant \dim M - 2p$ such that*

(1) $\mathcal{J}^0 = \operatorname{span}\{da^K,\ 1 \leqslant K \leqslant p\}$,

(2) $\alpha = da^K \wedge db_K$,

(3) *the set of functions $\{a^K, b_K,\ 1 \leqslant K \leqslant p\}$ is a complete set of independent first integrals of the characteristic distribution $\mathcal{D}$ of α.*

Note that the condition (1) says that (U, ϕ) is an adapted chart to $\mathcal{J}$ and (2) means that it is a Darboux chart of α.

Proof. Let $x \in M$ and (x^J, a^K) be adapted coordinates to $\mathcal{J}$ in a neighborhood of x. Let ρ be such that $\alpha = d\rho$ in a neighborhood of x; in the adapted coordinates,

$$\rho = \rho_J\, dx^J + \hat{\rho}_K\, da^K. \tag{9.6}$$

Since by assumption $d\rho = 0$ on the leaves of $\mathcal{J}$ we get

$$\frac{\partial \rho_J}{\partial x^I} - \frac{\partial \rho_I}{\partial x^J} = 0,$$

which means that locally there is a function $f(x^J, a^K)$ such that

$$\rho_J = \frac{\partial f}{\partial x^J}. \tag{9.7}$$

Now, we have

$$\rho = \frac{\partial f}{\partial x^J} dx^J + \hat{\rho}_K \, da^K = df + \left(\hat{\rho}_K - \frac{\partial f}{\partial a^K}\right) da^K. \tag{9.8}$$

Put

$$b_K = \hat{\rho}_K - \frac{\partial f}{\partial a^K}. \tag{9.9}$$

Since $\operatorname{rank} d\rho = 2p$, we can see that the rank of the matrix $(\partial b_K / \partial x^J)$ is maximal and equals p. This means that the one-forms da^K, db_K, $1 \leqslant K \leqslant p$, are linearly independent, i.e., the functions a^K, b_K can be, in a neighborhood of x, completed to a chart (a^K, b_K, y^S). This chart is obviously a Darboux chart of α, proving that the system $\{a^K, b_K, 1 \leqslant K \leqslant p\}$ is a complete set of independent first integrals of $\mathcal{D}$. □

Using Theorem 9.2.6 we have the following obvious reformulation of the above theorem:

9.2.8. Generalized Liouville Theorem. *Let α be a closed two-form of constant rank $2p$ on M. Let a^K, $1 \leqslant K \leqslant p$, be independent first integrals in involution of the characteristic distribution $\mathcal{D}$ of α, defined on an open subset U of M. Then the system of functions $\{a^K, b_K, 1 \leqslant K \leqslant p\}$, where b_K are defined by* (9.9), *is a complete set of independent first integrals of $\mathcal{D}$.*

Notice that the explicit formulas for the functions ρ_J, $\hat{\rho}_K$ and f can be easily found using the Poincaré Lemma. This means that the first integrals b_K can be found explicitly by means of integration and differentiation (i.e., "by quadratures"). To a given closed two-form

$$\alpha = g_{JK} \, dx^J \wedge da^K + h_{LK} \, da^L \wedge da^K$$

we find a suitable one-form ρ by the standard formulas

$$\begin{aligned} \rho_J &= -a^L \int_0^1 (g_{JL} \circ \psi) \, u \, du \\ \hat{\rho}_K &= x^J \int_0^1 (g_{JK} \circ \psi) \, u \, du + 2a^L \int_0^1 (h_{LK} \circ \psi) \, u \, du, \end{aligned} \tag{9.10}$$

where ψ is defined by $\psi(u, x^J, a^K) = (ux^J, ua^K)$. Knowing ρ, we obtain a suitable f easily by putting

$$f = x^J \int_0^1 (\rho_J \circ \mu) \, dv + h(a^K) \tag{9.11}$$

where

$$\mu(v, x^J, a^K) = (vx^J, a^K),$$

and $h(a^K)$ is an arbitrary (fixed) function (note that h is a first integral of the distribution $\mathcal{J}$). Another direct consequence of Theorem 9.2.7 is the following "*Coordinate free Liouville Theorem*":

9.2.9. Corollary. *Let $\mathcal{J}_1$ be a complete integral of a closed two-form α of a constant rank on M. Then to each point x of* dom $\mathcal{J}_1$ *there exists a neighborhood U and a complete integral $\mathcal{J}_2$ of α on U such that $\mathcal{J}_1 \cap \mathcal{J}_2$ is the characteristic distribution $\mathcal{D}$ of α on U.*

Theorem 9.2.7 can be reformulated also in the following way which in a sense generalizes the classical *Jacobi Theorem.*

9.2.10. Corollary. *Let $\mathcal{J}$ be a complete integral of a closed two-form α of a constant rank on M. Then to each point x of* dom $\mathcal{J}$ *there exists a neighborhood U and a one-form $\bar{\rho}$ on U such that $\alpha = d\bar{\rho}$ and $\bar{\rho}$ belongs to $\mathcal{J}^0$ (i.e., $\bar{\rho}$ vanishes on the maximal integral manifolds of $\mathcal{J}$).*

By the above corollary, given a (local) one-form ρ such that $\alpha = d\rho$, there *exists* a function f such that

$$\rho - df \in \mathcal{J}^0. \tag{9.12}$$

In a chart adapted to $\mathcal{J}$ (such that $\mathcal{J}^0 = \text{span}\{da^K,\ 1 \leqslant K \leqslant p\}$), the condition (9.12) reads

$$\bar{\rho} = \rho - df = b_K\, da^K, \tag{9.13}$$

The meaning of the form $\bar{\rho}$ for the solution of the integration problem is now clear: we have

$$b_K = i_{\partial/\partial a^K}\, \bar{\rho},$$

which obviously are independent first integrals of $\mathcal{D}$. In this sense we can speak of (9.12) or (9.13) as of a *generalized Hamilton-Jacobi equation.*

Equation (9.13) represents a generalization of two equations obtained by Z. Oziewicz and W. Gruhn [1], as a modification of Jacobi equation (cf. the "Noether splitting" and the "Jacobi splitting").

Let us now consider instead of a general manifold M, a prolongation $J^{s-1}Y$, $s \geqslant 1$, of a fibered manifold $\pi : Y \to X$. Let α be a closed two-form on $J^{s-1}Y$. This form need not represent a Lagrangean system. However, directly from the definition of a Lepagean two-form we get the following characterization of Lagrangean systems, which is a kind of *the inverse problem of the calculus of variations*:

9.2.11. Proposition. *Let α be a closed two-form of constant rank $= 2p$ on $J^{s-1}Y$, let (U, χ), $\chi = (a^K, b_K, y^S)$, $1 \leqslant K \leqslant p$, be a Darboux chart of α such that $U \subset V_{s-1}$, where (V, ψ), $\psi = (t, q^\sigma)$ is a fiber chart on Y.*

(1) *α is Lepagean if and only if*

$$\frac{da^K}{dt}\frac{\partial b_K}{\partial q^\sigma_j} - \frac{db_K}{dt}\frac{\partial a^K}{\partial q^\sigma_j} = 0, \quad 1 \leqslant j \leqslant s-1. \tag{9.14}$$

(2) *If α is Lepagean then the corresponding Euler-Lagrange expressions read*

$$E_\sigma = \frac{db_K}{dt}\frac{\partial a^K}{\partial q^\sigma} - \frac{da^K}{dt}\frac{\partial b_K}{\partial q^\sigma}, \quad 1 \leqslant \sigma \leqslant m, \tag{9.15}$$

and the Lagrangians are given by the formula

$$L = -b_K \frac{da^K}{dt} + \frac{df}{dt}, \tag{9.16}$$

where f is an arbitrary function.

(3) *If α is Lepagean and if the characteristic distribution of α is weakly horizontal, then*

$$\mathcal{D} = \Delta = \text{span}\left\{\frac{\partial}{\partial y^J}\,,\; 1 \leqslant J \leqslant ms + 1 - 2p\right\} \approx \text{span}\{da^K,\, db_K\}\,.$$

Note that the Lagrangian $L = -b_K\, da^K/dt$ (corresponding to the Lepagean one-form $\theta = b_K\, da^K$) *is not* a minimal-order Lagrangian; it is generally of order s.

The inverse problem of the calculus of variations in the formulation for closed two forms, namely, *under which conditions to a closed two-form of a constant rank there exists a Lagrangian*, has been studied in L. Klapka [1].

Now, let α_E be a *semiregular Lagrangean system* of order $s-1$, $s \geqslant 1$, denote rank $\alpha_E = 2p$. Recall that by Theorem 5.4.1 the Euler-Lagrange distribution Δ_E coincides with the characteristic distribution $\mathcal{D}_E$ of α_E, and is weakly horizontal, and by Cartan Theorem 2.3.2 it is completely integrable. This means that the Euler-Lagrange distribution Δ_E is a *flat generalized connection* of corank $2p$ on the phase space. Consequently, the above results concerning closed two-forms of a constant rank apply to semiregular higher-order Lagrangean systems, and can be used to find the foliation of the phase space related to the Euler-Lagrange distribution. In this case however, the following terminology will be used:

9.2.12. Definition. Let α_E be a semiregular Lagrangean system, let Δ_E be its Euler-Lagrange distribution.

A $(ms + 1 - p)$-dimensional (immersed) submanifold Q of the phase space will be called a *Lagrangean submanifold* of the Lagrangean system α_E if Q is foliated by the leaves of Δ_E.

If $\mathfrak{I}$ is a complete integral of α_E then the leaves of the foliation defined by $\mathfrak{I}$ are Lagrangean submanifolds; therefore, a complete integral of a Lagrangean system will be also called a *field of Lagrangean submanifolds* and the corresponding foliation of the phase space will be called a *Lagrangean foliation.*

By the definition, every leaf of the Lagrangian foliation is foliated by the leaves of the Euler-Lagrange distribution.

Notice that every complete integral of a semiregular Lagrangean system is a *flat generalized connection on the phase space.*

The above concepts of a Lagrangean submanifold and a field of Lagrangean submanifolds for semiregular Lagrangean systems have been introduced in O. Krupková [4]. Analogous objects are considered within the symplectic geometry (and refer to autonomous Lagrangians regular in the standard sense). For more details the reader can consult eg. S. Benenti [1,2], W. M. Tulczyjew [4], or P. Libermann and Ch.-M. Marle [1].

9.3. Jacobi complete integrals and the Hamilton-Jacobi integration method

By the previous section, given a complete integral of a closed two-form α, one is able to find Darboux charts, computing first integrals of α "by quadratures". In this section, we shall show that for *Lagrangean systems* on fibered manifolds there is a possibility to simplify the procedure described by the Liouville theorem. This possibility is closely related with the existence of the *canonical form* of a Lepagean two-form. As a result, we obtain a generalization of the classical *Jacobi Theorem*, and we shall see that this theorem is in fact a *particular case of the Liouville Theorem.*

We start with a theorem proving for closed two-forms of a constant rank on *fibered manifolds* the existence of complete integrals of a *special kind.*

Consider a fixed fibered manifold $\pi : Y \to X$ ($\dim X = 1$), let $s \geqslant 1$. Let c be defined as in 4.5.1.

9.3.1. Theorem. *Let α be a closed two-form of constant rank $2p$ on $J^{s-1}Y$. Let $x \in J^{s-1}Y$ be a point. There exists a complete integral $\mathfrak{I}$ of α defined in a neighborhood of x, and a chart (W, χ), $\chi = (t, q^\sigma, \dots, q^\sigma_{c-1}, a^1, \dots, a^p, u^1, \dots, u^{ms-mc-p})$ at x such that*

(1) $\mathfrak{I} \approx \mathrm{span}\{da^1, \dots, da^p\}$,

(2) $(\pi_{s-1,c-1}W, \psi_{c-1})$, *where* $\psi_{c-1} = (t, q^\sigma, \dots, q^\sigma_{c-1})$, *is a chart on* $J^{c-1}Y$, *associated with a fiber chart* (V, ψ), $\psi = (t, q^\sigma)$ *on* Y,

(3) *it holds*

$$\mathrm{rank}\Big(\frac{\partial a^K}{\partial q^\nu_j}\Big) \text{ is maximal,} \tag{9.17}$$

where in the above matrix, K labels rows, $1 \leqslant K \leqslant p$, and (ν, j) label columns, $1 \leqslant \nu \leqslant m$, $c \leqslant j \leqslant s-1$.

Proof. Let $\mathfrak{I}'$ be a complete integral of α defined in a neighborhood of x. Let (U, ϕ), $\phi = (a^K, b_K, y^J)$, $1 \leqslant K \leqslant p$, $1 \leqslant J \leqslant \dim J^{s-1}Y - 2p$, be a chart at x, satisfying the conditions of Theorem 9.2.7. Let (V, ψ), $\psi = (t, q^\sigma)$ be a fiber chart on Y such that $x \in V_{s-1} \cap U$, consider the overlap mapping $(t, q^\sigma, \dots, q^\sigma_{s-1}) \to (a^K, b_k, y^J)$. The matrix

$$\begin{pmatrix} \dfrac{\partial a^K}{\partial t} & \dfrac{\partial a^K}{\partial q^\sigma_i} & \dfrac{\partial a^K}{\partial q^\nu_j} \\ \dfrac{\partial b_K}{\partial t} & \dfrac{\partial b_K}{\partial q^\sigma_i} & \dfrac{\partial b_K}{\partial q^\nu_j} \end{pmatrix} \tag{9.18}$$

where K labels rows, $1 \leqslant K \leqslant p$, and (σ, i), (ν, j) label columns, $0 \leqslant i \leqslant c-1$, $c \leqslant j \leqslant s-1$, has maximal rank $2p$.

If the condition (9.17) is satisfied then the functions $t, q^\sigma, \dots, q^\sigma_{c-1}, a^K$, $1 \leqslant K \leqslant p$, can be completed to coordinates in a neighborhood of x, and da^K, $1 \leqslant K \leqslant p$, define a desired complete integral.

Now suppose that the condition (9.17) is not satisfied. Then it must hold

$$\mathrm{rank}\Big(\frac{\partial b_K}{\partial q^\nu_j}\Big) = \text{maximal} = p, \text{ where } 1 \leqslant \nu \leqslant m,\ c \leqslant j \leqslant s-1. \tag{9.19}$$

Put $\bar{a}^K = \delta^{KL} b_L$. Then the functions $\bar{a}^K$, $1 \leqslant K \leqslant p$, satisfy the condition (9.17), the distribution $\mathfrak{I} \approx \mathrm{span}\{d\bar{a}^K,\ 1 \leqslant K \leqslant p\}$ is a desired complete integral, and

$t, q^\sigma, \ldots, q^\sigma_{c-1}, \bar{a}^K$, $1 \leqslant K \leqslant p$, can be completed to coordinates in a neighborhood of x. This completes the proof. □

9.3.2. Definition. A complete integral $\mathfrak{I}$ of α and a chart (W, χ), $\chi = (t, q^\sigma_j, a^K, u^L)$, $1 \leqslant j \leqslant c-1$, $1 \leqslant K \leqslant p$, $1 \leqslant L \leqslant ms - mc - p$, satisfying the conditions of Theorem 9.3.1 will be called a *Jacobi complete integral* and *Jacobi chart* of α, respectively.

Jacobi complete integrals arc canonically associated with the prolongation structure arising on $J^{s-1}Y$. More precisely, their geometric meaning is the following:

9.3.3. Proposition. *Let α be a closed two-form of a constant rank on $J^{s-1}Y$. Every Jacobi complete integral of α is a $\pi_{s-1,s-c}$-horizontal distribution, hence a (flat) generalized connection on an open subset of $J^{s-1}Y$. It is a connection (i.e., a local section $J^{s-1}Y \to J^1\pi_{s-1,c-1}$ of the fibered manifold $\pi_{s-1,c-1} : J^{s-1}Y \to J^{c-1}Y$) if and only if s is even ($s = 2c$) and $p = mc$ (i.e., $r = \operatorname{rank} \mathfrak{D} = 1$).*

Proof. If (W, ψ), $\psi = (t, q^\sigma_i, a^K, u^L)$, $0 \leqslant i \leqslant c-1$, $1 \leqslant K \leqslant p$, $1 \leqslant L \leqslant ms - mc - p$, is a Jacobi chart and $\mathfrak{I} \approx \operatorname{span}\{da^K, 1 \leqslant K \leqslant p\}$ is the corresponding Jacobi complete integral then obviously

$$\mathfrak{I} = \operatorname{span}\left\{\frac{\partial}{\partial t}, \frac{\partial}{\partial q^\sigma_i}, \frac{\partial}{\partial u^L}, 0 \leqslant i \leqslant c-1, 1 \leqslant L \leqslant ms - mc - p\right\}, \tag{9.20}$$

i.e., $\mathfrak{I}$ is complementary to a subbundle of the $\pi_{s-1,s-c}$-vertical bundle (spanned by the $\pi_{s-1,s-c}$-vertical vector fields $\partial/\partial a^K$, $1 \leqslant K \leqslant p$).

If $\mathfrak{I}$ is a connection on $\pi_{s-1,s-c}$ then

$$\mathfrak{I} = \operatorname{span}\left\{\frac{\partial}{\partial t}, \frac{\partial}{\partial q^\sigma_i}, 0 \leqslant i \leqslant c-1\right\}. \tag{9.21}$$

Hence, $ms - mc - p = 0$. If $s = 2c$, we have $p = mc$ and $r = \dim J^{s-1}Y - 2p = 1 + 2mc - 2mc = 1$. If $s = 2c+1$, we get $p = mc + m = m(c+1)$; however, by assumption, $2p < \dim J^{s-1}Y = m(2c+1)+1$, leading to $m < 1$ which is a contradiction.

Conversely, if $s = 2c$ and $p = mc$ then every Jacobi chart is of the form (t, q^σ_i, a^K), $0 \leqslant i \leqslant c-1$, $1 \leqslant K \leqslant mc$.

This completes the proof. □

From now on, let α_E be a *semiregular Lagrangean system*.

By Proposition 9.3.3, Jacobi complete integrals of α_E are *connections* (with respect to the projection $\pi_{s-1,s-c}$) if and only if the Lagrangean system is *regular and odd-order* (i.e., the corresponding variational equations are regular and even order). In all other situations, Jacobi complete integrals are *flat generalized connections* on the phase space, complementary to a subbundle of the $\pi_{s-1,s-c}$-vertical bundle.

Considering a semiregular Lagrangean system, we can make use of the existence of local Lepagean equivalents of *minimal-order Lagrangians*. In this case, taking in the Liouville Theorem a *Jacobi* complete integral, the theorem takes the following simpler form which is a *generalization of the classical Jacobi Theorem*:

9.3.4. Jacobi Theorem for even-order variational equations. *Let α_E be a semiregular Lagrangean system of rank $2p$ on $J^{s-1}Y$, suppose that $s = 2c$. Let $\mathfrak{J}$ be a Jacobi complete integral of α_E, and let (W, χ), $\chi = (t, q_i^\sigma, a^K, u^L)$, $0 \leqslant i \leqslant c-1$, $1 \leqslant K \leqslant p$, $1 \leqslant L \leqslant mc - p$, be a related Jacobi chart. Then in a neighborhood of every point of W there exists a function S such that the family of functions $\{a^K, b_K\}$, $1 \leqslant K \leqslant p$, where b_K are defined by*

$$b_K = \frac{\partial S}{\partial a^K}, \quad 1 \leqslant K \leqslant p, \tag{9.22}$$

is a complete set of independent first integrals of the Euler-Lagrange distribution Δ_E of α_E.

For the b_K, and every Hamiltonian and the related momenta of α_E it holds

$$\frac{\partial b_K}{\partial u^M} = 0, \quad \frac{\partial H}{\partial u^M} = 0, \quad \frac{\partial p_\sigma^i}{\partial u^M} = 0, \tag{9.23}$$

where $1 \leqslant K \leqslant p$, $1 \leqslant M \leqslant mc - p$, $1 \leqslant \sigma \leqslant m$, $0 \leqslant i \leqslant c-1$.

The function S is a solution of the equations

$$H(t, q_j^\nu, a^K) = -\frac{\partial S}{\partial t}, \quad p_\sigma^i(t, q_j^\nu, a^K) = \frac{\partial S}{\partial q_i^\sigma}, \quad \frac{\partial S}{\partial u^L} = 0, \tag{9.24}$$

where $1 \leqslant \sigma \leqslant m$, $0 \leqslant i \leqslant c-1$, $1 \leqslant L \leqslant mc - p$, and $H, p_\sigma, \dots, p_\sigma^{c-1}$ are a Hamiltonian and the related momenta of α_E.

At each point of their domain of definition, the first integrals b_K, $1 \leqslant K \leqslant p$, satisfy the condition

$$\operatorname{rank}\left(\frac{\partial b_K}{\partial q_i^\sigma}\right) = maximal = p. \tag{9.25}$$

9.3.5. Jacobi Theorem for odd-order variational equations. *Let α_E be a semiregular Lagrangean system of rank $2p$ on $J^{s-1}Y$, suppose that $s = 2c+1$. Let $\mathfrak{J}$ be a Jacobi complete integral of α_E, and let (W, χ), $\chi = (t, q_i^\sigma, a^K, u^L)$, $0 \leqslant i \leqslant c-1$, $1 \leqslant K \leqslant p$, $1 \leqslant L \leqslant m(c+1) - p$, be a related Jacobi chart. Then in a neighborhood of every point of W there exists a function S such that the family of functions $\{a^K, b_K\}$, $1 \leqslant K \leqslant p$, where b_K are defined by*

$$b_K = \frac{\partial S}{\partial a^K} - p_\rho^c \frac{\partial q_c^\rho}{\partial a^K}, \quad 1 \leqslant K \leqslant p, \tag{9.26}$$

is a complete set of independent first integrals of the Euler-Lagrange distribution Δ_E of α_E.

The function S is a solution of the equations

$$\begin{aligned} (H - p_\rho^c \frac{\partial q_c^\rho}{\partial t})(t, q_j^\nu, a^K, u^M) &= -\frac{\partial S}{\partial t}, \\ (p_\sigma^i + p_\rho^c \frac{\partial q_c^\rho}{\partial q_i^\sigma})(t, q_j^\nu, a^K, u^M) &= \frac{\partial S}{\partial q_i^\sigma}, \\ \left(p_\rho^c \frac{\partial q_c^\rho}{\partial u^L}\right)(t, q_j^\nu, a^K, u^M) &= \frac{\partial S}{\partial u^L}, \end{aligned} \tag{9.27}$$

where $1 \leqslant \sigma \leqslant m$, $0 \leqslant i \leqslant c-1$, $1 \leqslant L \leqslant m(c+1) - p$, and the H, $p_\sigma, \dots, p_\sigma^{c-1}$, p_σ^c are a Hamiltonian and the related momenta of α_E.

At each point of their domain of definition, the first integrals b_K, $1 \leqslant K \leqslant p$, satisfy the condition

$$\operatorname{rank}\left(\frac{\partial b_K}{\partial q_i^\sigma}, \frac{\partial b_K}{\partial u^L}\right) = \textit{maximal} = p. \tag{9.28}$$

Proof of the Jacobi Theorems. The assertions follow from the Liouville theorem if we take for a local one-form ρ such that $\alpha_E = d\rho$ the Lepagean equivalent of a minimal-order Lagrangian.

(1) Let $s = 2c$. Expressing this Lepagean equivalent in the Jacobi chart we get

$$\rho = -H(t, q_j^\nu, a^K, u^M)\, dt + \sum_{i=0}^{c-1} p_\sigma^i(t, q_j^\nu, a^K, u^M)\, dq_i^\sigma = dS - b_K\, da^K \tag{9.29}$$

This means that (9.22) and (9.24) are satisfied. Since $\partial S/\partial u^M = 0$, we get from (9.24) and (9.22) that the relations (9.23) hold.

Finally, we shall prove (9.25). We have

$$\Delta_E \approx \operatorname{span}\{da^K,\ db^K,\ 1 \leqslant K \leqslant p\},$$

where (in the Jacobi chart)

$$db_K = \frac{\partial b_K}{\partial t}\, dt + \frac{\partial b_K}{\partial q_i^\sigma}\, dq_i^\sigma + \frac{\partial b_K}{\partial a^L}\, da^L.$$

Δ_E is equivalently spanned by the vector fields $\partial/\partial u^L$, $1 \leqslant L \leqslant mc - p$, and

$$\zeta = \zeta^0 \frac{\partial}{\partial t} + \zeta_j^\nu \frac{\partial}{\partial q_j^\nu},$$

where the functions ζ^0, ζ_j^ν satisfy the equations

$$\zeta^0 \frac{\partial b_K}{\partial t} + \zeta_i^\sigma \frac{\partial b_K}{\partial q_i^\sigma} = 0.$$

These equations must posses rank $\Delta_E - (mc - p) = mc - p + 1$ linearly independent solutions ζ, which means that

$$\operatorname{rank}\left(\frac{\partial b_K}{\partial t}, \frac{\partial b_K}{\partial q_i^\sigma}\right) = p.$$

Since Δ_E is π_{s-1}-horizontal, i.e., there exists a solution with $\zeta^0 \neq 0$, we get that

$$\operatorname{rank}\left(\frac{\partial b_K}{\partial q_i^\sigma}\right) = \operatorname{rank}\left(\frac{\partial b_K}{\partial t}, \frac{\partial b_K}{\partial q_i^\sigma}\right),$$

i.e., the condition (9.25) is satisfied.

(2) For $s = 2c + 1$ the proof is similar. Taking the Lepagean equivalent of a minimal-order Lagrangian and expressing it in the Jacobi chart we get

$$\begin{aligned}\rho =& \Big(-H + p_\rho^c \frac{\partial q_c^\rho}{\partial t}\Big)(t, q_j^\nu, a^K, u^M)\, dt + \sum_{i=0}^{c-1}\Big(p_\sigma^i + p_\rho^c \frac{\partial q_c^\rho}{\partial q_i^\sigma}\Big)(t, q_j^\nu, a^K, u^M)\, dq_i^\sigma \\ &+ p_\rho^c \frac{\partial q_c^\rho}{\partial a^K}\, da^K + p_\rho^c \frac{\partial q_c^\rho}{\partial u^L}\, du^L = dS - b_K\, da^K,\end{aligned} \tag{9.30}$$

hence we obtain (9.26) and (9.27). The Euler-Lagrange distribution Δ_E is spanned by the vector fields

$$\zeta = \zeta^0 \frac{\partial}{\partial t} + \zeta_j^\nu \frac{\partial}{\partial q_j^\nu} + \hat{\zeta}^L \frac{\partial}{\partial u^L},$$

where the functions ζ^0, ζ_j^ν, $\hat{\zeta}^L$, satisfy the equations

$$\zeta^0 \frac{\partial b_K}{\partial t} + \zeta_i^\sigma \frac{\partial b_K}{\partial q_i^\sigma} + \hat{\zeta}^M \frac{\partial b_M}{\partial u^M} = 0.$$

By similar arguments as above we get (9.28). □

For semiregular Lagrangean systems the above Jacobi theorems represent a powerful *integration method*. Clearly, to apply the theorem, one needs to know a *Jacobi complete integral*, i.e., p independent first integrals of Δ_E satisfying the condition (9.17). Such first integrals, however, can be found by means of *symmetries* of the closed two-form α_E (cf. Chapter 8).

9.3.6. Remark. Notice that if the Lagrangean system α_E is *regular* then (for both odd and even-order Lagrangean systems) the function S has a nice geometrical interpretation. Namely, in this case, the one-form dS gives rise to a *completely integrable distribution of corank* 1 which is *complementary* to the Euler-Lagrange distribution Δ_E. Actually, if ζ is a semispray spanning Δ_E then

$$i_\zeta \theta_{\lambda_{\min}} = i_\zeta (b_K \, da^K) + i_\zeta \, dS,$$

and since $i_\zeta \, da^K = 0$ for all K, we get

$$i_\zeta \, dS = L_{\min} \neq 0,$$

proving that ζ does not belong to the distribution annihilated by span$\{dS\}$. This means that the leaves of the foliation associated with dS have the meaning of *wave fronts* whereas the motion in the phase space proceeds transversally in the form of *rays*.

In the semiregular (singular) case the distribution annihilated by span$\{dS\}$ is no more complementary to Δ_E.

Consider a *regular odd-order Lagrangean system*. Then momenta are independent, and the Jacobi theorem directly leads to the following corollary:

9.3.7. Jacobi Theorem for regular even-order variational equations. *Let α_E be a regular Lagrangean system of order $2c-1$. Let H, p_σ^i, $1 \leqslant \sigma \leqslant m$, $0 \leqslant i \leqslant c-1$, be a Hamiltonian and the related momenta of α_E defined on an open subset U of the phase space, such that $U \subset V_{2c-1}$ where (V, ψ), $\psi = (t, q^\sigma)$ is a fiber chart on Y. In a neighborhood of every point of U there exists a function S depending on $t, q^\sigma, \ldots, q_{c-1}^\sigma$, and mc parameters a^K, $1 \leqslant K \leqslant mc$, such that*

$$H\left(t, q_j^\nu, \frac{\partial S}{\partial q_j^\nu}\right) = -\frac{\partial S}{\partial t}. \tag{9.31}$$

Then $(t, q^\sigma, \ldots, q_{c-1}^\sigma, a^1, \ldots, a^{mc})$ is a Jacobi chart of α_E, and the functions

$$a^K, \quad b_K = \frac{\partial S}{\partial a^K}, \quad 1 \leqslant K \leqslant mc \tag{9.32}$$

form a complete set of independent first integrals of the Euler-Lagrange distribution Δ_E of α_E.

The partial differential equation (9.31) for S is called the *Hamilton-Jacobi equation*.

9.3.8. Remark. The above assertion gives one a possibility to find a Jacobi complete integral *directly* by solving the Hamilton-Jacobi equation (9.31), without searching for p appropriate symmetries of the Lagrangean system. Simultaneously, it also provides one with the remaining first integrals b_K. However, this procedure can be applied *only* for *regular odd-order Lagrangean systems*, i.e., Lagrangean systems described by even-order variational equations, which are "solved" with respect to the highest derivatives.

The concept of a Jacobi complete integral and the generalization of the Jacobi theorem to semiregular higher-order Lagrangean systems, is due to O. Krupková [4], where also the relation with the Liouville theorem has been pointed out. A further generalization of the Jacobi theorem to arbitrary closed two-forms of a constant rank (non-variational equations) has been studied in O. Krupková [6].

We have seen, and we shall see also later, that the integration of a Lagrangean system on $J^{s-1}Y$ is naturally connected with the fibered manifold $\pi_{s-1,c-1} : J^{s-1}Y \to J^{c-1}Y$. In what follows, the base space $J^{c-1}Y$ of this fibered manifold (which is a manifold of dimension $mc + 1$) will be called the *quasi-configuration space*.

For $s = 2, 3$ the quasi-configuration space coincides with the configuration space Y.

9.4. Canonical transformations

Let α_E be a Lagrangean system, $J^{s-1}Y$, $s \geqslant 1$, the corresponding phase space.

9.4.1. Definition. Let M be a manifold locally diffeomorphic with the phase space $J^{s-1}Y$. Any local diffeomorphism $\phi : M \to J^{s-1}Y$ will be called a *canonical transformation*.

If α_E is a Lagrangean system and ϕ is a canonical transformation then the two-form $\phi^*\alpha_E$ is closed. Therefore, we can say that *canonical transformations carry Lepagean two-forms into closed two-forms*. Denoting

$$\phi^*\alpha_E = \alpha$$

we get (in a neighborhood of every point) the equality

$$\phi^*\theta = \rho + df \tag{9.33}$$

where $d\theta = \alpha_E$, $d\rho = \alpha$, and f is a function; such a function will be called a *generating function* of the canonical transformation ϕ.

Canonical transformations can be classified in the following way:

(1) General canonical transformations, transferring a Lepagean two-form α_E on the phase space into a closed two-form α on M which need not be Lepagean (i.e., transferring a Lagrangean = variational system into a system not necessarily variational).

(2) Canonical transformations, transferring a Lagrangean system α_E on the phase space $J^{s-1}Y$ of a fibered manifold $\pi : Y \to X$, to another Lagrangean system $\alpha_{E'}$ on the phase space $J^{s-1}Y'$ of a fibered manifold $\pi' : Y' \to X'$, i.e., satisfying

$$\phi^*\alpha_E = \alpha_{E'}.$$

(In this case obviously for every Lagrangian λ for E such that θ_λ is defined on the phase space, the form $\phi^*\theta_\lambda$ is a Lepagean one-form corresponding to E').

(3) Canonical transformations on the phase space, which are *invariant transformations* of the Lagrangean system α_E, i.e. satisfying

$$\phi^*\alpha_E = \alpha_E; \tag{9.34}$$

in this case (9.33) reads

$$\phi^*\theta = \theta + df \tag{9.35}$$

and the generating function f is defined on an open subset of the phase space.

(4) Canonical transformations on the phase space, which are *invariant transformations* of a Lepagean equivalent of a (local) Lagrangian λ, i.e., satisfying

$$\phi^*\theta_\lambda = \theta_\lambda; \tag{9.36}$$

in this case the generating functions are equal to zero.

9.4.2. Remark. Suppose that, in particular, a canonical transformation is of the form $J^{s-1}\phi : J^{s-1}Y \to J^{s-1}Y$, where ϕ is an *isomorphism* of the fibered manifold $\pi : Y \to X$. In this case, as we have seen in Section 4.7, $J^{s-1}\phi$ transfers a Lagrangean system α_E on $J^{s-1}Y$ into another *Lagrangean system* $\alpha_{E'}$ on $J^{s-1}Y$, and it holds $\alpha_{E'} = J^{s-1}\phi^*\alpha_E$, and $E' = J^s\phi^*E$.

Directly from the definition of a canonical transformation we get the following important theorem:

9.4.3. Theorem. *Let α_E be a Lagrangean system on $J^{s-1}Y$, let $\phi : M \to J^{s-1}Y$ be a canonical transformation. Denote $\alpha = \phi^*\alpha_E$, and by $\mathcal{D}_E$ (resp. $\mathcal{D}$) the characteristic distribution of α_E (resp. of α). It holds*

(1) $\operatorname{rank}\alpha_E = \operatorname{rank}\alpha$.

(2) $\mathcal{D} = \operatorname{span}\{\xi_\kappa, \kappa \in K\} \approx \operatorname{span}\{\eta_\iota, \iota \in I\}$ *if and only if* $\mathcal{D}_E = \operatorname{span}\{T\phi^{-1}\cdot\xi_\kappa, \kappa \in K\} \approx \operatorname{span}\{\phi^*\eta_\iota, \iota \in I\}$, *where I, K are suitable index sets.*

(3) *A local section δ of π_{s-1} is a Hamilton extremal of α_E if and only if $\phi^{-1}\circ\delta$ is an integral section of the characteristic distribution of $\phi^*\alpha_E$.*

By the above theorem, *canonical transformations can be used to simplify the integration of Lagrangean systems*. Practically, this means to find a suitable canonical transformation of the given Lagrangean system to another system, represented by a closed two-form such that the "new" equations would take a convenient form; to solve these equations, and finally, to transfer by the canonical transformation, the solutions to solutions of the original equations.

Note that the methods for integration of semiregular Lagrangean systems presented up to now, can be viewed as *based on particular canonical transformations*. Namely,

Legendre transformation is a canonical transformation of type (4) with the *zero generating function.* The Liouville method and the Hamilton-Jacobi method represent ways of searching for canonical transformations of type (3) with nonzero generating functions—the Hamilton-Jacobi equation then can be understood as an *equation for the generating function.* More precisely, we have:

9.4.4. General Liouville-Jacobi Theorem. *Let α be a closed two-form of a constant rank on M. Let ρ be defined on an open subset U of M and such that $d\rho = \alpha$ on U. Let $\mathcal{J}$ be a complete integral of α defined on U. Then there exists an α-preserving canonical transformation $\phi : U \to U$ such that*

$$\phi^* \rho \in \mathcal{J}. \tag{9.37}$$

In adapted coordinates to $\mathcal{J}$ such that $\mathcal{J} \approx \text{span}\{da^K,\ 1 \leqslant K \leqslant p\}$, the relation (9.37) means that there exists a function f such that

$$\phi^* \rho = \rho + df = b_K\, da^K; \tag{9.38}$$

in this equation we recognize the *generalized Hamilton-Jacobi equation* (9.13).

Notice that if we take in the above theorem a *Lagrangean system* α_E then the meaning of the equation (9.38) is the following: instead of a Lagrangian $\lambda = h\rho$ for α_E we are looking for an *equivalent* Lagrangian $\lambda' = h\rho + hdf$ such that its Lepagean equivalent belongs to $\mathcal{J}^0$. Note that in general, *the order of the Lagrangian λ' is higher* than the order of the original Lagrangian λ.

9.4.5. Remark. We can illustrate the above theorem on the most familiar situation of *classical mechanics.* Let us take a *regular* Lagrangean system α_E on J^1Y, and a (local) minimal order Lagrangian $\lambda_{\min}$ of α_E. Clearly, in this case, $p = m$. Then $\alpha_E = d\theta_{\lambda_{\min}}$ and

$$\theta_{\lambda_{\min}} = -H\, dt + p_\sigma\, dq^\sigma.$$

The problem is to find a canonical transformation ϕ on the phase space with a generating function S such that

$$\phi^* \theta_{\lambda_{\min}} \equiv -b_\sigma\, da^\sigma = \theta_{\lambda_{\min}} - dS.$$

If we suppose that the one-forms $da^1, \ldots, da^m$ span a *Jacobi* complete integral of α_E we can express the latter equation in Jacobi coordinates, and we get precisely the *classical Hamilton-Jacobi equation* for a generic function S.

Notice that since it is supposed that $b_k \neq 0$, i.e., $\partial S / \partial a^k \neq 0$, at least for one k, and since the form $\phi^* \theta_{\lambda_{\min}}$ is Lepagean, it is the Lepagean equivalent of a Lagrangian $\lambda' = L'\, dt$, where

$$L' = -b_k \frac{da^k}{dt}.$$

This Lagrangian is, however, *nontrivially of order two* (affine in the second derivatives). This shows us an interesting fact that *even within the range of classical mechanics one cannot avoid a nontrivial use of higher order Lagrangians.*

9.5. Fields of extremals and the generalized Van Hove Theorem

Let α_E be a Lagrangean system of order $s-1$. Recall that then the the manifold $J^{s-1}Y$ is the *phase space* and the manifold $J^{c-1}Y$ is the *quasi-configuration space* of α_E (cf. Sec. 9.3) . Denote by $\mathfrak{J}$ the differential ideal generated by α_E. Recall that a section φ is called an *integral section* of the ideal $\mathfrak{J}$ if

$$\varphi^*\alpha_E = 0.$$

In this section we shall study some special integral sections of the ideal $\mathfrak{J}$. Namely, we shall be interested in local sections (jet fields) of the fibered manifold $\pi_{s-1,c-1} : J^{s-1}Y \to J^{c-1}Y$ with the phase space as the total space and the quasi-configuration space as the base space. Sections of this kind will be shown to define an important structure on the quasi-configuration space, which geometrically describes the (local) structure of extremals.

Let $\varphi : J^{c-1}Y \to J^{s-1}Y$ be a jet field defined on an open set U. Recall that a section $\gamma : I \to Y$, defined on an open subset $I \subset X$ is called an *integral section* of the jet field φ if $\gamma(I) \subset U$ and

$$\varphi \circ J^{c-1}\gamma = J^{s-1}\gamma.$$

9.5.1. Theorem. *Let α_E be a Lagrangean system on $J^{s-1}Y$, $\mathfrak{J}$ the differential ideal generated by α_E. Let $\varphi : J^{c-1}Y \supset U \to J^{s-1}Y$ be an integral section of $\mathfrak{J}$. If a section γ of π is an integral section of φ then γ is an extremal of E.*

Proof. Let ξ be a π-projectable vector field on Y. Then by assumptions,

$$J^{s-1}\gamma^* i_{J^{s-1}\xi}\alpha_E = J^{c-1}\gamma^*\varphi^* i_{J^{s-1}\xi}\alpha_E = J^{c-1}\gamma^* i_{J^{c-1}\xi}\,\varphi^*\alpha_E = 0,$$

proving our assertion. □

9.5.2. Definition. Let α_E be a Lagrangean system on $J^{s-1}Y$, $\mathfrak{J}$ the corresponding differential ideal, and let $J^{c-1}Y$ be the corresponding quasi-configuration space. In accordance with Theorem 9.5.1, any integral section $\varphi : J^{c-1}Y \supset U \to J^{s-1}Y$ of $\mathfrak{J}$ will be called a *field of extremals* of α_E on U. The equation

$$\varphi^*\alpha_E = 0 \tag{9.39}$$

for fields of extremals will be called the *Hamilton-Jacobi equation* (*for fields of extremals*).

Note that the Hamilton-Jacobi equation (9.39) implies that for every (local) Lagrangian λ such that θ_λ is of order $s-1$ and defined on U, the form $\varphi^*\theta_\lambda$ is closed, i.e.,

$$\varphi^*\theta_\lambda = dS \tag{9.40}$$

for a function S on an open subset of the quasi-configuration space. Expressing α_E in the canonical form we get the Hamilton-Jacobi equation (9.40) in the following form:

$$H \circ \varphi = -\frac{\partial S}{\partial t}, \quad p^k_\nu \circ \varphi = \frac{\partial S}{\partial q^\nu_k}, \tag{9.41}$$

where $0 \leqslant k \leqslant c-1$, $1 \leqslant \nu \leqslant m$, if $s = 2c$, and

$$H \circ \varphi - (p^c_\nu \circ \varphi)\frac{\partial(q^\nu_c \circ \varphi)}{\partial t} = -\frac{\partial S}{\partial t}, \quad p^k_\nu \circ \varphi + (p^c_\nu \circ \varphi)\frac{\partial(q^\nu_c \circ \varphi)}{\partial q^\nu_k} = \frac{\partial S}{\partial q^\nu_k}, \tag{9.42}$$

where $0 \leqslant k \leqslant c-1$, $1 \leqslant \nu \leqslant m$, if $s = 2c+1$, respectively.

If E is *regular* and *even-order* then using the fact that momenta are independent, we can rewrite the equations (9.41) in a form of a system, consisting of one partial differential equation

$$H\Big(t, q_k^\nu, \frac{\partial S}{\partial q_k^\nu}\Big) = -\frac{\partial S}{\partial t} \tag{9.43}$$

for a function S on the quasi-configuration space, and equations for the components of the jet field φ,

$$p_\nu^k \circ \varphi = \frac{\partial S}{\partial q_k^\nu}. \tag{9.44}$$

In analogy with classical mechanics the equation (9.43) will be called *eikonal equation*, or *Hamilton-Jacobi equation.*

Notice that (by Sec. 2.3) the vector fields belonging to the characteristic distribution $\mathcal{D}$ of α_E, are *Cauchy characteristic vector fields* of the ideal $\mathfrak{J}$. If the Lagrangean system α_E is *semiregular*, then the characteristic distribution coincides with the Euler-Lagrange distribution, hence the Cauchy characteristic vector fields coincide with the Hamilton vector fields. Moreover, in this case, $\mathcal{D}$ is completely integrable, and the corresponding foliation is of dimension $\geqslant 1$. Recall that, in the terminology of the theory of differential ideals, the leaves are called *Cauchy characteristics*. As we know from Chapter 5 and 7, the leaves in this case represent solutions of the generalized Hamilton equations (i.e., equations for integral sections of the distribution $\mathcal{D}$), since every Hamilton extremal is a mapping from the base into a leaf of this foliation, and conversely, every section of π_{s-1}, whose image lies in a leaf, is a Hamilton extremal. Thus, for *semiregular* Lagrangean systems, instead of the generalized Hamilton equations one can *equivalently* work with the equations for the Cauchy characteristics of the differential ideal $\mathfrak{J}$, generated by the Lepagean two-form α_E. In the *regular* case the situation is even more simple: the Hamilton equations are just the equations for the Cauchy characteristics of the ideal $\mathfrak{J}$.

With the help of the Hamilton-Jacobi equation (9.39), resp. (9.40) we can look for jet fields which are "built from extremals". However, there arises a question: can *every* extremal be represented as an integral section of a field of extremals? We shall show now, that for semiregular Lagrangean systems, every extremal can be "locally" embedded in a field of extremals. This result generalizes the well-known *Van Hove Theorem* from regular first order variational theory (L. Van Hove [1]).

9.5.3. Embedding Theorem (Generalized Van Hove Theorem). *Let α_E be a semiregular Lagrangean system of order $s-1$, let γ be an extremal of α_E. Let $x \in \operatorname{dom}\gamma$ be a point. There exists a neighborhood W of $J_x^{s-1}\gamma$ and a field of extremals φ defined on an open subset $U \subset J^{c-1}Y$ such that $\varphi(U) \subset W$ and γ is an integral section of φ.*

Proof. Let $\operatorname{rank}\alpha_E = 2p$. Consider a Jacobi chart $\chi = (t, q_j^\nu, a^K, u^L)$ of α_E at the point $J_x^{s-1}\gamma$ and the related Jacobi complete integral $\mathfrak{I} \approx \operatorname{span}\{da^1, \dots, da^p\}$. Since $J^{s-1}\gamma$ is a Hamilton extremal of E, there exists a leaf N of $\mathfrak{I}$ such that $J^{s-1}\gamma(I) \subset N$ for some open set $I \ni x$. $J^{s-1}\gamma(I)$ is a submanifold of N, hence there exists a chart (z, w^J), $1 \leqslant J \leqslant ms-p$ on N at $J_x^{s-1}\gamma$ adapted to $J^{s-1}\gamma(I)$ (i.e., such that the equations of $J^{s-1}\gamma(I)$ are $w^J = \text{const}.$). Denote by ξ a vector field along $J^{s-1}\gamma(I)$, tangent to $J^{s-1}\gamma(I)$. Since $T\pi_{s-1}.\xi \neq 0$, we get that the functions (z, q^ν_j, u^L), defined in a neighborhood W of $J^{s-1}_x\gamma$ in N, form another chart at $J^{s-1}_x\gamma$ adapted to the submanifold $J^{s-1}\gamma(I)$. Consider the $(ms+1)$-dimensional submanifold M of W, defined by the equations $u^L = \text{const}$. Then for some open set I_0 of I, $I_0 \ni x$, the $J^{s-1}\gamma(I_0)$ lies in M. Denote by τ the projection $\pi_{s-1,c-1}$ restricted to M. Then $\tau(M)$ is an open submanifold of the quasi-configuration space $J^{c-1}Y$ and $T_yM \to T_{\tau(y)}\tau(M)$ is an isomorphism of vector spaces for all $y \in M$, proving that τ is a local diffeomorphism. Putting $\varphi = \tau^{-1}$ we get a jet field in a neighborhood of $J^{c-1}_x\gamma$. Now, γ is locally embedded in φ and $\varphi^*\alpha_E = 0$, since $\operatorname{Im}\varphi \subset N$. □

By the Embedding Theorem, a semiregular variational problem can be solved completely if one knows *all fields of extremals*. Notice that this integration method leads to an essential *order-reduction* of the integration problem. Namely, if one knows a field of extremals φ, the corresponding extremals are found as solutions of $m(s-c)$ ODE *of order* c, while the original Euler-Lagrange equations are m ODE of order s. Geometrically, a field of extremals represents a collection of extremals with "partially prescribed initial conditions" (more precisely, with prescribed "higher velocities" of order c, $c+1, \dots, s-1$, resp. with prescribed momenta).

9.5.4. Example (Fields of extremals of first-order and second-order Lagrangean systems). For *first-order Lagrangean systems*, equivalently represented by second-order variational equations ($s = 2$, $c = 1$), a field of extremals is a (local) section $Y \to J^1Y$, i.e., from the configuration space to the phase space; notice that in this case, the phase space is also the space where the minimal-order Lagrangians live. In other words, fields of extremals of first order Lagrangean systems are (local) *connections on the configuration space* Y. A field of extremals then represents a collection of extremals with "prescribed velocities". Since every connection is in correspondence with a horizontal distribution, we can see that the extremals embedded in a field of extremals φ are integral sections of the distribution

$$\Delta_\varphi = \operatorname{span}\left\{\frac{\partial}{\partial t} + \varphi^\sigma(t, q^\nu)\frac{\partial}{\partial q^\sigma}\right\},$$

where $\varphi^\sigma = \dot{q}^\sigma \circ \varphi$ are the components of the jet field φ. In other words, knowing a field of extremals (which is found as a solution of the Hamilton-Jacobi equation), we can find the corresponding extremals as solutions of m *first-order ODE*. The distribution Δ_φ is called *Hamilton-Jacobi distribution* of the Lagrangean system α_E, related to φ. This distribution is of rank one, hence completely integrable. This means that fields of extremals of first-order Lagrangean systems have a nice geometric structure: through every point of dom Δ_φ, which is an open subset of the configuration space, there passes exactly one (maximal) extremal, embedded in φ.

Consider a *second-order Lagrangean system* (which is equivalently represented by m *third-order* variational equations); in this case $s = 3$, $c = 1$, and the phase space again is the space where the minimal-order Lagrangians live. Fields of extremals are (local) sections from the configuration space Y (which coincides with the quasi-configuration space) to the phase space J^2Y, i.e., they *are not connections*. In a field of extremals, all extremals with prescribed velocities and accelerations are collected. The corresponding horizontal distribution is of the form

$$\Delta_\varphi = \operatorname{span}\left\{\frac{\partial}{\partial t} + \varphi^\sigma(t, q^\nu)\frac{\partial}{\partial q^\sigma} + \varphi_1^\sigma(t, q^\nu)\frac{\partial}{\partial \dot{q}^\sigma}\right\},$$

where $\varphi^\sigma = \dot{q}^\sigma \circ \varphi$, and $\varphi_1^\sigma = \ddot{q}^\sigma \circ \varphi$. In this way we get the corresponding extremals as solutions of $2m$ *first-order ODE* of the form

$$\frac{d\gamma^\sigma}{dt} = \varphi^\sigma(t, \gamma^\nu), \quad \frac{d^2\gamma^\sigma}{dt^2} = \varphi_1^\sigma(t, \gamma^\nu);$$

the first set of these equations represent equations for γ, the second one represents compatibility conditions.

On the contrary to the first-order situation, Δ_φ is a distribution on J^1Y (not on the configuration space), i.e., its integral sections are local sections from X to J^1Y (not extremals). The above mentioned compatibility conditions clearly mean that all integral sections of Δ_φ are of the form of prolongations of extremals.

For *higher-order Lagrangean systems*, fields of extremals carry too many compatibility conditions. This, however, is inconvenient from the practical point of view. We shall show that the number of the compatibility conditions can be minimized if we take another structure related to fields of extremals; this structure will have similar properties as are those of fields of extremals of first and second order Lagrangean systems. Consequently, this will lead to a further reduction of the integration problem.

9.5.5. Definition. Let $s \geqslant 4$. Let φ be a field of extremals of a Lagrangean system α_E on $J^{s-1}Y$. We set

$$\hat{\varphi} = \pi_{s-1,s-c} \circ \varphi \tag{9.45}$$

and call the jet-field $\hat{\varphi}$ a *Hamilton-Jacobi field* of the Lagrangean system α_E. The corresponding distribution will be called the *Hamilton-Jacobi distribution* of α_E and will be denoted by $\Delta_{\hat{\varphi}}$.

Clearly, $\hat{\varphi}$ is a local section of the fibered manifold $J^{s-c}Y \to J^{c-1}Y$ over the quasi-configuration space. Notice that the total space is the manifold where minimal-order Lagrangians live. Further notice that for $s = 2c$, $\hat{\varphi}$ is a local section $J^{c-1}Y \to J^cY$, i.e., a *semispray connection*; this means that the corresponding Hamilton-Jacobi distribution is a *semispray distribution*, defined on an open subset of the quasi-configuration space $J^{c-1}Y$. We have

$$\begin{aligned}\Delta_{\hat{\varphi}} &= \operatorname{span}\left\{\frac{\partial}{\partial t} + \sum_{i=0}^{c-2} q_{i+1}^\sigma \frac{\partial}{\partial q_i^\sigma} + \varphi_{c-1}^\sigma \frac{\partial}{\partial q_{c-1}^\sigma}\right\} \\ &\approx \operatorname{span}\{\omega_i^\sigma, 0 \leqslant i \leqslant c-2, \ dq_{c-1}^\sigma - \varphi_{c-1}^\sigma\, dt\},\end{aligned}$$

where $\varphi_{c-1}^\sigma = q_c^\sigma \circ \varphi$. For $s = 2c+1$, $\hat{\varphi}$ is a local section $J^{c-1}Y \to J^{c+1}Y$, i.e., it is *not* a semispray connection; the corresponding Hamilton-Jacobi distribution is not a semispray distribution and it is defined on an open subset of J^cY (i.e., not on the quasi-configuration space). In this case,

$$\Delta_{\hat{\varphi}} = \text{span}\left\{\frac{\partial}{\partial t} + \sum_{i=0}^{c-2} q_{i+1}^{\sigma}\frac{\partial}{\partial q_i^{\sigma}} + \varphi_{c-1}^{\sigma}\frac{\partial}{\partial q_{c-1}^{\sigma}} + \varphi_c^{\sigma}\frac{\partial}{\partial q_c^{\sigma}}\right\}$$
$$\approx \text{span}\{\omega_i^{\sigma}, 0 \leqslant i \leqslant c-2,\ dq_{c-1}^{\sigma} - \varphi_{c-1}^{\sigma}\, dt,\ dq_c^{\sigma} - \varphi_c^{\sigma}\, dt\},$$

where $\varphi_{c-1}^{\sigma} = q_c^{\sigma} \circ \varphi$, $\varphi_c^{\sigma} = q_{c+1}^{\sigma} \circ \varphi$.

It is easy to see that if γ is an integral section of φ, it is an integral section of $\hat{\varphi}$; this means that all the extremals belonging to the field of extremals φ, belong also to the Hamilton-Jacobi field $\hat{\varphi}$. However, does $\hat{\varphi}$ possess some other integral sections? Let us turn to this question now.

9.5.6. Lemma. *Let α_E be a Lagrangean system on $J^{s-1}Y$, let φ be a jet field on the fibered manifold $\pi_{s-1,c-1}$. Put $\hat{\varphi} = \pi_{s-1,s-c} \circ \varphi$. If γ is an integral section of $\hat{\varphi}$ then for every π_{s-1}-vertical vector field ξ on $J^{s-1}Y$*

$$J^{c-1}\gamma^*\varphi^*\, i_{\xi}\alpha_E = J^{c-1}\gamma^*\, i_{\zeta}\, \varphi^*\alpha_E, \tag{9.46}$$

where ζ is a vector field on $J^{c-1}Y$ defined by

$$\zeta(J_x^{c-1}\gamma) = T\pi_{s-1,c-1}.\xi(\varphi(J_x^{c-1}\gamma)).$$

Proof. Let $x \in \operatorname{dom}\gamma$ be a point, $\xi_0 \in T_xX$ a tangent vector. We have

$$J^{c-1}\gamma^*\varphi^*\, i_{\xi}\alpha_E(x)(\xi_0) = J^{c-1}\gamma^*\, i_{\zeta}\, \varphi^*\alpha_E(x)(\xi_0)$$
$$+\, \alpha_E(\varphi(J_x^{c-1}\gamma))(\xi(\varphi(J_x^{c-1}\gamma)) - T\varphi.\zeta(J_x^{c-1}\gamma), T\varphi.TJ^{c-1}\gamma.\xi_0).$$

Now, one can check by a direct calculation that the second term on the right-hand side vanishes (see D. Krupka [6], Lemma in Sec. 7 for details), proving (9.46). □

9.5.7. Theorem. *Let $\hat{\varphi}$ be a Hamilton-Jacobi field of a Lagrangean system α_E on $J^{s-1}Y$. If γ is an integral section of $\hat{\varphi}$ then $\varphi \circ J^{c-1}\gamma$ is a Hamilton extremal of α_E. If α_E is regular then γ is an integral section of φ, hence an extremal.*

Proof. Computing $(\varphi \circ J^{c-1}\gamma)^*\, i_{\xi}\alpha_E$ for every π_{s-1}-vertical vector-field ξ on $J^{s-1}Y$ and using Lemma 9.5.6 we get $(\varphi \circ J^{c-1}\gamma)^*\, i_{\xi}\alpha_E = 0$, since $\varphi^*\alpha_E = 0$ by assumption. If E is regular, then $\varphi \circ J^{c-1}\gamma = J^{s-1}\bar{\gamma}$ for an extremal $\bar{\gamma}$, hence $\pi_{s-1,s-c} \circ (\varphi \circ J^{c-1}\gamma) = J^{c-1}\bar{\gamma} = J^{c-1}\gamma$. □

By the above theorem, Hamilton-Jacobi fields generally possess more solutions than fields of extremals, since they are built from *Hamilton extremals* which are *holonomic up to the order* $c-1$; we shall call such Hamilton extremals *quasi-holonomic*. On the other hand, compared with the integration method based on fields of extremals, Theorem 9.5.7 provides the possibility of *lowering the number of equations* in higher-order integration problems. Instead of solving $m(s-c)$ equations for fields of extremals, one can consider equations for Hamilton-Jacobi fields, representing m equations for $\hat{\varphi}$ if s is even, and $2m$ equations (consisting of m equations for $\hat{\varphi}$ and m compatibility conditions) if s is odd, respectively.

Hamilton-Jacobi distributions are of rank one, i.e., they are completely integrable. This means that through every point of dom $\Delta_{\hat{\varphi}}$ there passes exactly one (maximal) quasi-holonomic Hamilton extremal embedded in $\hat{\varphi}$. In this way we get the original variational problem represented by means of a lower-order geometric structure, namely by a family of local one-dimensional foliations of the quasi-configuration space (resp. of J^cY) if $s = 2c$ (resp. $s = 2c + 1$).

The meaning of Hamilton-Jacobi fields for practical integration of Lagrangean systems increases if the Lagrangean system is *regular*. Then all integral sections of a Hamilton-Jacobi distribution are *prolongations of extremals*. Applying the Embedding Theorem we can see that *a regular variational problem can be completely solved if one knows all Hamilton-Jacobi fields*. Practically, compared with the integration method using fields of extremals, this means an essential *reduction of both the order and the number of equations* to be solved.

Summarizing the results, for *regular Lagrangean systems*, Hamilton-Jacobi distributions provide an important information on the *structure of solutions* of the Euler-Lagrange equations. Moreover, from the point of view of the integration problem, instead of looking for integral sections of the Euler-Lagrange distribution (ODE of order $2c$, resp. $2c + 1$), extremals can be found as integral sections of the Hamilton-Jacobi distributions (ODE of order c, resp. $c+1$). In this sense, *Hamilton-Jacobi distributions of a regular Lagrangean system have a geometric meaning of integrals of the Euler-Lagrange distribution.*

9.5.8. Remark. The results of this section, obtained in O. Krupková [4], are a further development and generalization to not necessarily regular, higher-order Lagrangians of the *theory of fields of extremals*, initiated by L. Van Hove [1] and geometrically interpreted by H. Goldschmidt and S. Sternberg [1]. (The Goldschmidt-Sternberg theory has been generalized also to regular Lagrangians in higher-order field theory (D. Krupka [6])). The concepts of a Hamilton-Jacobi field and a Hamilton-Jacobi distribution have been introduced and studied in O. Krupková [3] for regular first-order Lagrangean systems on fibered manifolds. To higher-order singular Lagrangean systems the theory has been generalized in O. Krupková [4].

9.6. Hamilton-Jacobi distributions for regular odd-order Lagrangean systems

Now, we shall be interested in an important example of Lagrangean systems, namely in *regular odd-order Lagrangean systems* (i.e., represented by *regular even-order* Euler-Lagrange equations). This means that throughout this section we shall consider $s = 2c$ and $p = 2mc$.

It is an important feature of these systems that the main structures (jet fields, generalized connections) related to the corresponding variational equations reduce to *connections*. Namely, the Euler-Lagrange distribution coincides with the characteristic distribution, and it is a semispray distribution on the phase space (i.e., a semispray connection on the fibered manifold π_{2c-1}). Every complete integral is a completely integrable distribution of rank $mc+1$, and (prolonged) extremals passing in its domain define a one-dimensional subfoliation of its related $(mc + 1)$-dimensional foliation. In particular, Jacobi complete integrals are connections (horizontal distributions) on the fibered manifold $\pi_{2c-1,c-1}$, and they can be found as solutions of the Hamilton-Jacobi equation (9.31). Hamilton-Jacobi fields are semispray connections on the fibered manifold π_{c-1}, i.e., Hamilton-Jacobi distributions are rank one semispray distributions on the quasi-configuration space; their integral sections are $(c-1)$-prolongations of extremals.

Notice that Jacobi charts are of the form (t, q_j^ν, a^K), $0 \leqslant j \leqslant c-1$, $1 \leqslant K \leqslant mc$, where the a^K's are first integrals of the Euler-Lagrange distribution, satisfying the condition

$$\det\left(\frac{\partial a^K}{\partial q_i^\sigma}\right) \neq 0, \quad \text{where } 1 \leqslant \sigma \leqslant m,\ c \leqslant i \leqslant 2c-1.$$

Immediately from the Jacobi Theorem we get conditions for a chart on the phase space to be a Jacobi chart:

9.6.1. Proposition. *Let $y \in J^{2c-1}Y$ be a point, (W, χ), $\chi = (\bar{t}, \bar{q}_j^\nu, a^K)$ where $0 \leqslant j \leqslant c-1$, $1 \leqslant K \leqslant mc$, a chart at y. Let α_E be a Lagrangean system on $J^{2c-1}Y$. The chart (W, χ) is a Jacobi chart of α_E if and only if for every Legendre chart (U, φ), $\varphi = (t, q_j^\nu, p_\nu^j)$, $0 \leqslant j \leqslant c-1$, of α_E at y there exists a function $S(\bar{t}, \bar{q}_j^\nu, a^K)$ on $W \cap U$ such that the overlap mapping is defined by the formulas*

$$t = \bar{t}, \quad q_j^\nu = \bar{q}_j^\nu, \quad p_\nu^j = \frac{\partial S}{\partial q_j^\nu}$$

for all ν, j, and for the Hamiltonian associated with the momenta p_ν^j it holds

$$H(\bar{t}, \bar{q}_j^\nu, a^K) = -\frac{\partial S}{\partial t}.$$

It is easy to see that if $\varphi : J^{c-1}Y \supset U \to J^{2c-1}Y$ is a field of extremals of a regular odd-order Lagrangean system then $\varphi(U)$ is a *Lagrangean submanifold.* Conversely, if we have a *Jacobi complete integral* and N is a leaf of the corresponding foliation then every point $y \in N$ has a neighborhood M in N such that $\tau = \pi_{2c-1,c-1}|_M$ is a local diffeomorphism and $\varphi = \tau^{-1}$ is a field of extremals. In this way, a Jacobi complete integral can be locally represented by a suitable family of fields of extremals, and, in particular, by a suitable family of *Hamilton-Jacobi distributions*. More precisely, we have

9.6.2. Theorem. *Let α_E be a regular odd-order Lagrangean system on $J^{2c-1}Y$. Let (W, χ), $\chi = (\bar{t}, \bar{q}_i^\sigma, a^K)$, $1 \leqslant i \leqslant c-1$, $1 \leqslant K \leqslant mc$, be a Jacobi chart of α_E, and suppose that $W \subset V_{2c-1,c-1}$, where (V, ψ), $\psi = (t, q^\sigma)$ is a fiber chart on Y such that $t = \bar{t}$ and $q_i^\sigma = \bar{q}_i^\sigma$ for $0 \leqslant i \leqslant c-1$. Consider the vector field*

$$\zeta = \frac{\partial}{\partial \bar{t}} + \sum_{i=0}^{c-2} \bar{q}_{i+1}^\sigma \frac{\partial}{\partial \bar{q}_i^\sigma} + q_c^\sigma(\bar{t}, \bar{q}_j^\nu, a^K) \frac{\partial}{\partial \bar{q}_{c-1}^\sigma} \tag{9.47}$$

on W. The vector field ζ is tangent to the foliation the leaves N_a of which are defined by $a = (a^1, \ldots, a^{mc}) =$ const. For every $a =$ const $\in R^{mc}$ denote by ζ_a the restriction of ζ to N_a and put

$$\xi_a = T\pi_{2c-1,c-1} \cdot \zeta_a. \tag{9.48}$$

Then ξ_a is a vector field on $\pi_{2c-1,c-1}(W)$, spanning a Hamilton-Jacobi distribution Δ_a of α_E. If $\gamma : U \to Y$ is an extremal of α_E such that $J^{2c-1}\gamma(U) \subset W$ then there exists a *parameter $a \in R^{mc}$ and a Hamilton-Jacobi distribution Δ_a such that $J^{c-1}\gamma$ is an integral section of Δ_a.*

Proof. Proceeding similarly as in the proof of the Embedding Theorem we get for every fixed a a field of extremals φ_a mapping an open subset of the quasi-configuration space onto an open subset of the leaf N_a. The equations of φ_a are

$$(t, q^\sigma, \dots, q^\sigma_{c-1}, q^\sigma_c, \dots, q^\sigma_{2c-1}) \circ \varphi_a = (t, q^\sigma, \dots, q^\sigma_{c-1}, q^\sigma_c|_{N_a}, \dots, q^\sigma_{2c-1}|_{N_a}).$$

The corresponding Hamilton-Jacobi field is then defined by the equations

$$(t, q^\sigma, \dots, q^\sigma_{c-1}, q^\sigma_c) \circ \hat{\varphi}_a = (t, q^\sigma, \dots, q^\sigma_{c-1}, q^\sigma_c|_{N_a}).$$

Hence, we get the associated Hamilton-Jacobi distribution Δ_a spanned by the vector field ξ_a (9.48).

If γ is an extremal of α_E such that $J^{2c-1}\gamma$ passes through W then it is embedded in N_a for some a, and by the above construction, $J^{c-1}\gamma$ is an integral section of Δ_a. □

In correspondence with the above theorem we say that a Jacobi complete integral on the phase space *is equivalent with an mc-parameter family of Hamilton-Jacobi distributions* on the quasi-configuration space; such a family will be denoted by $\{\Delta_a\}$. Clearly, $\{\Delta_a\}$ determines the complete solution of the Euler-Lagrange distribution on W, and in this way this structure provides *complete local information* about extremals.

Moreover, using the Jacobi Theorem, we now easily obtain

9.6.3. Theorem. *Let α_E be a regular odd-order Lagrangean system on $J^{2c-1}Y$, let $\{\Delta_a\}$ be an mc-parameter family of Hamilton-Jacobi distributions defined on an open subset W of $J^{c-1}Y$, equivalent with a Jacobi complete integral of α_E. Then in a neighborhood of every point of W there exists a function S such that for every parameter a, the functions*

$$\beta_K = \frac{\partial S}{\partial a^K} \circ \varphi_a, \quad 1 \leqslant K \leqslant mc, \tag{9.49}$$

are a complete set of independent first integrals of Δ_a. The function S is an mc-parameter solution of the Hamilton-Jacobi equation (9.31).

By Theorems 9.6.2 and 9.6.3, *regular odd-order Lagrangean systems* possess important properties which lead to an *essential simplification of the Hamilton-Jacobi integration method.* First of all, to find a local complete solution of the Euler-Lagrange equations it is not necessary to find *all* Hamilton-Jacobi fields, but it is sufficient to find an appropriate mc-parameter family $\{\hat{\varphi}_a\}$ and then to integrate the corresponding family $\{\Delta_a\}$ of Hamilton-Jacobi distributions. Secondly, by Theorem 9.6.3, the integration of $\{\Delta_a\}$ reduces to a trivial procedure, since one gets immediately adapted charts using the formula (9.49). Hence, summarizing, we have the following *integration procedure*:

(1) Find a solution S of the Hamilton-Jacobi equation (9.31), depending on mc parameters $a = (a^1, \dots, a^{mc})$,
(2) construct the functions β_a according to (9.49),
(3) solve the equations

$$\beta_K \circ J^{c-1}\gamma = \text{const}, \quad 1 \leqslant K \leqslant mc. \tag{9.50}$$

Notice that in case of $s = 2$ (classical mechanics), equations (9.50) are *m implicit functional equations* for the *extremals* γ, and the above described integration method is the familiar *classical Hamilton-Jacobi method* of integration of variational equations coming from regular first-order Lagrangians. For $s \geqslant 4$ equations (9.50) represent mc ordinary *differential* equations of order $c-1$ for extremals *in an implicit form.*

We shall close this section with an interesting property of (both regular and non-regular) *odd-order* Lagrangean systems.

By solving the Hamilton-Jacobi equation (9.39) one obtains *local* fields of extremals. However, one can ask about the existence of a *global* solution of this equation. The following theorem shows the relation between the existence of global fields of extremals and the existence of global minimal-order Lagrangians.

9.6.4. Theorem. *Let α_E be a Lagrangean system on $J^{2c-1}Y$, $\varphi : J^{c-1}Y \to J^{2c-1}Y$ a (global) jet field. The following two conditions are equivalent:*

(1) *φ is a field of extremals of α_E,*

(2) *there exists a unique minimal-order Lagrangian λ_φ such that $E_{\lambda_\varphi} = E$ and $\varphi^*\theta_{\lambda_\varphi} = 0$.*

Proof. Since α_E is an odd-order Lagrangean system, every point in J^cY has a neighborhood W such that, on W, there exists a minimal-order Lagrangian $\lambda_{\min}$ for α_E. Suppose that $\varphi : J^{c-1}Y \to J^{2c-1}Y$ is a field of extremals of α_E. Then the one-form $\varphi^*\theta_{\lambda_{\min}}$ on $\pi_{c,c-1}(W) \subset J^{c-1}Y$ is closed. Put

$$\lambda_\varphi = \lambda_{\min} - h\,\varphi^*\theta_{\lambda_{\min}}. \tag{9.51}$$

λ_φ is a minimal-order Lagrangian, equivalent with $\lambda_{\min}$, such that $\varphi^*\theta_{\lambda_\varphi} = \varphi^*\theta_{\lambda_{\min}} - \varphi^*\theta_{\lambda_{\min}} = 0$.

Let us show that λ_φ is a global Lagrangian for α_E. Denote by $\{W_\iota\}$ an open covering of J^cY such that for every ι there exists a Lagrangian λ_ι on W_ι satisfying $\varphi^*\theta_{\lambda_\iota} = 0$. Let $W_\iota \cap W_\kappa$ be nonempty. Then on the intersection we have $\lambda_\iota = \lambda_\kappa + h\rho$, where ρ is a closed one-form on $\pi_{c,c-1}(W_\iota \cap W\kappa)$. Since $\varphi^*\theta_{\lambda_\iota} = \varphi^*\theta_{\lambda_\kappa} = 0$, we get $\varphi^*\rho = \rho = 0$, i.e., $\lambda_\iota = \lambda_\kappa$. This, however, implies both the existence and uniqueness of the desired Lagrangian λ_φ. □

9.6.5. Corollary. *Let α_E be a Lagrangean system on $J^{2c-1}Y$. If there does not exist a global minimal-order Lagrangian then there exists no global field of extremals.*

Since $\varphi^*\theta_{\lambda_{\min}} = dS$, we have for the Lagrangian (9.51) the following fiber-chart expression:

$$L_\varphi = L_{\min} - \frac{dS}{dt} = -H + \sum_{i=0}^{c-1} p^i_\sigma q^\sigma_{i+1} - \frac{dS}{dt}. \tag{9.52}$$

9.7. Geodesic distance in a field of extremals

Consider an *odd-order Lagrangean system* α_E. Let φ be a field of extremals of α_E. In this section we shall study the meaning of the closed one-form $\varphi^*\theta_{\lambda_{\min}}$ where $\lambda_{\min}$ is a minimal-order Lagrangian for α_E. We shall see that the situation becomes interesting namely in the case when α_E is *regular*.

9.7.1. Theorem. *Let α_E be a Lagrangean system on $J^{2c-1}Y$. Let U be an open set in the quasi-configuration space $J^{c-1}Y$, and $\varphi : U \to J^{2c-1}Y$ a field of extremals of α_E. Denote by $\hat{\varphi} : U \to J^cY$ the associated Hamilton-Jacobi field. Let $\lambda_{\min}$ be a minimal-order Lagrangian for α_E, defined on an open subset of J^cY over U. If*

$$\hat{\varphi}^*\lambda_{\min} \neq 0 \tag{9.53}$$

then the one-form $\varphi^*\theta_{\lambda_{\min}}$ defines a completely integrable distribution $\mathcal{C}_{\lambda_{\min}}$ of rank mc on U, complementary to the Hamilton-Jacobi distribution $\Delta_{\hat{\varphi}}$ of α_E.

Proof. If $\lambda_{\min}$ satisfies (9.53) then (with the notation of Theorem 4.5.8)

$$\varphi^*\theta_{\lambda_{\min}} = ((L_{\min} \circ \hat{\varphi}) - (p_\sigma^{c-1} \circ \varphi)\varphi_{c-1}^\sigma)\, dt + \sum_{i=0}^{c-2}(p_\sigma^i \circ \varphi)\, \omega_i^\sigma + (p_\sigma^{c-1} \circ \varphi)\, dq_{c-1}^\sigma \neq 0;$$

actually, if $\varphi^*\theta_{\lambda_{\min}}$ would be equal to zero then $p_\sigma^{c-1} \circ \varphi = 0$, and consequently, $L_{\min} \circ \hat{\varphi} = 0$, which contradicts to our assumption. This means that $\varphi^*\theta_{\lambda_{\min}}$ defines a distribution of rank mc on U. This distribution is completely integrable, since the one-form $\varphi^*\theta_{\lambda_{\min}}$ is closed. It holds rank $\mathcal{C}_{\lambda_{\min}}$ + rank $\Delta_{\hat{\varphi}} = mc + 1 = \dim J^{c-1}Y$. If ξ is a semispray spanning the Hamilton-Jacobi distribution $\Delta_{\hat{\varphi}}$, we have

$$i_\xi \varphi^*\theta_{\lambda_{\min}} = (L_{\min} \circ \hat{\varphi}) \neq 0$$

proving the complementarity of $\Delta_{\hat{\varphi}}$ and $\mathcal{C}_{\lambda_{\min}}$. □

By Theorem 9.7.1 we have on U complementary distributions $\mathcal{C}_{\lambda_{\min}} \approx \operatorname{span}\{\varphi^*\theta_{\lambda_{\min}}\}$ and $\Delta_{\hat{\varphi}}$. This means that at each point of U there is a chart (W, χ), $\chi = (u, x^J)$, $1 \leqslant J \leqslant mc$, such that

$$\mathcal{C}_{\lambda_{\min}} = \operatorname{span}\left\{\frac{\partial}{\partial x^J},\ 1 \leqslant J \leqslant mc\right\} \approx \operatorname{span}\{du\},$$

$$\Delta_{\hat{\varphi}} = \operatorname{span}\left\{\frac{\partial}{\partial u}\right\} \approx \operatorname{span}\{dx^J,\ 1 \leqslant J \leqslant mc\}.$$

Consequently, $\varphi^*\theta_{\lambda_{\min}} = f\, du$ for a nowhere zero function f. On the other hand, the one-form $\varphi^*\theta_{\lambda_{\min}}$ is closed, i.e., locally,

$$\varphi^*\theta_{\lambda_{\min}} = dS.$$

This means that also (S, x^J) are local coordinates, adapted to the complementary distributions $\mathcal{C}_{\lambda_{\min}}$ and $\Delta_{\hat{\varphi}}$.

Theorem 9.7.1 has the following important consequence

9.7.2. Corollary. *Let $\varphi : U \to J^{2c-1}Y$ be a field of extremals of a Lagrangean system α_E on $J^{2c-1}Y$, let $\lambda_{\min}$ be a minimal order Lagrangian of α_E over U such that $\hat{\varphi}^*\lambda_{\min} \neq 0$. Let $\iota : Q \to U$ be an integral manifold of the Hamilton-Jacobi distribution $\Delta_{\hat{\varphi}}$. Then $\iota^*\varphi^*\theta_{\lambda_{\min}}$ is a line element on Q.*

Proof. Let $\iota : Q \to U$ be an integral manifold of $\Delta_{\hat{\varphi}}$. We have to show that $\iota^*\varphi^*\theta_{\lambda_{\min}}$ is a nowhere zero one-form on Q. Let $x \in Q$ and $\xi \in T_xQ$. Then

$$\iota^*\varphi^*\theta_{\lambda_{\min}}(\xi)(x) = \varphi^*\theta_{\lambda_{\min}}(T\iota \cdot \xi)(\iota(x)) \neq 0,$$

since the vector $T\iota \cdot \xi(\iota(x))$ is tangent to Q, and hence does not belong to $\mathcal{C}_{\lambda_{\min}}(\iota(x))$. □

Notice the difference between the regular and non-regular case: In general, $\iota^*\varphi^*\theta_{\lambda_{\min}}$ is a line element along a *quasi-holonomic Hamilton extremal*. In case that the considered Lagrangean system is regular, this form is a line element along the $(c-1)$-prolongation of an *extremal*.

9.7.3. Definition. Let $\iota : Q \to U \subset J^{c-1}Y$ be an integral manifold of the Hamilton-Jacobi distribution $\Delta_{\hat{\varphi}}$. The one-form $\iota^*\varphi^*\theta_{\lambda_{\min}}$ on Q will be called a *line element induced by the minimal-order Lagrangian* $\lambda_{\min}$.

Let $y \in U$ be a point, $\iota : Q \to U$ an integral manifold of $\Delta_{\hat{\varphi}}$ such that $y \in \iota(Q)$. For a fixed integral manifold $\kappa : M \to U$ of the distribution $\mathcal{C}_{\lambda_{\min}}$ such that $\iota(Q) \cap \kappa(M) \neq \emptyset$ we set

$$A(y) = \int_{y_0}^{y} \iota^*\varphi^*\theta_{\lambda_{\min}}, \tag{9.54}$$

where $y_0 = \iota(Q) \cap \kappa(M)$. There arises a real function $A : J^{c-1}Y \supset U \to R$ which will be called the *geodesic distance* of y from the hypersurface $\kappa(M)$ *in the Hamilton-Jacobi field* $\hat{\varphi}$, *induced by the Lagrangian* $\lambda_{\min}$. If, in particular, the Lagrangean system α_E is *regular* then A will be called the *geodesic distance* of y from the hypersurface $\kappa(M)$ *in the field of extremals* φ, *induced by the Lagrangian* $\lambda_{\min}$.

9.7.4. Proposition.

(1) *The function $A : J^{c-1}Y \supset U \to R$ defined by* (9.54) *is a solution of the Hamilton-Jacobi equation*

$$H \circ \varphi = -\frac{\partial A}{\partial t}, \quad p_\nu^k \circ \varphi = \frac{\partial A}{\partial q_k^\nu}, \quad 0 \leqslant k \leqslant c-1,$$

where H and p_ν^k are the Hamiltonian and momenta associated to $\lambda_{\min}$.

(2) *If α_E is regular then the function $A : J^{c-1}Y \supset U \to R$ defined by* (9.54) *is a solution of the eikonal equation*

$$\frac{\partial A}{\partial t} + H\left(t, q^\sigma, \dots, q^\sigma_{c-1}, \frac{\partial A}{\partial q^\sigma}, \dots, \frac{\partial A}{\partial q^\sigma_{c-1}}\right) = 0,$$

where H is the Hamiltonian associated to $\lambda_{\min}$.

Proof. Let f be an integral mapping of $\Delta_{\hat{\varphi}}$ defined on an interval $[0, t]$ in R such that $f([0,t]) \subset \iota(Q)$, and $f(0) = y_0$, $f(t) = y$. Since f is an embedding, we have along Q,

$$A(y) = \int_{y_0}^{y} \iota^*\varphi^*\theta_{\lambda_{\min}} = \int_{y_0}^{y} \iota^* dS = \int_0^t f^*\iota^* dS = S(y) - S(y_0),$$

where $S(y_0)$ is a constant. By assumption, S is a solution of the Hamilton-Jacobi equation (9.41), resp. of the eikonal equation (9.43) in the regular case, hence A satisfies the same equation. □

We have the following easy corollaries which show relations among φ, $\lambda_{\min}$ and S for odd-order Lagrangean systems.

9.7.5. Corollary.

(1) *If $S(y_0) = 0$ then $S(y)$ is the geodesic distance of y from the hypersurface $S = 0$ in the Hamilton-Jacobi field $\hat{\varphi}$, induced by λ_{min}.*

(2) *If α_E is regular and $S(y_0) = 0$ then $S(y)$ is the geodesic distance of y from the hypersurface $S = 0$ in the field of extremals φ, induced by λ_{min}.*

9.7.6. Corollary. *Let α_E be a regular Lagrangean system on $J^{2c-1}Y$, λ_{min} its local minimal-order Lagrangian. Let $S : U \to R$ be a solution of the eikonal equation*

$$\frac{\partial S}{\partial t} + H\left(t, q^\sigma, \dots, q^\sigma_{c-1}, \frac{\partial S}{\partial q^\sigma}, \dots, \frac{\partial S}{\partial q^\sigma_{c-1}}\right) = 0,$$

where H is the Hamiltonian associated to λ_{min}. There exists a field of extremals φ such that $\varphi^*\theta_{\lambda_{\mathrm{min}}} = dS$. If $\varphi^*\lambda_{\mathrm{min}} \neq 0$, and $0 \in S(U)$ then for every $y \in U$, $S(y)$ is the geodesic distance of y from the hypersurface $S = 0$ in the field of extremals φ.

Proof. We get an appropriate field of extremals φ by putting $p^i_\sigma \circ \varphi = \partial S / \partial q^\sigma_i$, $1 \leqslant i \leqslant c-1$. □

9.7.7. Corollary. *Let α_E be a Lagrangean system on $J^{2c-1}Y$. Let $U \subset J^{c-1}Y$ be be an open set, $S : U \to R$ a function. Let φ be a field o extremals of α_E, defined on U. Then there exists a unique minimal-order Lagrangian λ_{min} such that $\varphi^*\theta_{\lambda_{\mathrm{min}}} = dS$. If $0 \in S(U)$, and*

$$\partial_\xi S \neq 0$$

for a generator ξ of the Hamilton-Jacobi distribution $\Delta_{\hat{\varphi}}$ then $S(y)$ is the geodesic distance of y from the hypersurface $S = 0$ in the Hamilton-Jacobi field $\hat{\varphi}$, induced by the Lagrangian λ_{min}. If α_E is regular then $S(y)$ is the geodesic distance of y from the hypersurface $S = 0$ in the field of extremals φ, induced by the Lagrangian λ_{min}.

Proof. By Theorem 9.6.4 there exists a unique minimal-order Lagrangian λ_φ of α_E on U such that $\varphi^*\theta_{\lambda_\varphi} = 0$. Put

$$\lambda_{\mathrm{min}} = \lambda_\varphi + h\, dS.$$

Then λ_{min} is the desired Lagrangian. Finally, if $\Delta_{\hat{\varphi}}$ is spanned by

$$\xi = f\left(\frac{\partial}{\partial t} + \sum_{i=0}^{c-2} q^\sigma_{i+1} \frac{\partial}{\partial q^\sigma_i} + \varphi^\sigma_{c-1} \frac{\partial}{\partial q^\sigma_{c-1}}\right),$$

where f is a nowhere zero function then by the Hamilton-Jacobi equation (9.41),

$$\partial_\xi S = f\left(-H \circ \varphi + \sum_{i=0}^{c-1} (p^i_\sigma q^\sigma_{i+1}) \circ \varphi\right) = f\,(L_{\mathrm{min}} \circ \varphi).$$

Now, if $\partial_\xi S \neq 0$ then $L \circ \varphi \neq 0$, i.e., the condition (9.53) is satisfied, and we get our assertion using Corollary 9.7.5. □

We shall close this section with a theorem which is *inverse* to Theorem 9.7.1, saying that *every* distribution complementary to a Hamilton-Jacobi distribution comes from a minimal-order Lagrangian such that (9.53) is satisfied:

9.7.8. Theorem. *Let α_E be a Lagrangean system on $J^{2c-1}Y$, $\varphi : U \to J^{2c-1}Y$ a field of extremals of α_E, defined on an open subset U of the quasi-configuration space. Let $\Delta_{\hat{\varphi}}$ be the Hamilton-Jacobi distribution associated with φ. Let $\mathcal{C}$ be a distribution on U, complementary to $\Delta_{\hat{\varphi}}$. Then to each point of $\pi^{-1}_{c,c-1}(U)$ there exists a neighborhood W and a minimal-order Lagrangian $\lambda_{\min}$ of α_E defined on W, such that*

$$\varphi^*\lambda_{\min} \neq 0, \quad \textit{and} \quad \mathcal{C}^0 = \operatorname{span}\{\varphi^*\theta_{\lambda_{\min}}\}.$$

Proof. By assumption, U can be covered by local charts of the form (W, χ), $\chi = (u, x^J)$ where $1 \leqslant J \leqslant mc$, such that

$$\Delta_{\hat{\varphi}} = \operatorname{span}\left\{\frac{\partial}{\partial u}\right\}, \quad \mathcal{C} = \operatorname{span}\left\{\frac{\partial}{\partial x^J}, \ 1 \leqslant J \leqslant mc\right\} \approx \operatorname{span}\{du\}.$$

Let λ_φ be the minimal-order Lagrangian of α_E on U such that $\varphi^*\theta_{\lambda\varphi} = 0$. Putting

$$\lambda_{\min} = \lambda_\varphi - h\,du$$

we get a minimal-order Lagrangian for α_E on W such that $\varphi^*\theta_{\lambda_{\min}} = du$, i.e., $\mathcal{C}^0 = \operatorname{span}\{\varphi^*\theta_{\lambda_{\min}}\}$. Moreover,

$$\varphi^*\lambda_{\min} = \varphi^* h\,du = \left(\frac{du}{dt} \circ \varphi\right) dt = \partial_\xi u,$$

where ξ is the semispray spanning $\Delta_{\hat{\varphi}}$. However, $\xi = f(\partial/\partial u)$ for a nowhere zero function f, which means that $\varphi^*\lambda_{\min} \neq 0$, as desired. □

9.8. Two illustrative examples

9.8.1. Consider the Lagrangian

$$L = \tfrac{1}{2}(\ddot{q}^2 - \dot{q}^2) \tag{9.55}$$

on the fibered manifold $\pi : R \times R \to R$ (L. Klapka [1]). This Lagrangian describes the one-dimensional motion of a "black box" in which a harmonic oscillator is hidden (a system of units is chosen in such a way that the angle frequence of oscillations is a unit one).

Using results of Chapter 4, we get that the Lagrangian (9.55) defines the following *Lagrangean system of order* 3:

$$\alpha_E = -d\left(\tfrac{1}{2}\ddot{q}^2 - \tfrac{1}{2}\dot{q}^2 - \dot{q}\,\dddot{q}\right) \wedge dt - d(\dot{q} + \dddot{q}) \wedge dq + d\ddot{q} \wedge d\dot{q}.$$

The corresponding Euler-Lagrange form is

$$E = (\ddddot{q} + \ddot{q})\,dq \wedge dt,$$

which, by (6.2), means that the Lagrangean system is *regular*. (Notice that the Lagrangian (9.55) is regular also within the standard understanding of regulatity).

We have $s = 4$ and $c = 2$, i.e., the Euler-Lagrange equations are of order 4 and L is a minimal-order Lagrangian. The configuration space is $R \times R$, the phase space is $J^3(R \times R) = R \times R^4$, and the quasi-configuration space is $J^1(R \times R) = R \times R^2$.

The Euler-Lagrange distribution coincides with the characteristic distribution and it is spanned by a vector field

$$\zeta = \frac{\partial}{\partial t} + \dot{q}\frac{\partial}{\partial q} + \ddot{q}\frac{\partial}{\partial \dot{q}} + \dddot{q}\frac{\partial}{\partial \ddot{q}} - \ddot{q}\frac{\partial}{\partial \dddot{q}}.$$

The Hamiltonian related with the minimal-order Lagrangian (9.55) is

$$H = \tfrac{1}{2}\ddot{q}^2 - \tfrac{1}{2}\dot{q}^2 - \dot{q}\,\dddot{q}, \tag{9.56}$$

and since the Lagrangean system is *autonomous*, this is the unique time-independent Hamiltonian (Proposition 4.5.10). Legendre transformation with respect to the minimal-order Lagrangian (9.55) is a diffeomorphism of the phase space,

$$(t, q, \dot{q}, \ddot{q}, \dddot{q}) \to (t, q, \dot{q}, p_0, p_1),$$

where

$$p_0 = -(\dot{q} + \dddot{q}), \quad p_1 = \ddot{q}. \tag{9.57}$$

Hence, in the Legendre coordinates,

$$H = \tfrac{1}{2} p_1^2 + p_0\dot{q} + \tfrac{1}{2}\dot{q}^2,$$

and

$$\zeta = \frac{\partial}{\partial t} + \dot{q}\frac{\partial}{\partial q} + p_1\frac{\partial}{\partial \dot{q}} - (\dot{q} + p_0)\frac{\partial}{\partial p_1}.$$

According to (6.37), the Hamilton canonical equations read

$$\frac{d(p_0 \circ \delta)}{dt} = 0, \quad \frac{d(p_1 \circ \delta)}{dt} = -(\dot{q} + p_0)\circ\delta,$$
$$\frac{d(q \circ \delta)}{dt} = \dot{q}\circ\delta, \quad \frac{d(\dot{q} \circ \delta)}{dt} = p_1 \circ \delta.$$

Since by (8.1) the vector fields $\partial/\partial t$ and $\partial/\partial q$ are symmetries of the Lagrangian (9.55), Noether Theorem gives us that the functions H and p_0 are *first integrals* of the Lagrangean system α_E (H is the total energy and p_0 is the momentum of the center of mass). Checking the condition (9.17) we get

$$\det\begin{pmatrix} \dfrac{\partial H}{\partial \ddot{q}} & \dfrac{\partial H}{\partial \dddot{q}} \\ \dfrac{\partial p_0}{\partial \ddot{q}} & \dfrac{\partial p_0}{\partial \dddot{q}} \end{pmatrix} = \det\begin{pmatrix} \ddot{q} & -\dot{q} \\ 0 & -1 \end{pmatrix} = -\ddot{q},$$

i.e., this matrix is regular on the open subset $U = R \times R^4 - \{\ddot{q} = 0\}$ of the phase space. Applying Theorem 9.2.6, we can see that the first integrals H and p_0 are on U independent and in involution, i.e., they define the *complete integral*

$$\mathcal{J} \approx \operatorname{span}\{dH, dp_0\},$$

which, by Definition 9.3.2, is a *Jacobi complete integral* of α_E. Applying the Jacobi Theorem 9.3.4 we can compute other two first integrals b_1, b_2 which together with H and p_0 form a complete set of independent first integrals of α_E on U. Expressing the form θ_λ in the Jacobi coordinates $(t, q, \dot{q}, H, p_0)$ on U (cf. 9.29), we get

$$\theta_\lambda = -H\,dt + p_0\,dq + \sqrt{2H - \dot{q}^2 - 2\dot{q}\,p_0}\,d\dot{q} = dS - b_1\,dH - b_2\,dp_0,$$

which, with the help of (9.11), gives us

$$S(t, q, \dot{q}, H, p_0) = -Ht + p_0 q + I\dot{q}$$

where

$$I = \int_0^1 \sqrt{2H - \dot{q}^2 v^2 - 2\dot{q}\, p_0 v}\, dv\,.$$

Now, by (9.22),

$$b_1 = \frac{\partial S}{\partial H} = -t + \frac{\partial I}{\partial H}, \quad b_2 = \frac{\partial S}{\partial p_0} = q + \frac{\partial I}{\partial p_0}$$

are desired first integrals. We obtain a chart (t, H, p_0, b_1, b_2) which is a *Darboux chart* of α_E, hence, adapted to the Euler-Lagrange distribution; in this chart,

$$\alpha_E = dH \wedge db_1 + dp_0 \wedge db_2$$

and

$$\zeta = \frac{\partial}{\partial t}.$$

Since our Lagrangean system is regular and odd-order, we can use Theorem 9.6.3 to get the extremals. By (9.49) we have

$$\beta_1 = b_1|_{\{H=\text{const}, p_0=\text{const}\}}, \quad \beta_2 = b_2|_{\{H=\text{const}, p_0=\text{const}\}},$$

which are implicit equations for $q(t)$, $\dot{q}(t)$, giving us extremals $\gamma = (t, q(t), \dot{q}(t))$. Notice that in this way we get extremals assorted into *fields of extremals* (each of which contains extremals corresponding to the same (constant) energy, and center of mass).

9.8.2. Let us turn to an example of a *regular even-order Lagrangean system*.

Consider the fibered manifold $pr_1 : R \times R^2 \to R$ with the canonical coordinates (t, x, y) and the two-form

$$E = -(m_1\dddot{x} + \dddot{y})\, dx \wedge dt + (-m_2\dddot{y} + \dddot{x})\, dy \wedge dt \tag{9.58}$$

on $J^3(R \times R^2)$ (O. Krupková [2]). E is a dynamical form satisfying (4.9), i.e., it is *variational*, and defines a *second-order Lagrangean system* α_E. Hence, we have $s = 3$, $c = 1$; the phase space is $J^2(R \times R^2)$, the quasi-configuration space coincides with the configuration space $R \times R^2$.

Let us compute a family of momenta and the related Hamiltonian. By (4.36), we get

$$f = \tfrac{1}{2}(\ddot{x}y - \ddot{y}x) + \phi(t, x, y, \dot{x}, \dot{y}),$$

where ϕ is a gauge function. Let us choose ϕ in the form

$$\phi = -\tfrac{1}{2}(m_1\dot{x}x + m_2\dot{y}y). \tag{9.59}$$

Then we get from (4.37), (4.40)

$$\begin{aligned} p_x^1 &= -\tfrac{1}{2}y, \quad p_x^0 = \ddot{y} + m_1\dot{x}, \\ p_y^1 &= \tfrac{1}{2}x, \quad p_y^0 = -\ddot{x} + m_2\dot{y}, \\ H &= \tfrac{1}{2}m_1\dot{x}^2 + \tfrac{1}{2}m_2\dot{y}^2 + \dot{x}\ddot{y} - \dot{y}\ddot{x}. \end{aligned} \tag{9.60}$$

By (6.2), the given Lagrangean system is *regular*. Since the Jacobi matrix of the mapping $(t, x, y, \dot{x}, \dot{y}, \ddot{x}, \ddot{y}) \to (t, x, y, p_x^0, p_y^0, p_x^1, p_y^1)$ is regular, we get that the mapping is a Legendre transformation. The inverse Legendre transformation is defined by

$$\dot{x} = 2p_y^1, \quad \dot{y} = -2p_x^1,$$
$$\ddot{x} = -p_y^0 - 2m_2 p_x^1, \quad \ddot{y} = p_x^0 - 2m_1 p_y^1.$$

In the Legendre coordinates,

$$H = 2p_x^0 p_y^1 - 2p_y^0 p_x^1 - 2m_1 (p_y^1)^2 - 2m_2 (p_x^1)^2,$$

and the Hamilton canonical equations by (6.43) take the form

$$\begin{aligned} \frac{dx}{dt} &= 2p_y^1, \quad \frac{dy}{dt} = -2p_x^1, \\ \frac{dp_x^0}{dt} &= 0, \quad \frac{dp_y^0}{dt} = 0, \\ \frac{dp_x^1}{dt} &= -\tfrac{1}{2}p_x^0 + m_1 p_y^1, \quad \frac{dp_y^1}{dt} = -\tfrac{1}{2}p_y^0 - m_2 p_x^1. \end{aligned} \tag{9.61}$$

Notice that by (6.41) we have for α_E,

$$\alpha_E = -dH \wedge dt + dp_x^0 \wedge dx + dp_y^0 \wedge dy + 4\, dp_x^1 \wedge dp_y^1.$$

The Euler-Lagrange distribution is spanned by a semispray

$$\zeta = \frac{\partial}{\partial t} + \dot{x}\frac{\partial}{\partial x} + \dot{y}\frac{\partial}{\partial y} + \ddot{x}\frac{\partial}{\partial \dot{x}} + \ddot{y}\frac{\partial}{\partial \dot{y}} + m_2 \ddot{y}\frac{\partial}{\partial \ddot{x}} - m_1 \ddot{x}\frac{\partial}{\partial \ddot{y}},$$

which in the Legendre coordinates reads

$$\zeta = \frac{\partial}{\partial t} + 2p_y^1\frac{\partial}{\partial x} - 2p_x^1\frac{\partial}{\partial y} - (\tfrac{1}{2}p_x^0 - m_1 p_y^1)\frac{\partial}{\partial p_x^1} - (\tfrac{1}{2}p_y^0 + m_2 p_x^1)\frac{\partial}{\partial p_y^1}.$$

It is clear that another choice of a gauge function ϕ than (9.59) could lead to another Legendre coordinates of the given Lagrangean system, hence to another form of the Hamilton canonical equations than is that of (9.61).

It is worth to compute the minimal-order Lagrangian corresponding to the choice of ϕ in the form (9.59): using (4.42) we get

$$L_{\min} = \tfrac{1}{2}m_1\dot{x}^2 + \tfrac{1}{2}m_2\dot{y}^2 + \tfrac{1}{2}(\dot{x}\ddot{y} - \dot{y}\ddot{x}). \tag{9.62}$$

Notice that to get a family of momenta and the corresponding Hamiltonian for our Lagrangean system, we could proceed also in a "more standard way": first we could find a minimal-order Lagrangian for E, applying the formulas (4.6) and (4.42) for the Vainberg-Tonti Lagrangian and minimal-order Lagrangian, respectively, and then we could compute the corresponding p's and H according to (4.49).

Now, let us apply to our Lagrangean system the Hamilton-Jacobi integration method. It is obvious that the functions H, p_x^0, p_y^0 are *first integrals in involution*. (Note that H, p_x^0, and p_y^0 correspond to the symmetries $\partial/\partial t$, $\partial/\partial x$, and $\partial/\partial y$ of the minimal-order

Lagrangian (9.62), respectively). The condition (9.17) gives us

$$\operatorname{rank}\begin{pmatrix} \dfrac{\partial H}{\partial \dot x} & \dfrac{\partial H}{\partial \dot y} & \dfrac{\partial H}{\partial \ddot x} & \dfrac{\partial H}{\partial \ddot y} \\ \dfrac{\partial p_x^0}{\partial \dot x} & \dfrac{\partial p_x^0}{\partial \dot y} & \dfrac{\partial p_x^0}{\partial \ddot x} & \dfrac{\partial p_x^0}{\partial \ddot y} \\ \dfrac{\partial p_y^0}{\partial \dot x} & \dfrac{\partial p_y^0}{\partial \dot y} & \dfrac{\partial p_y^0}{\partial \ddot x} & \dfrac{\partial p_y^0}{\partial \ddot y} \end{pmatrix}$$

$$= \operatorname{rank}\begin{pmatrix} m_1\dot x + \ddot y & m_2\dot y - \ddot x & -\dot y & \dot x \\ m_1 & 0 & 0 & 1 \\ 0 & m_2 & -1 & 0 \end{pmatrix} = \operatorname{rank}\begin{pmatrix} 1 & 0 & 0 & m_1 \\ 0 & -1 & m_2 & 0 \\ 0 & 0 & -\ddot x & \ddot y \end{pmatrix} = 3$$

on the open subset $W = R \times R^6 - \{\ddot x = \ddot y = 0\}$ of the phase space. On this set, the first integrals H, p_x^0, p_y^0 are independent, and (by Theorem 9.2.6 and Definition 9.3.2)

$$\mathfrak{I} \approx \operatorname{span}\{dH,\ dp_x^0,\ dp_y^0\}$$

is a *Jacobi complete integral* of α_E. Let us choose some Jacobi coordinates. Put $U = \{\ddot x \neq 0\}$ and consider the mapping

$$(t, x, y, \dot x, \dot y, \ddot x, \ddot y) \to (t, x, y, H, p_x^0, p_y^0, \dot x).$$

U is an open subset of W, and the above mapping is a coordinate transformation on U, hence $(U, (t, x, y, H, p_x^0, p_y^0, \dot x))$ is a *Jacobi chart.* Putting

$$g = \sqrt{(p_y^0)^2 - 2m_2H - m_1m_2\dot x^2 + 2m_2\dot x p_x^0},$$

we get the inverse transformation formulas in the form

$$\dot y = \frac{1}{m_2}(p_y^0 + g), \quad \ddot x = g, \quad \ddot y = p_x^0 - m_1\dot x.$$

Now, we can apply Jacobi Theorem 9.3.5. The equations for S are by (9.27)

$$\frac{\partial S}{\partial t} = -H, \quad \frac{\partial S}{\partial x} = p_x^0, \quad \frac{\partial S}{\partial y} = p_y^0, \quad \frac{\partial S}{\partial \dot x} = -\frac{1}{2m_2}(p_y^0 + g) + \frac{1}{2g}\dot x(p_x^0 - m_1\dot x),$$

hence

$$S = -Ht + p_x^0 x + p_y^0 y + \dot x \int_0^1 F(H, p_x^0, p_y^0, v\dot x)\, dv,$$

where

$$F = -\frac{1}{2m_2}(p_y^0 + g) + \frac{1}{2g}\dot x(p_x^0 - m_1\dot x).$$

Computing first integrals according to (9.26), we get

$$b_1 = \frac{\partial S}{\partial H} - \tfrac{1}{2}\dot x\frac{\partial \dot y}{\partial H} = \frac{\partial S}{\partial H} + \frac{1}{g}\dot x,$$

$$b_2 = \frac{\partial S}{\partial p_x^0} - \tfrac{1}{2}\dot x\frac{\partial \dot y}{\partial p_x^0} = \frac{\partial S}{\partial p_x^0} - \frac{1}{2g}\dot x^2,$$

$$b_3 = \frac{\partial S}{\partial p_y^0} - \tfrac{1}{2}\dot x\frac{\partial \dot y}{\partial p_y^0} = \frac{\partial S}{\partial p_y^0} - \frac{1}{2m_2}\dot x\,(1 + g^{-1}p_y^0).$$

The functions H, p_x^0, p_y^0, b_1, b_2, b_3 are a complete set of independent first integrals of the given Lagrangean system on U.

It should be emphasized that the Lagrangian (9.62) is not regular in the standard sense; this means that it cannot be investigated by methods of the standard (higher-order) Hamilton-Jacobi theory.

9.9. A few remarks on integration of non-variational equations

Differential equations which can be interpreted as equations for integral mappings of a completely integrable *distribution* can be solved if one knows a set of independent first integrals, the existence of which is ensured by the Frobenius theorem. The problem is to find such a family; to this purpose certain methods have been developed, based on relations between first integrals and symmetries of the corresponding distribution (S. V. Duzhin and V. V. Lychagin [1], V. V. Lychagin [1]). The application of these methods is based on finding a family of symmetries with appropriate properties; the number of such symmetries equals to the corank of the distribution in question.

We have seen that if the equations are *variational*, there are also other methods which enable one to find the solution. To apply them, it is sufficient to know appropriate symmetries the number of which equals $1/2$ of the corank of the distribution. Some of these integration methods have been recently generalized also to *non-variational equations* (though their concrete application in the case when a Lagrangian is known is easier and more straightforward than in the general case).

Notice that in fact the range for application of the Liouville integration method in Sec. 9.2 and the method of canonical transformations in Sec. 9.3 is wider than merely variational equations: they can be applied to any distribution (on a general manifold) which can be interpreted as a characteristic distribution of (possibly locally) defined closed 2-form (this includes, in particular, the case of a vector field or of a "higher-order differential equation vector field").

The integration method based on fields of extremals can be generalized to any semispray distribution on a fibered manifold (cf. the references in the introduction to this chapter). This extension covers both systems of regular ordinary differential equations (the so called "higher-order equation vector fields") and systems of partial differential equations which are "solved" with respect to the highest derivatives.

Since for variational equations the application of concrete integration methods turns out to be more simple than in the case when a Lagrangian does not exist, it is possible in principle to combine the integration problem with the inverse problem of the calculus of variations (the problem under which conditions given equations are equivalent with some variational equations—see Chapter 7 for more details). Then if one would find the equivalent Euler-Lagrange form, one could apply some of the exact integration methods for variational equations, and, in this way, to solve the original integration problem. Such a procedure is quite advantageous namely if one regular second-order differential equation has to be solved (as we know, in this case a variational integrating factor always exists and can be found by means of integration of a relatively simple equation (cf. Chapter 1).

Chapter 10.

LAGRANGEAN SYSTEMS ON $\pi : R \times M \to R$

10.1. Introduction

There is a rich bibliography of books and papers dealing with (in the stadard sense) *regular*, and *autonomous* Lagrangians (symplectic geometry, Finsler geometry). To study other kinds of Lagrangians, different authors use different geometric structures and constructions, which are difficult to be compared with each other. In particular, often it is not possible to transfer in a straightforward way constructions from $T^r M$ to fibered manifolds $R \times T^r M \to R$, and then to $Y \to R$, or finally to $Y \to X$. Moreover, it may be not clear at a first sight how these results fit in our general scheme.

In this book, we have presented an opposite approach to build a general theory: not to generalize known results from classical Lagrangians to more general ones, but better to build a systematic theory of variational ODE, covering all the possibilities, and then to study its consequences for particular cases of variational problems. We obtained a theory which applies to any Lagrangean system on any fibered manifold over an one-dimensional base. Now, we want to touch some important particular cases in more detail to show that, on the one hand, the theory of Lepagean two-forms covers standard approaches to Lagrangean systems and represents in fact their universal generalization to all kinds of Lagrangians, and on the other hand, it helps to reveal limitations and reasons for ambiguities in standard approaches when non-classical Lagrangians are considered.

Sections 10.2 and 10.3 in this chapter are devoted to relations of the theory of Lepagean two-forms with the symplectic/presymplectic geometry, and Section 10.4 with Finsler geometry and its generalizations.

10.2. Lagrangean systems on $R \times M \to R$

First we shall discuss some of the additional geometric properties of Lagrangean systems which appear in the case when the underlying fibered manifold is simplified to a product over R.

More precisely, throughout this chapter we shall consider a fibered manifold $\pi : R \times M \to R$, where π is the first canonical projection, and M is a manifold of dimension m. In particular, this structure is suitable for classical or relativistic mechanics. In classical mechanics, the base R represents time, and M is the manifold of configurations. In relativistic mechanics, the base R represents a space of parameters, and M is the space-time which usually is considered to be R^4 with the Minkowski metric.

We shall denote the global coordinate on R by t and adopt the most frequent terminology calling t the *time* (although for relativistic mechanics, this terminology is not suitable). On $R \times M \to R$ we shall utilize fibered coordinates of the form (t, q^σ), where (q^σ) are local coordinates on M. Working with an atlas of this kind is natural and convenient, since the transformation formulas are simplified to

$$\bar{t} = t, \quad \bar{q}^\sigma = \bar{q}^\sigma(q^1, \ldots, q^m), \quad 1 \leqslant \sigma \leqslant m,$$

i.e., the "time" and "spacc" coordinates transform independently. For simplicity, wc shall call an atlas of this kind *adapted* to the product structure $R \times M$ of the total space.

Recall that the r-jet prolongation $J^r(R \times M)$ identifies with $R \times T^r M$.

The vector field

$$\xi = \frac{\partial}{\partial t}$$

on $R \times M$ is the global generator of the *time translation* on the configuration space. It defines a *connection* on $R \times M$. For every $r \geqslant 1$ it is a global generator of an everywhere non-vertical distribution on $R \times T^r M$. Consequently, every vector field on $R \times M$, or on $R \times T^r M$ uniquely splits into two parts – the horizontal part, belonging to the distribution $\operatorname{span}\{\partial/\partial t\}$, and the vertical part, which is a π_r-vertical vector field.

Now, let us turn to some interesting geometrical properties of Lagrangean systems, which are connected with the special product structure $R \times M$ of the configuration space, and which do not remain valid in more general fibered manifolds.

Consider on $R \times M$ an adapted atlas. Let α_E be a Lagrangean system of order $s - 1 \geqslant 0$ (i.e, defined on $R \times T^{s-1} M$). From the Euler-Lagrange form E there arises a global *one-form*

$$-i_{\partial/\partial t} E = E_\sigma \, dq^\sigma$$

which is often used to describe the Euler-Lagrange equations instead of the Euler-Lagrange form.

Let H, and p^k_ν, $1 \leqslant \nu \leqslant m$, $0 \leqslant k \leqslant s - c$, be a Hamiltonian and momenta of α_E (c is defined by 4.5.1). The Lepagean two-form

$$\alpha_E = -dH \wedge dt + \sum_{k=0}^{s-c-1} dp^k_\nu \wedge dq^\nu_k$$

splits into two *invariant* (global) two-forms: $-dH \wedge dt$, and

$$\omega = \sum_{k=0}^{s-c-1} dp^k_\nu \wedge dq^\nu_k. \tag{10.1}$$

Notice that the rank of ω generally depends both on the q's and on time (i.e., it may be different on different fibers).

Similarly, the Lepagean equivalent (Cartan form) of a minimal-order Lagrangian $\lambda_{\min}$,

$$\theta_{\lambda_{\min}} = -H \, dt + \sum_{k=0}^{s-c-1} p^k_\nu \, dq^\nu_k$$

splits into two *invariant* parts: the one-forms $-H \, dt$, and

$$\eta = \sum_{k=0}^{s-c-1} p^k_\nu \, dq^\nu_k. \tag{10.2}$$

For the H and p's we get the following simplified transformation formulas:

$$\bar{H} = H, \quad \bar{p}^i_\sigma = p^k_\nu \frac{\partial q^\nu_k}{\partial \bar{q}^\sigma_i}.$$

A *global* Lagrangian λ of order r can be represented by a *global function* $L : R \times T^r M \to R$; this is a very frequent definition of a Lagrangian on $R \times T^r M$.

For a global minimal-order Lagrangian we get a global one-form $H\,dt$, and consequently, a *global Hamiltonian* H, defined on the phase space $R \times T^{s-1} M$.

Hamilton equations of a Lagrangean system α_E on $R \times T^{s-1} M$ are of the form

$$\delta^* i_\xi \alpha_E = 0, \tag{10.3}$$

for every π_{s-1}-vertical vector field ξ. Recall that they are equations for *sections* δ *of the fibered manifold* π_{s-1} (Hamilton extremals). If α_E is *semiregular* (i.e., if the Euler-Lagrange distribution is of a *constant rank and weakly horizontal*) then these equations are *equivalent* to the following equations for *vector fields* ζ *on the phase space*

$$i_\zeta \alpha_E = 0, \quad T\pi_{s-1} \cdot \zeta = \frac{\partial}{\partial t}, \tag{10.4}$$

i.e., to *equations for non-vertical vector fields belonging to the Euler-Lagrange* (resp. to the characteristic) *distribution*. We stress that for more general Lagrangean systems (i.e., possessing dynamical constraints), equations (10.3) and (10.4) are no more equivalent. In other words, the dynamics of constrained systems generally *is not* obtained completely by finding "all vector fields ζ satisfying the equations (10.4)". As we have seen in Chapter 7, in this case one can get a complete dynamical picture only by *applying the constraint algorithm to the Euler-Lagrange distribution.*

10.2.1. Remark. Lagrangean systems on a fibered manifold $R \times M \to R$ are also called *precosymplectic* Lagrangean systems. Regular precosymplectic Lagrangean systems are usually called *cosymplectic*.

10.3. Autonomous Lagrangean systems: Higher-order symplectic and presymplectic systems

Consider a fibered manifold $R \times M \to R$ and its prolongation $R \times T^{s-1} M$, where $s \geqslant 1$ (in this notation we identify $T^0 M$ with M).

Taking into account results of Chapter 8 we can define autonomous Lagrangean systems as follows:

10.3.1. Definition. A Lagrangean system α_E will be called *time-independent*, or *autonomous* if the generator of the time translation is a symmetry of the Euler-Lagrange form E, or, *equivalently*, of the Lepagean two-form α_E. Autonomous Lagrangean systems on $R \times M \to R$ are also called *presymplectic* Lagrangean systems.

Expressing the above requirement in coordinates (using the Noether–Bessel-Hagen equation) we get that a Lagrangean system is autonomous if and only if its Euler-Lagrange expressions do not depend on time explicitly, i.e.,

$$\frac{\partial E_\sigma}{\partial t} = 0 \quad \text{for all } \sigma.$$

From the formula for local Lagrangians we can see immediately that Vainberg-Tonti Lagrangians of an autonomous Lagrangean system do not depend on time explicitly. This means that every autonomous Lagrangean system possesses (local) *time-independent minimal-order Lagrangians*. Consequently, every autonomous Lagrangean system possesses *time-independent momenta*, and a *unique up to a constant, time-independent Hamiltonian* H, called the *total energy* of this Lagrangean system (see Proposition 4.5.10, (4.51)). Notice that, however, for a time-independent Lagrangean system there exist time-dependent minimal-order Lagrangians, momenta and Hamiltonians, as well. From Chapter 8 we also get that the vector field $\partial/\partial t$ is a symmetry of the Lepagean equivalents $\theta_{\lambda_{\min}}$ of all time-independent minimal-order Lagrangians, and that the total energy H is a *first integral* of the characteristic distribution $\mathcal{D}_E$. Consequently,

$$H = -i_{\partial/\partial t}\,\theta_{\lambda_{\min}}.$$

There arises a unique global *exact* one-form on the phase space

$$\varepsilon = i_{\partial/\partial t}\,\alpha_E = dH,$$

where H is the total energy, called the *energy one-form*. Notice that although for a time-independent Lagrangean system one generally does not have a distinguished Lagrangian, the system possesses a unique global energy one form (even in case that there does *not* exist a global Lagrangian), and a unique up to a constant total energy.

From (10.1) we can see that an autonomous Lagrangean system gives rise to a global time-independent closed two form

$$\omega = \alpha_E + dH \wedge dt = \sum_{k=0}^{s-c-1} dp_\nu^k \wedge dq_k^\nu. \tag{10.5}$$

Clearly (within an obvious convention), this form can be understood as a closed two-form on $T^{s-1}M$, i.e., in accordance with Definition 2.3.9, as a *presymplectic form*.

Thus, any autonomous Lagrangean system of order $s-1 \geqslant 0$ on a fibered manifold $R \times M \to R$ gives rise to a presymplectic structure on $T^{s-1}M$. In particular, if α_E is of order zero then (M, ω) is a presymplectic manifold.

We can see that every *autonomous* Lagrangean system of order $s-1 \geqslant 0$ can be *uniquely* determined by a presymplectic form ω and a (global) Hamilton function H on $T^{s-1}M$ (this is in a full correspondence with the presymplectic geometry).

For an autonomous Lagrangean system defined by a *global* minimal-order Lagrangian, the form (10.2) is a *Liouville form*. Notice, however, that if the Lagrangean system is not globally variational, we do not have a global Liouville form.

Concerning the *Hamilton equations*, for autonomous Lagrangean systems we have the same arguments as in Sec. 10.2, i.e., a complete and geometrically clear dynamics can be obtained directly from the equations (10.3). In the *semiregular* case one can equivalently use equations (10.4). However, for autonomous Lagrangean systems, these equations can be expressed in a simpler form: Since α_E is time-independent, one can restrict in (10.4) to vector fields ζ which are of the form

$$\zeta = \frac{\partial}{\partial t} + \zeta_0,$$

where ζ_0 is a time-independent vertical vector field (i.e., a vector field tangent to the fibers which is "the same" for all the fibers). Computing now $i_\zeta \alpha_E = 0$, and using that H is a first integral of the Euler-Lagrange distribution Δ_E, i.e., that $i_\zeta dH = 0$ for every vector field belonging to Δ_E, we get

$$i_{\zeta_0}\omega = -dH. \tag{10.6}$$

Equation (10.6) will be called the *presymplectic Hamilton equation*.

Summarizing, we have obtained that the Hamilton equation (10.3) for a *semiregular autonomous* Lagrangean system is *equivalent* with the equation (10.6), which is an equation for vector fields ζ_0 *on* $T^{s-1}M$. In accordance with the definition, semiregularity now means that ω is of a constant rank on $T^{s-1}M$, *and* the conditions of weak horizontality of Δ_E are satisfied (Theorem 5.4.1).

For autonomous Lagrangean systems with dynamical constraints equations (10.4), or even (10.6) are not appropriate. Indeed, often there exist Hamilton extremals which are not integral mappings of a vector field satisfying (10.4)—it may even happen that although there exist Hamilton extremals, the equation (10.4) has no solution at all. To give this equation an interpretation, one has to consider it not as an equation for vector fields on the phase space, but rather as an equation for vector fields along certain submanifolds of the phase space (which have to be determined!, and, as we know from Chapter 7, their union can be quite a complicated subset of $J^{s-1}Y$). Moreover, to get a picture of the proper dynamics, one has to pick up only those solutions which are semisprays along certain submanifolds (cf. the (higher-order) semispray problem discussed in Chapter 7). Of course, such an interpretation gives a sense to the equation (10.4), and the procedure to solve this equation is described by the *constraint algorithm* (see Sec. 7.3).

Also, one should notice that equation (10.3) has a straightforward clear dynamical meaning (being the equation for integral sections of the characteristic distribution of α_E, which describes the dynamics), while the dynamical content of the equation (10.6) is not so straightforward.

Let α_E be a time-independent Lagrangean system of order $s-1$, ω the corresponding closed 2-form on $T^{s-1}M$ (10.5). Then any local automorphism ϕ of $T^{s-1}M$, leaving the form ω *invariant* will be called a *contact transformation*. Since (in the notations of (10.2)) we have $\omega = d\eta$ in a neighborhood of every point in $T^{s-1}Y$, we get

$$\phi^*\eta = \eta + df, \tag{10.7}$$

where f is a function; it will be called a *generating function* of the contact transformation ϕ. Contact transformations can be studied in a way similar to that of canonical transformations (Sec. 9.4). Notice that in local coordinates, equation (10.7) reads

$$\sum_{k=0}^{s-c-1} P_\nu^k \, dQ_k^\nu = \sum_{k=0}^{s-c-1} p_\nu^k \, dq_k^\nu + df,$$

where $P_\nu^k = p_\nu^k \circ \phi$, and $Q_k^\nu = q_k^\nu \circ \phi$, $0 \leqslant k \leqslant s-c-1$. If, in particular, the Lagrangean system α_E is regular and (p_ν^k, q_k^ν) are its Legendre coordinates then the latter relations represent the equations for the components of the automorphism ϕ.

Let us turn to discuss some particular cases of presymplectic Lagrangean systems.

10.3.2. Definition. Let α_E be a Lagrangean system of order $s-1 \geqslant 0$ on a fibered manifold $\pi : Y \to X$. The Lagrangean system α_E will be called *symplectic* if the fibered manifold is of the form $Y = R \times M$ and $X = R$, and the Lagrangean system is *autonomous*, and *regular* (i.e., rank α_E is maximal).

As we have seen so far, for Lagrangean systems possessing the above properties the Lepagean 2-form α_E gives rise to the (global) closed two-form

$$\omega = \sum_{k=1}^{s-c-1} dp_\sigma^k \wedge dq_k^\sigma$$

on $T^{s-1}M$, such that rank $\omega = \dim T^{s-1}M$. This means, however, that $\dim T^{s-1}M$ is even, i.e., ω is a *symplectic form*. In other words, every autonomous regular Lagrangean system on $R \times M \to R$ gives rise to a symplectic structure on $T^{s-1}M$. In particular, if the Lagrangean system α_E is of order zero then (M, ω) is a symplectic manifold.

Notice that, by 6.3.3, if $s = 2c$, then ω is a symplectic form on $T^{2c-1}M$ if and only if (any) minimal-order Lagrangian for α_E satisfies the regularity condition

$$\det\left(\frac{\partial^2 L_{\min}}{\partial q_c^\sigma \partial q_c^\nu}\right) \neq 0 \tag{10.8}$$

(cf. (6.20)). Similarly, if $s = 2c + 1$, then ω is a symplectic form on $T^{2c}M$ if and only if (any) minimal order Lagrangian for the Lagrangean system α_E satisfies the condition

$$\det\left(\frac{\partial^2 L_{\min}}{\partial q_{c+1}^\sigma \partial q_c^\nu} - \frac{\partial^2 L_{\min}}{\partial q_c^\sigma \partial q_{c+1}^\nu}\right) \neq 0 \tag{10.9}$$

(cf. (6.21)).

The Hamilton equation (10.3) is then *equivalent* with the equation (10.6), which has a *unique solution*—the dynamics is deterministic. We emphasize that in the context of symplectic geometry, by Hamilton equation one understands the equation (10.3); we shall call it the *symplectic Hamilton equation*.

Recall (Chapter 2) that the symplectic form ω gives rise to a one-to-one correspondence between vector fields and one-forms on $T^{s-1}M$, $\xi \to i_\xi \omega$. Consequently, one can define for functions on $T^{s-1}M$ the *Poisson bracket*

$$\{f, g\} = i_{\xi_f} i_{\xi_g} \omega = \partial_{\xi_f} g = -\partial_{\xi_g} f,$$

where ξ_f is given by $i_{\xi_f}\omega = df$.

A symplectic Lagrangean system is regular, i.e., *Legendre transformation* is a *local diffeomorphism* of $R \times T^{s-1}M$. Notice that since the Lagrangean system is autonomous, it possesses *time-independent momenta*. Recall that these are obtained either directly by means of the Euler-Lagrange expressions (formulas (4.37), (4.36), where ϕ is any time-independent gauge function), or equivalently, by means of (any) *time-independent minimal-order Lagrangian* of α_E according to (4.49). We stress that the Ostrogradskii momenta (4.52) of Lagrangians of a higher-order than the minimal one cannot be used as a part of Legendre coordinates (see the discussion surrounding the formulas (4.52)).

Restricting ourselves to *time-independent* momenta, we can consider the mapping $(q^\sigma, q_1^\sigma, \ldots, q_{s-1}^\sigma) \to (q^\sigma, \ldots, q_{c-1}^\sigma, p_\sigma, \ldots, p_{s-c-1}^\sigma)$, which apparently is a local diffeomorphism of $T^{s-1}M$. We shall call it a *Legendre transformation* on $T^{s-1}M$. If $s = 2c$, we get in the Legendre coordinates the expression

$$\zeta_0 = \sum_{k=0}^{c-1} \frac{\partial H}{\partial p_\nu^k} \frac{\partial}{\partial q_k^\nu} - \sum_{k=0}^{c-1} \frac{\partial H}{\partial q_k^\nu} \frac{\partial}{\partial p_\nu^k},$$

for the Hamilton vector field ζ_0 on $T^{s-1}M$, and

$$\{f, g\} = \sum_{k=0}^{c-1} \frac{\partial f}{\partial p_\nu^k} \frac{\partial g}{\partial q_k^\nu} - \sum_{k=0}^{c-1} \frac{\partial f}{\partial q_k^\nu} \frac{\partial g}{\partial p_\nu^k}$$

for the Poisson bracket. Hence, the evolution of a function f on $T^{s-1}M$ is given by

$$\partial_{\zeta_0} f = \{H, f\}.$$

(We suggest the reader to write down explicitly analogous formulas for the case $s = 2c+1$, and to compare them with (6.46)–(6.48)).

As we have seen so far, for a geometric description of autonomous regular Lagrangean systems on $\pi : R \times M \to R$ one can utilize both the tools of the theory of Lepagean 2-forms, and of the symplectic geometry. Now, let us investigate the (frequently studied) case of *first-order presymplectic* Lagrangean systems, satisfying the condition $\operatorname{rank} \omega =$ const. In terms of the corresponding Euler-Lagrange expressions, resp. time-independent minimal-order (=first-order) Lagrangians this means that

$$\operatorname{rank}(B_{\sigma\nu}) = \operatorname{rank}\Big(\frac{\partial^2 L}{\partial \dot{q}^\sigma \partial \dot{q}^\nu}\Big) = \text{const},$$

i.e., such systems *admit generalized Legendre transformations*.

(Notice that many authors understand by presymplectic systems just those satisfying the condition $\operatorname{rank} \omega =$ const; our more general understanding is in correspondence with that of P. Libermann and Ch.-M. Marle [1]. As we have seen above, the restrictive condition $\operatorname{rank} \omega =$ const. means in fact a restriction to those Lagrangean systems which admit generalized Legendre transformations).

Let us turn to investigate the dynamics of such Lagrangean systems in more detail.

Let α_E be a presymplectic system, L its (local) first-order time-independent Lagrangian, and H and p_σ, $1 \leqslant \sigma \leqslant m$, the corresponding Hamiltonian and momenta. Suppose that

$$\operatorname{rank}(B_{\sigma\nu}) = \operatorname{rank}\Big(\frac{\partial^2 L}{\partial \dot{q}^\sigma \partial \dot{q}^\nu}\Big) = \operatorname{rank}\Big(\frac{\partial p_\sigma}{\partial \dot{q}^\nu}\Big) = K \geqslant 1. \tag{10.10}$$

This condition means that K of the momenta are independent; without the loss of generality we can suppose that $p_1, \dots, p_K$ are the independent momenta. Now, (in a neighborhood of every point) we can introduce *generalized Legendre coordinates*; for simplicity, we can add to the q's and the independent p's, $m - K$ of appropriate velocities $\dot{q}$'s; we denote these velocities by $v^{K+1}, \dots, v^m$. Hence, let us consider a generalized Legendre transformation in the form:

$$(q^1, \dots, q^m, \dot{q}^1, \dots, \dot{q}^m) \to (q^1, \dots, q^m, p_1 \dots, p_K, v^{K+1}, \dots v^m). \tag{10.11}$$

Consider a fixed generalized Legendre chart (U, φ), $\varphi = (q^i, p_i, v^j)$, where $1 \leqslant i \leqslant K$,

$K+1 \leqslant j \leqslant m$, on TM. For every two functions f, g on U denote

$$\{f, g\}_{(U,\varphi)} = \sum_{i=1}^{K}\Big(\frac{\partial f}{\partial p_i}\frac{\partial g}{\partial q^i} - \frac{\partial f}{\partial q^i}\frac{\partial g}{\partial p_i}\Big). \tag{10.12}$$

Further denote the remaining momenta $p_{K+1}, \ldots, p_m$ by $\phi_{K+1}, \ldots, \phi_m$. Taking into account the assumption (10.10), we easily get for the remaining momenta and the Hamiltonian

$$\frac{\partial \phi_j}{\partial v^l} = 0, \quad \frac{\partial H}{\partial v^l} = 0, \quad K+1 \leqslant j, l \leqslant m. \tag{10.13}$$

Expressing our Lagrangean system in the generalized Legendre coordinates we get

$$\begin{aligned}\alpha_E = &-\Big(\sum_{\sigma=1}^{m}\frac{\partial H}{\partial q^\sigma}dq^\sigma + \sum_{i=1}^{K}\frac{\partial H}{\partial p_i}dp_i\Big)\wedge dt\\ &+\sum_{i=1}^{K}dp_i\wedge dq^i + \sum_{j=K+1}^{m}\Big(\sum_{\sigma=1}^{m}\frac{\partial \phi_j}{\partial q^\sigma}dq^\sigma + \sum_{i=1}^{K}\frac{\partial \phi_j}{\partial p_i}dp_i\Big)\wedge dq^j.\end{aligned}$$

Hence, the *Euler-Lagrange distribution* Δ_E is annihilated by the one-forms

$$\begin{aligned}&-\frac{\partial H}{\partial q^i}dt - dp_i + \sum_{l=K+1}^{m}\frac{\partial \phi_l}{\partial q^i}dq^l,\\ &-\frac{\partial H}{\partial p_i}dt + dq^i + \sum_{l=K+1}^{m}\frac{\partial \phi_l}{\partial p_i}dq^l,\\ &-\frac{\partial H}{\partial q^j}dt - \sum_{i=1}^{K}\frac{\partial \phi_j}{\partial q^i}dq^i - \sum_{l=K+1}^{m}\Big(\frac{\partial \phi_j}{\partial q^l} - \frac{\partial \phi_l}{\partial q^j}\Big)dq^l - \sum_{i=1}^{K}\frac{\partial \phi_j}{\partial p_i}dp_i,\\ &\quad 1 \leqslant i \leqslant K, \quad K+1 \leqslant j \leqslant m.\end{aligned} \tag{10.14}$$

and the *characteristic distribution* $\mathcal{D}_E$ is annihilated by (10.14), and the energy one-form

$$dH = \sum_{\sigma=1}^{m}\frac{\partial H}{\partial q^\sigma}dq^\sigma + \sum_{i=1}^{K}\frac{\partial H}{\partial p_i}dp_i. \tag{10.15}$$

As we have seen in Chapter 7, the dynamics is characterized by the relation between these two distributions, and by their properties.

Let us compute *the ranks* of the characteristic distribution $\mathcal{D}_E$ and of the Euler-Lagrange distribution Δ_E. We have $\operatorname{rank}\mathcal{D}_E = \dim R \times TM - \operatorname{rank}\alpha_E$. By (5.22),

$$\begin{aligned}\operatorname{rank}\alpha_E &= \operatorname{rank}\begin{pmatrix} 2F^{00}_{\sigma\nu} & B_{\sigma\nu} & A_\sigma \\ -B_{\sigma\nu} & 0 & 0\end{pmatrix}\\ &= \operatorname{rank}\begin{pmatrix} \frac{1}{2}\Big(\frac{\partial A_\sigma}{\partial q^\nu_1} - \frac{\partial A_\nu}{\partial q^\sigma_1}\Big) & B_{\sigma\nu} & A_\sigma \\ -B_{\sigma\nu} & 0 & 0\end{pmatrix} \geqslant 2K,\end{aligned}$$

since we suppose (10.10). Hence, $\operatorname{rank}\mathcal{D}_E \leqslant 2(m-K)+1$. On the other hand, by (10.14), (10.15), there are at most $2K + m - K + 1 = K + m + 1$ independent one-forms spanning $\mathcal{D}^0_E$, i.e., $\operatorname{rank}\mathcal{D}_E \geqslant 2m + 1 - K - m - 1 = m - K$. Summarizing, we get

$$m - K \leqslant \operatorname{rank}\mathcal{D}_E \leqslant 2(m-K)+1, \tag{10.16}$$

and consequently,

$$m - K + 1 \leqslant \operatorname{rank} \Delta_E \leqslant 2(m - K) + 1.$$

Now, let us find *the primary constraint set* (Definition 5.4.4). By definition, it is the subset of the phase space where $\Delta_E \neq \mathcal{D}_E$. Notice that this is a subset of the phase space where the Hamiltonian is not a first integral of Δ_E. Computing this condition using the above generators of the distributions Δ_E and $\mathcal{D}_E$ we get that the primary constraint set for a presymplectic system is the set of all points of the phase space where the ranks of the matrices

$$\begin{aligned} \mathrm{P} &= \left(\{\phi_l, \phi_j\}_{(U,\varphi)} + \frac{\partial \phi_l}{\partial q_j} - \frac{\partial \phi_j}{\partial q_l} \right), \\ \mathrm{P}' &= \left(\{\phi_l, \phi_j\}_{(U,\varphi)} + \frac{\partial \phi_l}{\partial q_j} - \frac{\partial \phi_j}{\partial q_l} \,\middle|\, \{H, \phi_j\}_{(U,\varphi)} + \frac{\partial H}{\partial q_j} \right), \end{aligned} \tag{10.17}$$

where j labels rows and l labels columns, are different.

It is interesting to investigate the case when the Lagrangean system is *semiregular*. Recall that this means that there are *no dynamical constraints*, i.e., $\operatorname{rank} \mathrm{P} = \operatorname{rank} \mathrm{P}'$ at each point, and $\operatorname{rank} \alpha_E$ *is a constant*. Then it is possible to find a system of *non-vertical vector fields* on the phase space, belonging to $\mathcal{D}_E = \Delta_E$.

Denote

$$\zeta = \zeta^0 \frac{\partial}{\partial t} + \sum_{\sigma=1}^{m} \zeta^\sigma \frac{\partial}{\partial q^\sigma} + \sum_{i=1}^{K} \zeta_i \frac{\partial}{\partial p_i} + \sum_{j=K+1}^{m} \hat{\zeta}^j \frac{\partial}{\partial v^j}.$$

Then we get from (10.14) for the components of vector fields ζ belonging to the Euler-Lagrange distribution the following equations

$$\begin{aligned} & -\frac{\partial H}{\partial q^i} \zeta^0 - \zeta_i + \sum_{l=K+1}^{m} \frac{\partial \phi_l}{\partial q^i} \zeta^l = 0, \quad -\frac{\partial H}{\partial p_i} \zeta^0 + \zeta^i + \sum_{l=K+1}^{m} \frac{\partial \phi_l}{\partial p_i} \zeta^l = 0, \\ & \quad 1 \leqslant i \leqslant K, \\ & -\frac{\partial H}{\partial q^j} \zeta^0 - \sum_{i=1}^{K} \frac{\partial \phi_j}{\partial q^i} \zeta^i - \sum_{l=K+1}^{m} \left(\frac{\partial \phi_j}{\partial q^l} - \frac{\partial \phi_l}{\partial q^j} \right) \zeta^l - \sum_{i=1}^{K} \frac{\partial \phi_j}{\partial p_i} \zeta_i = 0, \\ & \quad K + 1 \leqslant j \leqslant m, \end{aligned} \tag{10.18}$$

i.e.,

$$\begin{aligned} & \zeta_i = -\frac{\partial H}{\partial q^i} \zeta^0 + \sum_{l=K+1}^{m} \frac{\partial \phi_l}{\partial q^i} \zeta^l, \quad \zeta^i = \frac{\partial H}{\partial p_i} \zeta^0 - \sum_{l=K+1}^{m} \frac{\partial \phi_l}{\partial p_i} \zeta^l, \\ & \quad 1 \leqslant i \leqslant K, \end{aligned} \tag{10.19}$$

$$\begin{aligned} & \sum_{l=K+1}^{m} \left(\{\phi_l, \phi_j\}_{(U,\varphi)} + \frac{\partial \phi_l}{\partial q_j} - \frac{\partial \phi_j}{\partial q_l} \right) \zeta^l = \left(\{H, \phi_j\}_{(U,\varphi)} + \frac{\partial H}{\partial q_j} \right) \zeta^0, \\ & \quad K + 1 \leqslant j \leqslant m. \end{aligned} \tag{10.20}$$

Solving these equations we get generators of Δ_E in the form (7.26) and (7.27) if $\operatorname{rank} \mathrm{P} = 0$ and $\operatorname{rank} \mathrm{P} \neq 0$, respectively.

Notice that for different semiregular presymplectic systems with the same rank of ω, the rank of $\Delta_E = \mathcal{D}_E$ can be different and depends on the rank of P; more precisely, $m - K + 1 \leqslant \operatorname{rank} \Delta_E \leqslant 2(m-K)$.

Recall that the Euler-Lagrange distributions of semiregular Lagrangean systems are *completely integrable*. This means that maximal integral manifolds define a *foliation* of the phase space, and the motion proceeds from an initial point within the leaf passing through this point.

The system of vector fields (7.26) (resp. (7.27)) describes completely the *extended* (= Hamiltonian) dynamics of semiregular autonomous Lagrangean systems. To find the *proper dynamics* one has to compute the *semispray constraints* and to apply the *constraint algorithm* (Chapter 7).

Other presymplectic systems with relatively easy dynamics are *weakly regular* autonomous Lagrangean systems, i.e., satisfying the condition $\operatorname{rank} \mathrm{B} = \operatorname{rank}(\mathrm{B} \mid \mathrm{A})$ (cf. Chapter 7). In this case we can find a subdistribution of Δ_E spanned by $m+1-K$ *semisprays*, describing the proper dynamics completely. Unfortunately, this distribution is not completely integrable. On the other hand, the proper motion in this case is not restricted to a submanifold of the phase space (we have a "global dynamics"). If, additionally, a weakly regular autonomous Lagrangean system is semiregular, then the semispray subdistribution is spanned by an appropriate subsystem of vector fields (7.26) (resp. (7.27)).

It is important to note that the condition (10.10) does not mean that rank α_E is constant. Practically this implies that the assumption (10.10) does not lead to any simplification in the procedure of searching for the dynamics. If there are *no dynamical constraints*, but the distribution $\Delta_E = \mathcal{D}_E$ is *not semiregular* (since it has not a constant rank), it is not possible to span it by a system of continuous vector fields. In other words, one cannot obtain the dynamics expressed in the form analogous to (7.26) (resp. (7.27)). In this case one has to apply to Δ_E the *constraint algorithm*. The same procedure has to be applied also in the more general case when there are dynamical constraints. Similarly, the *semispray problem* has to be solved by means of the constraint algorithm. Notice that in concrete calculations one can utilize either fibered (adapted) coordinates, or generalized Legendre coordinates.

10.3.3. Hamiltonian formalism. We have seen above, that for symplectic Lagrangean systems one has the possibility to define Legendre transformation as a coordinate transformation on $T^{s-1}M$. Now, we want to show that with help of the Legendre transformation, *odd-order symplectic Lagrangean systems* can be canonically represented by means of a symplectic structure on a *cotangent bundle*.

Let α_E be a symplectic Lagrangean system of order $2c-1$ on a fibered manifold $R \times M \to R$. Then the Legendre transformation on $T^{2c-1}M$ can be viewed as a mapping $\mathfrak{Leg} : T^{2c-1}M \to T^*T^{c-1}M$, in adapted coordinates $(q^\sigma, \dots, q^\sigma_{c-1}, p_\sigma, \dots, p^\sigma_{c-1})$ on $T^*T^{c-1}M$ defined by

$$p^k_\nu = \sum_{j=0}^{c-k-1} (-1)^j \frac{d^j}{dt^j} \frac{\partial L_{\min}}{\partial q^\nu_{k+1+j}}, \quad 1 \leqslant \nu \leqslant m,\ 0 \leqslant k \leqslant c-1, \tag{10.21}$$

where $L_{\min}$ is an autonomous minimal-order Lagrangian for α_E, and called the *Legendre mapping*. Since the Lagrangean system is symplectic, the mapping (10.21) has the maximal rank, i.e., it is a local diffeomorphism of $T^{2c-1}M$ onto $T^*T^{c-1}M$. If ω is the symplectic form on $T^{2c-1}M$, associated with α_E, and ω_0 is the canonical symplectic form on $T^*T^{c-1}M$ then obviously,

$$\mathfrak{Leg}^*\omega_0 = \omega.$$

The symplectic manifold $(T^*T^{c-1}M, \omega_0)$ is called a *cotangent representation* of the odd-order symplectic Lagrangean system α_E. The Legendre transformation (10.21) represents a one-to-one correspondence between the two symplectic structures $(T^{2c-1}M, \omega)$ and $(T^*T^{c-1}M, \omega_0)$, therefore it is possible instead of $(T^{2c-1}M, \omega)$ *equivalently* deal with its representation $(T^*T^{c-1}M, \omega_0)$. This procedure is often called the *Hamiltonian formalism*. Notice that the Hamiltonian formalism does not depend on a choice of equivalent minimal-order Lagrangians; this means, in particular, that even in case that there does not exist a *global* minimal-order Lagrangian for α_E, we get a unique (global) representation of α_E by means of the symplectic structure $(T^*T^{c-1}M, \omega_0)$.

Notice that the manifold $R \times T^{c-1}M$ is the *quasi-configuration space* (defined in Sec. 9.3) for the Lagrangean system α_E. If $c = 1$, we have $T^{c-1}M = M$, i.e., $R \times T^{c-1}M$ coincides with the configuration space $R \times M$. Thus, if $c = 1$ (first-order symplectic Lagrangean systems), the Legendre transformation $\mathfrak{Leg}$ (10.21) is a local diffeomorphism $TM \to T^*M$.

It should be emphasized that for *even-order symplectic Lagrangean systems* a cotangent representation by means of an equivalent canonical symplectic structure is no longer possible. To get a cotangent representation for an even-order symplectic Lagrangean system α_E of order $s - 1 = 2c \geqslant 0$, we can proceed as follows. Consider the manifold T^*T^cM with the canonical symplectic structure endowed by the canonical symplectic form ω_0; recall that in the adapted coordinates $(q^\sigma, q_1^\sigma, \ldots, q_c^\sigma, p_\sigma, \ldots, p_\sigma^c)$ on T^*T^cM,

$$\omega_0 = \sum_{i=0}^{c} dp_\sigma^c \wedge dq_c^\sigma.$$

Next, consider the *Legendre mapping* $\mathfrak{Leg} : T^{2c}M \to T^*T^cM$, defined by

$$p_\nu^k = \sum_{j=0}^{c-k} (-1)^j \frac{d^j}{dt^j} \frac{\partial L_{\min}}{\partial q_{k+1+j}^\nu}, \quad 1 \leqslant \nu \leqslant m, \; 0 \leqslant k \leqslant c, \tag{10.22}$$

where $L_{\min}$ is a time-independent minimal-order Lagrangian for α_E. Since $\operatorname{rank}\omega$ is maximal, the mapping $\mathfrak{Leg}$ has the maximal rank, i.e., it is an *immersion* of $T^{2c}M$ into T^*T^cM. Obviously,

$$\mathfrak{Leg}^*\omega_0 = \omega.$$

Hence, for an even-order Lagrangean system we have a cotangent representation as an *immersed symplectic submanifold* $(T^{2c}M, \omega)$ of the canonical symplectic manifold (T^*T^cM, ω_0). This submanifold is also referred to as a *final constraint submanifold* (terminology in the sense of the Dirac theory of constrained systems).

An extension of the Hamiltonian formalism to general *odd-oder* presymplectic systems now consists in an idea to represent via the Legendre map (10.22) the presymplectic system $(T^{2c-1}M, \omega)$ as a presymplectic system (N, ω_N), where N is a submanifold of $T^*T^{c-1}M$ (the so called *final constraint submanifold*), and $\omega_N = \iota^*\omega_0$ is the restricted to

N canonical symplectic form ω_0 on $T^*T^{c-1}M$. Similarly, an extension of the Hamilton formalism to *even-oder* presymplectic systems means to represent via the Legendre map (10.22) the presymplectic system $(T^{2c}M, \omega)$ as a presymplectic system (N, ω_N), where N is a submanifold of T^*T^cM, and $\omega_N = \iota^*\omega_0$ is the restricted to N canonical symplectic form ω_0 on T^*T^cM. However, such a representation is *not possible in general*—it has a sense only if the rank of the Legendre map is (locally) constant. This means, that, *in principle*, only those presymplectic Lagrangean systems admit a Hamiltonian formalism which *admit generalized Legendre transformations* (see Sec. 7.7). The cotangent representation of α_E is then constructed as follows: The Legendre map (which is supposed to have a constant rank) maps $T^{s-1}M$ onto a *submanifold* of $T^*T^{c-1}M$ (resp. T^*T^cM); equations defining this submanifold are called *primary constraints*. The symplectic form ω_0 restricted to this submanifold then defines a presymplectic system, which, however, is *not equivalent* with the original one, $(T^{s-1}M, \omega)$, i.e., the "new dynamics" is not in one-to-one correspondence with the original dynamics. Accordingly, (as it is standard in presymplectic geometry) if one utilizes the cotangent representation for study the dynamics of $(T^{s-1}M, \omega)$, one is faced with many complications, arising from this non-equivalence (e.g., equivalence problems between constraints on the motion on both sides, equivalence problems between the true and new dynamics, validity of the "Dirac's conjecture", problems with the meaning of symmetries and first integrals, difficulties with a generalization of the Hamilton-Jacobi equation and its interpretation, ambiguous or even contradictory results in quantization, etc.). There is a rich bibliography on the presymplectic Hamiltonan formalism, where the ideas are developed and arising complications are discussed—we refer the reader to the Bibliography to the present book, and references therein. On the contrary, the theory of Lepagean 2-forms provides us with a "direct Hamilton theory" (Hamiltonian, momenta, Hamilton equations, etc.), applicable to *all* Lagrangean systems. It can be used to investigate the dynamics studying the properties of the corresponding distributions, or to find the corresponding quantum mechanical description directly, without additional complications arising form the cotangent representation of the dynamics (this advantage becomes even more transparent in case of Lageangean systems which do not admit a cotangent representation).

10.3.4. Remark. The attempts to generalize techniques and results of classical mechanics to *autonomous first-order singular Lagrangians* come back to P. A. M. Dirac [1] (for the Dirac's theory of constrained systems see also eg. the books by E. C. G. Sudarshan and N. Mukunda [1], or K. Sundermeyer [1]). Since that time, there appeared a plenty of papers dealing with different aspects of the theory proposed by Dirac, and with its correct mathematical formulation (based usually on the above mentioned cotangent representation of presymplectic Lagrangean systems). Recently, there has been growing an interest in generalizing the theory to a wider class of Lagrangians (time-dependent, higher-order), however, only particular problems have been studied, and not all kinds of Lagrangians have been considered (cf. papers listed in the Bibliography and references therein).

Also, within the range of the standard presymplectic geometry, for singular Lagrangians one does not have satisfactory generalizations of classical integration methods (based on a Liouville theorem, Hamilton-Jacobi equation, theory of canonical transformations, or theory of fields of extremals). There have been obtained only some particular results for certain special kinds of Lagrangians (see e.g. D. Dominici and J. Gomis [1,2], D. Dominici, G. Longhi, J. Gomis and J. M. Pons [1], J. Gomis [1], J. Llosa and N. Román [1], C. M. Marle [1]).

10.4. Metric structures associated with regular first-order Lagrangean systems

The dynamics of a first-order regular time-dependent mechanical system on a manifold M is described by a semispray (a "second order vector field") on a fibered manifold $R \times M \to R$, or, equivalently, by a *semispray connection* on $R \times TM$ (i.e. a section $R \times TM \to R \times T^2M$). Locally it is represented by a regular system of second order differential equations for sections of the fibered manifold $R \times M \to R$. In case that the manifold M is endowed with a Riemannian metric g, it carries a *canonical* semispray connection Γ such that the geodesics of Γ coincide with the graphs of geodesics of the Levi-Civita connection ∇ of g; a similar situation occurs in the case of a Finsler manifold. Moreover we know that in both the Riemannian and the Finsler geometry the equations for geodesics are variational (i.e., they are the Euler-Lagrange equations of a Lagrangian called the *kinetic energy* of the Riemannian and Finsler structure, respectively). Hence, the geodesics in the Riemannian and Finsler geometries can be viewed as geodesics (paths) of semispray connections describing the dynamics of the Riemannian and Finsler *free particle*, respectively. These two important particular cases of mechanical systems suggest us an idea to investigate the structure of semispray connections on $R \times TM$, and to search for all (semispray) connections describing the dynamics of "free particles". Naturally, it is required these connections be *variational*.

In classical Finsler geometry, a Finsler manifold is a manifold M endowed with a Finsler metric g on TM which is a regular symmetric fibered morphism $g : TM \to T^0_2M$ over id_M (where T^0_2M denotes the bundle of all tensors of type (0, 2) over M), satisfying the following two conditions:

$$\frac{\partial g_{ij}}{\partial \dot{x}^k} = \frac{\partial g_{ik}}{\partial \dot{x}^j} \quad (\text{"integrability"}), \qquad \frac{\partial g_{ij}}{\partial \dot{x}^k}\dot{x}^k = 0 \quad (\text{"homogeneity"}).$$

For basic theory on Finsler structures we refer to the book of M. Matsumoto [1].

Omitting the "integrability condition" one obtains a class of metrics which is studied in generalized Finsler geometry (cf. e.g. H. Shimada [1], and the references therein).

Another possibility is to consider regular symmetric fibered morphisms $g : R \times TM \to T^0_2M$ over id_M (time–dependent metrics on TM), which satisfy the above "integrability" condition, but not necessarily the "homogeneity" condition. This will lead to another generalization of the concept of a Finsler manifold, which is close to that of a *Lagrange space* (see P. L. Antonelli and R. Miron [1], R. Miron [1], R. Miron and G. Atanasiu [1–4], R. Miron and M. Anastasiei [1], O. Krupková [8]). Consequently, we shall see that regular first-order Lagrangean systems give rise to *metric structures* on $R \times TM$, resp. in the autonomous case, to metric structures on TM, or directly on M.

Denote by T^0_2M the bundle of all tensors of type (0, 2) over M. Let $g : R \times TM \to T^0_2M$ be a fibered morphism over the identity of M. Recall that g is said to be a *metric on* $R \times TM$ if it is regular and symmetric.

10.4.1. Definition. We shall say that a metric g on $R \times TM$ is *variational* if there exists a regular locally variational form E on $R \times T^2M$ such that in every adapted chart on $R \times M$, the components of g and E satisfy

$$g_{\sigma\nu} = -\frac{\partial E_\sigma}{\partial \ddot{q}^\nu}, \quad 1 \leqslant \sigma, \nu \leqslant m. \tag{10.23}$$

10.4.2. Proposition. *A metric g on $R \times TM$ is variational if and only if the components of g satisfy in each adapted chart on $R \times M$ the conditions*

$$\frac{\partial g_{\sigma\nu}}{\partial \dot{q}^\rho} = \frac{\partial g_{\sigma\rho}}{\partial \dot{q}^\nu}, \quad 1 \leqslant \sigma, \nu, \rho \leqslant m. \tag{10.24}$$

Proof. If g is a variational metric then the relations (10.24) follow from the Helmholtz conditions (1.8).

We shall prove the converse. Consider an open ball $W \subset R^m$ with the center at the origin, and denote by (q^σ) the canonical coordinates on W. Let g be a metric on $R \times TW$ satisfying (10.24). Define a mapping $\bar{\chi} : [0,1] \times (R \times TW) \to R \times TW$ setting

$$\bar{\chi}(v, (t, q^\sigma, \dot{q}^\sigma)) = (t, q^\sigma, v\dot{q}^\sigma), \tag{10.25}$$

and put

$$T = \dot{q}^\sigma \dot{q}^\nu \int_0^1 \left(\int_0^1 (g_{\sigma\nu} \circ \bar{\chi})\, dv \right) \circ \bar{\chi}\; v\, dv. \tag{10.26}$$

Then $T\,dt$ is a Lagrangian on the fibered manifold $\pi : R \times W \to R$ satisfying

$$g_{\sigma\nu} = \frac{\partial^2 T}{\partial \dot{q}^\sigma \partial \dot{q}^\nu} = -\frac{\partial E_\sigma(T)}{\partial \ddot{q}^\nu} = -B_{\sigma\nu},$$

where $E_\sigma(T)$, $1 \leqslant \sigma \leqslant m$, are the Euler-Lagrange expressions of the Lagrangian $T\,dt$.

Now, let $\pi : R \times M \to R$ be a fibered manifold, g a metric on $R \times TM$, satisfying the conditions (10.24). Then there exists an open covering $\mathcal{O}$ of $R \times TM$ such that on every open set of $\mathcal{O}$ there is defined the Lagrangian $T\,dt$ (10.26). From the transformation properties of the components $g_{\sigma\nu}$, $1 \leqslant \sigma, \nu \leqslant m$, of g and of the coordinates $\dot{q}^\sigma$, $1 \leqslant \sigma \leqslant m$, it is easy to see that the local Lagrangians $T\,dt$ give rise to a (global) Lagrangian λ_g on $R \times TM$ such that for each $U \in \mathcal{O}$, $\lambda_g|_U = T\,dt$. For the Euler-Lagrange form E_g of the Lagrangian λ_g we have (10.23), i.e., the metric g is variational. □

10.4.3. Definition. If g is a variational metric on $R \times TM$ then the (global) Lagrangian λ_g of g defined in the proof of Proposition 10.4.2 will be called the *kinetic energy* of the metric g. The Euler-Lagrange form E_g of the kinetic energy λ_g will be called the *canonical dynamical two-form* of the metric g.

A manifold M endowed with a variational metric g will be called a *semi-finslerian manifold.*

10.4.4. Theorem. *Let g be a variational metric on $R \times TM$. There exists a unique semispray connection $\Gamma_g : R \times TM \to R \times T^2M$ such that the geodesics of Γ_g coincide with the extremals of the kinetic energy λ_g, i.e., coincide with the solutions of the Euler-Lagrange equations*

$$\frac{\partial T}{\partial q^\sigma} - \frac{d}{dt}\frac{\partial T}{\partial \dot{q}^\sigma} = 0. \tag{10.27}$$

This connection is defined by the relation

$$\Gamma_g^* E_g = 0, \tag{10.28}$$

and its adapted chart components $\Gamma^\sigma(g)$, $1 \leqslant \sigma \leqslant m$, *are given by the following formulas:*

$$\ddot{q}^\sigma \circ \Gamma_g \equiv \Gamma^\sigma(g) = g^{\sigma\nu}\Gamma_\nu(g), \tag{10.29}$$

where $(g^{\sigma\nu})$ *is the inverse matrix to* $(g_{\sigma\nu})$, *and the functions* $\Gamma_\nu(g)$, $1 \leqslant \nu \leqslant m$, *are given by*

$$-\Gamma_\nu(g) = \Gamma_{\nu\mu\rho}(g)\,\dot{q}^\mu\dot{q}^\rho + \dot{q}^\mu \int_0^1 \left(\frac{\partial g_{\nu\mu}}{\partial t} \circ \bar{\chi}\right) dv, \tag{10.30}$$

where

$$\Gamma_{\nu\mu\rho}(g) = \tfrac{1}{2}\int_0^1 \left(\frac{\partial g_{\nu\mu}}{\partial q^\rho} + \frac{\partial g_{\nu\rho}}{\partial q^\mu} - 2\frac{\partial g_{\mu\rho}}{\partial q^\nu}\right) \circ \bar{\chi}\, dv + \int_0^1 \left(\frac{\partial g_{\mu\rho}}{\partial q^\nu} \circ \bar{\chi}\right) v\, dv\,. \tag{10.31}$$

Proof. Let E_g be the canonical dynamical two-form of the variational metric g, and consider sections $\Gamma : R \times TM \to R \times T^2M$. Since E_g is *regular*, we get that the equation $\Gamma^* E_g = 0$ has a unique solution $\Gamma = \Gamma_g$, and the components of Γ_g are of the form

$$\ddot{q}^\sigma \circ \Gamma_g \equiv \Gamma_g^\sigma = -B^{\sigma\nu}A_\nu = g^{\sigma\nu}A_\nu,$$

where we have used the notation $E_\sigma = A_\sigma + B_{\sigma\nu}\ddot{q}^\nu$ for the components of E_g.

Let us show that $A_\nu = \Gamma_\nu(g)$. Since E_g comes form the Lagrangian (10.26), we have

$$A_\nu = \frac{\partial T}{\partial q^\nu} - \frac{\partial^2 T}{\partial t\, \partial q^\nu} - \frac{\partial^2 T}{\partial q^\rho \partial \dot{q}^\nu}\dot{q}^\nu. \tag{10.32}$$

Now,

$$\begin{aligned}
\frac{\partial T}{\partial q^\nu} &= \dot{q}^\sigma\dot{q}^\rho \int_0^1 \left(\int_0^1 \left(\frac{\partial g_{\sigma\rho}}{\partial q^\nu} \circ \bar{\chi}\right) dv\right) \circ \bar{\chi}\, v\, dv \\
&= \dot{q}^\sigma\dot{q}^\rho \left(\int_0^1 \left(\frac{\partial g_{\sigma\rho}}{\partial q^\nu} \circ \bar{\chi}\right) dv - \int_0^1 \left(\frac{\partial g_{\sigma\rho}}{\partial q^\nu} \circ \bar{\chi}\right) v\, dv\right), \\
\frac{\partial T}{\partial \dot{q}^\nu} &= 2\dot{q}^\rho \int_0^1 \left(\int_0^1 (g_{\nu\rho} \circ \bar{\chi})\, dv\right) \circ \bar{\chi}\, v\, dv + \dot{q}^\sigma\dot{q}^\rho \int_0^1 \left(\int_0^1 \left(\frac{\partial g_{\sigma\rho}}{\partial \dot{q}^\nu} \circ \bar{\chi}\right) v\, dv\right) \circ \bar{\chi}\, v^2\, dv \\
&= \dot{q}^\rho \left(\int_0^1 \left(2\int_0^1 (g_{\nu\rho} \circ \bar{\chi})\, dv + \dot{q}^\sigma \int_0^1 \left(\frac{\partial g_{\sigma\rho}}{\partial \dot{q}^\nu} \circ \bar{\chi}\right) v\, dv\right) \circ \bar{\chi}\, v\, dv \right. \\
&= \dot{q}^\rho \int_0^1 \left(\int_0^1 (g_{\nu\rho} \circ \bar{\chi})\, dv\right) \circ \bar{\chi}\, v\, dv + \dot{q}^\rho \int_0^1 (g_{\nu\rho} \circ \bar{\chi}) v\, dv = \dot{q}^\rho \int_0^1 (g_{\nu\rho} \circ \bar{\chi})\, dv,
\end{aligned}$$

where we have used (10.24), and the following identities:

$$\begin{aligned}
\int_0^1 \left(\int_0^1 (F \circ \bar{\chi})\, dv\right) \circ \bar{\chi}\, v\, dv &= \int_0^1 (F \circ \bar{\chi})\, dv - \int_0^1 (F \circ \bar{\chi})\, v\, dv, \\
F &= \int_0^1 (F \circ \bar{\chi})\, dv + \dot{q}^\sigma \int_0^1 \left(\frac{\partial F}{\partial \dot{q}^\sigma} \circ \bar{\chi}\right) v\, dv\,.
\end{aligned}$$

Substituting into (10.32) we get $A_\nu = \Gamma_\nu(g)$. This completes the proof. □

10.4.5. Definition. The connection Γ_g defined by (10.28) will be called the *canonical connection* associated with the metric g.

The concepts of variational metric, the associated canonical connection and semi-finslerian manifold as a manifold endowed with a variational metric are due to O. Krupková [8].

According to the above theorem, on every semi-finslerian manifold (M, g) there exists a unique canonical dynamical two-form E_g, and a unique canonical semispray connection Γ_g.

Conversely, let λ be a *regular* Lagrangian on $R \times TM$. Then putting

$$g_{\sigma\nu} = \frac{\partial^2 L}{\partial \dot{q}^\sigma \partial \dot{q}^\nu} \tag{10.33}$$

we get a variational metric g on $R \times TM$, i.e., (M, g) is a semi-finslerian manifold. In this sense, *every regular Lagrangian on $R \times TM$ defines a semi-finslerian structure* on M.

10.4.6. Particular cases of semi-finslerian structures.

(1) *Autonomous semi-finslerian structure.* Suppose that a variational metric g is time-independent (i.e., it is a metric on TM). Then we get an autonomous semi-finslerian structure on M. The corresponding kinetic energy is of the form (10.26), and the canonical connection Γ_g reads

$$-\Gamma_\nu(g) = \Gamma_{\nu\mu\rho}(g)\, \dot{q}^\mu \dot{q}^\rho,$$

where the $\Gamma_{\nu\mu\rho}(g)$ are given by (10.31). Since the equations for geodesics of Γ_g coincide with the Euler-Lagrange equations of the Lagrangian T, we get

$$\Gamma^\nu(g) = -\tilde{\Gamma}^\rho_\nu\, \dot{q}^\rho,$$

where $\tilde{\Gamma}^\nu_\rho$ are the components of the canonical *Grifone connection* for the energy T.

We remind the reader that a *Grifone connection* (J. Grifone [1]) is a vector-valued 1-form $\tilde{\Gamma}$ on TM satisfying the conditions $J\tilde{\Gamma} = J$, $\tilde{\Gamma}J = -J$, where J is the canonical almost tangent structure on TM. The equations for geodesics of a Grifone connection $\tilde{\Gamma}$ are of the form

$$\ddot{x}^i + \tilde{\Gamma}^i_k \dot{x}^k = 0,$$

where $\tilde{\Gamma}^i_k(x^j, \dot{x}^j)$ are the components of $\tilde{\Gamma}$. Grifone has shown that every manifold M endowed with a kinetic energy T caries a canonical Grifone connection such that the equations of geodesics of this connection coincide with the Euler-Lagrange equations of T.

Notice that according to (10.33), *every regular autonomous Lagrangian L on TM defines an autonomous semi-finslerian structure* on M; the pair (M, L) is then also called a *Lagrange space* (see R. Miron and M. Anastasiei [1] for more details).

(2) *Finslerian structure.* If a variational metric g is time-independent and satisfies the following *homogeneity condition*

$$\frac{\partial g_{\sigma\nu}}{\partial \dot{q}^\rho}\, \dot{q}^\rho = 0 \tag{10.34}$$

then (M, g) is a Finsler manifold. Let us compute the kinetic energy $\lambda_g = T\,dt$ and the canonical connection Γ_g of g. Using the identity (see Lemma 6.7.3)

$$F = \int_0^1 (F \circ \bar{\chi})\,dv + \dot{q}^\sigma \int_0^1 \left(\frac{\partial F}{\partial \dot{q}^\sigma} \circ \bar{\chi}\right) v\,dv = 2\int_0^1 (F \circ \bar{\chi}) v\,dv + \dot{q}^\sigma \int_0^1 \left(\frac{\partial F}{\partial \dot{q}^\sigma} \circ \bar{\chi}\right) v^2 dv$$

for the functions $F = g_{\sigma\nu}$ and $F = \partial g_{\sigma\nu}/\partial q^\rho$, respectively, and applying the homogeneity condition, the formulas (10.26), resp. (10.31), simplify to

$$T = \tfrac{1}{2} g_{\sigma\nu}\dot{q}^\sigma \dot{q}^\nu$$

and

$$\Gamma^\sigma(g) = -\Gamma^\sigma_{\mu\rho}(g)\,\dot{q}^\nu \dot{q}^\rho, \quad \Gamma^\sigma_{\mu\rho}(g) = \tfrac{1}{2} g^{\sigma\nu}\left(\frac{\partial g_{\nu\mu}}{\partial q^\rho} + \frac{\partial g_{\nu\rho}}{\partial q^\mu} - \frac{\partial g_{\mu\rho}}{\partial q^\nu}\right).$$

Since the metric g satisfies the homogeneity condition, we get $\Gamma^i_{jk}(g)\,\dot{x}^j \dot{x}^k = \gamma^i_{jk}\dot{x}^j \dot{x}^k$, where γ^i_{jk}, $1 \leqslant i, j, k \leqslant m$ are the components of the *Cartan connection*.

(3) *Riemannian structure*. Let g be a metric on M. Then g is trivially a variational metric, and the formulas (10.26), resp. (10.31), for the kinetic energy $\lambda_g = T\,dt$, resp. for the canonical connection Γ_g of g, simplify to the familiar formulas

$$T = \tfrac{1}{2} g_{\sigma\nu}\dot{q}^\sigma \dot{q}^\nu,$$

resp.

$$\Gamma^\sigma = -\Gamma^\sigma_{\mu\rho}\,\dot{q}^\mu \dot{q}^\rho,$$

where

$$\Gamma^\sigma_{\mu\rho} = \tfrac{1}{2} g^{\sigma\nu}\left(\frac{\partial g_{\nu\mu}}{\partial q^\rho} + \frac{\partial g_{\nu\rho}}{\partial q^\mu} - \frac{\partial g_{\mu\rho}}{\partial q^\nu}\right)$$

are the Christoffel symbols of the *Levi-Civita connection* ∇ of g. The geodesics of Γ coincide with the graphs of geodesics of ∇.

(4) *Time-dependent finslerian structure*. If g is a variational metric on $R \times TM$ satisfying the homogeneity condition (10.34) then g defines a time-dependent finslerian structure on M. Similarly as in (2) we compute that the corresponding kinetic energy and canonical connection are of the form

$$T = \tfrac{1}{2} g_{\sigma\nu}\dot{q}^\sigma \dot{q}^\nu$$

and

$$\Gamma^\sigma(g) = \left(\frac{\partial g_{\sigma\mu}}{\partial t} - \Gamma^\sigma_{\mu\rho}(g)\,\dot{q}^\rho\right)\dot{q}^\mu, \quad \Gamma^\sigma_{\mu\rho}(g) = \tfrac{1}{2} g^{\sigma\nu}\left(\frac{\partial g_{\nu\mu}}{\partial q^\rho} + \frac{\partial g_{\nu\rho}}{\partial q^\mu} - \frac{\partial g_{\mu\rho}}{\partial q^\nu}\right).$$

Notice that $\Gamma^\sigma_{\mu\rho}(g)$ may depend on $\dot{q}$'s.

(5) *Time-dependent riemannian structure*. If g is a time-dependent metric on M, then g defines a time-dependent riemannian structure on M, and we get similarly as above

$$T = \tfrac{1}{2} g_{\sigma\nu}\dot{q}^\sigma \dot{q}^\nu$$

and

$$\Gamma^\sigma(g) = \frac{\partial g_{\sigma\mu}}{\partial t}\dot{q}^\mu - \Gamma^\sigma_{\mu\rho}(g)\,\dot{q}^\rho \dot{q}^\mu, \quad \Gamma^\sigma_{\mu\rho}(g) = \tfrac{1}{2} g^{\sigma\nu}\left(\frac{\partial g_{\nu\mu}}{\partial q^\rho} + \frac{\partial g_{\nu\rho}}{\partial q^\mu} - \frac{\partial g_{\mu\rho}}{\partial q^\nu}\right),$$

where the $\Gamma^{\sigma}_{\mu\rho}(g)$ do not depend on $\dot{q}$'s. Notice that the semispray connection $\Gamma^{\sigma}(g)$ is a polynomial in the velocities.

10.4.7. Mechanical systems on semi-finslerian manifolds. Let (M, g) be a semi-finslerian manifold, E_1, E_2 dynamical forms on $R \times T^2M$, associated with g. Then obviously $E_1 - E_2$ is a first-order dynamical form, i.e., defined on $R \times TM$. Conversely, if E_1 is a second order dynamical form associated with g and F is a dynamical form on $R \times TM$ then $E_2 = E_1 + F$ is another second order dynamical form associated with g. This leads us to the following definition: A triple (M, g, F) will be called a *mechanical system in the force field* F if (M, g) is a semi-finslerian manifold and F is a dynamical form on $R \times TM$; we shall also say that F is a *force* on a semi-finslerian manifold (M, g). A mechanical system (M, g, F) is characterized by the dynamical 2-form $E = E_g + F$. Hence, its motion is described by sections γ of $R \times M \to R$ which are solutions to the "Euler-Lagrange equations for a nonconservative mechanical system"

$$\left[\frac{\partial T}{\partial q^{\sigma}} - \frac{d}{dt}\frac{\partial T}{\partial \dot{q}^{\sigma}} - F_{\sigma}\right] \circ J^2\gamma = 0,$$

where T and F_{σ} are the components of the kinetic energy λ_g and the force F, respectively. A mechanical system $(M, g, 0)$ will be also called a *free particle* for the semi-finslerian manifold (M, g).

10.4.8. Examples. Let us mention some examples of variational metrics in classical and relativistic mechanics.

Consider the manifold R^3 with the canonical global chart (x^i).

(1) Puting $g = m\delta$, where $\delta = \delta_{ij}\, dx^i \otimes dx^j$ is the Kronecker tensor and m is a positive constant, we get the semi-finslerian manifold $(R^3, m\delta)$, and the mechanical system $(R^3, m\delta, 0)$, which is a free particle of classical mechanics with mass m; in this case the canonical connection $\Gamma_g = 0$, and $E_g = (m\delta_{ij}\ddot{x}^i)\, dx^j \wedge dt$. Considering a force F on $(R^3, m\delta)$ we get a classical particle of mass m in the force field F, described by the equations of motion $m\ddot{x}^i = F^i(t, x, \dot{x})$.

(2) Put $g = f(t)\delta$, where f is a nowhere zero function. Then $(R^3, f(t)\delta)$ is a semi-finslerian manifold. Computing the components of the canonical connection according to (10.29)–(10.31) we get

$$\ddot{x}^i \circ \Gamma_g = -\frac{1}{f(t)}\delta^{ij}\dot{x}^k\delta_{jk}\frac{df}{dt} = -\frac{1}{f(t)}\frac{df}{dt}\dot{x}^i.$$

Hence, the equations of motion of the free particle on $(R^3, f(t)\delta)$ are the Newton equations of a classical free particle with nonconstant mass, i.e.,

$$f\ddot{x}^i + \frac{df}{dt}\dot{x}^i = 0.$$

Considering a force field F on this semi-finslerian manifold we get the mechanical system $(R^3, f(t)\delta, F)$ which is a classical particle with nonconstant mass moving in the force field F.

(3) Let us define a semi-finslerian metric $g = g_{ij}\, dx^i \otimes dx^j$ on R^3 by

$$g_{ij} = \frac{m\delta_{ij}}{(1-\frac{v^2}{c^2})^{1/2}} + \frac{m}{c^2}\frac{\delta_{ip}\dot{x}^p\,\delta_{kq}\dot{x}^q}{(1-\frac{v^2}{c^2})^{3/2}},$$

where m, c are positive constants, and $v^2 = \delta_{ij}\dot{x}^i\dot{x}^j$. Then $\Gamma_g = 0$, $E_g = (g_{ij}\ddot{x}^j)\, dx^i \wedge dt$, i.e. $(R^3, g, 0)$ is a free particle of the special relativity theory. Let F be the Lorentz force on the semi-finslerian manifold (R^3, g), $F = \delta(\vec{F}, \cdot)$, where $\vec{F} = e\vec{E} + \frac{em}{c}(\vec{v} \times \vec{H})$. Since $\Gamma_g = 0$, we get the components of the connection Γ of the mechanical system (R^3, g, F) in the form

$$\Gamma^i = g^{ij}F_j = \frac{e}{m}\sqrt{1-\frac{v^2}{c^2}}\left(E^i + \frac{1}{c}(\vec{v}\times\vec{H})^i - \frac{1}{c^2}v^i\,\vec{v}\vec{E}\right).$$

This connection obviously differs from that describing a classical particle in the Lorentz force field, i.e., the mechanical system $(R^3, m\delta, F)$; in this case we have

$$\Gamma^i = \frac{e}{m}\left(E^i + \frac{1}{c}(\vec{v}\times\vec{H})^i - \frac{1}{c^2}v^i\right).$$

We stress that, by definition, a mechanical system does not identify with its corresponding semispray connection. The difference between a mechanical system and the semispray connection describing the motion of this system can be demonstrated on the following easy example: the semispray connection $\Gamma^i = k\dot{x}^i$ can describe a mechanical system $(R^3, m\delta, F)$, where $F = \delta(\vec{F}, \cdot)$, $\vec{F} = (k\dot{x}^1, k\dot{x}^2, k\dot{x}^3)$, i.e., a classical particle of mass m moving in the dissipative force field, or a mechanical system $(R^3, e^{kt}\delta)$, i.e., a classical free particle accretting mass according to the rule $f(t) = me^{kt}$, or some other mechanical system (according to a choice of a semi-finslerian metric on R^3). For a discussion to this point see also Remark 6.7.9.

Recall that every $\pi_{1,0}$-vertical valued π_1-horizontal one-form on $R \times TM$ is called a *soldering form*. Soldering forms on $R \times TM$ can be roughly characterized as "differences of semispray connections." More precisely, if Γ, Γ' are two semispray connections on $R \times TM$ then the vector valued one-form s defined by

$$s = h_\Gamma - h_{\Gamma'}$$

is a soldering form; conversely, if s is a soldering form on $R \times TM$ then there exist semispray connections Γ, Γ' such that $s = h_\Gamma - h'_\Gamma$. We shall denote by $\mathcal{S}(R \times TM)$ the $\mathcal{F}$-module of all soldering forms on $R \times TM$. Further we denote by $\Omega^{1,1}_{R\times M}(R \times TM)$ the module of all dynamical forms on $R \times TM$.

Let (M, g) be a semi-finslerian manifold. Then there arises a canonical isomorphism

$$\tilde{g} : \mathcal{S}(R \times TM) \ni s \to \tilde{g}(s) = E \in \Omega^{1,1}_{R\times M}(R \times TM) \tag{10.35}$$

of modules. It is defined in each fiber chart (V, ψ), $\psi = (t, q^\sigma)$ on $R \times M$ where

$$s = s^\nu \frac{\partial}{\partial \dot{q}^\nu} \otimes dt, \quad E = E_\sigma\, dq^\sigma \wedge dt,$$

by the formula

$$E_\sigma = g_{\sigma\nu}s^\nu. \tag{10.36}$$

By this isomorphism, on a semi-finslerian manifold, forces can be identified with soldering forms. This means however, that a mechanical system (M, g, F) can be equivalently represented by the semispray connection Γ such that $h_\Gamma = h_{\Gamma_g} + s$, where $s = \tilde{g}^{-1}(F)$.

Now, let Γ be a semispray connection on a fibered manifold $R \times M \to R$. Note that if there is chosen a variational metric g on $R \times TM$, then Γ represents a *unique mechanical system* (M, g, F): it holds $F = \tilde{g}(s)$ where $s = h_\Gamma - h_{\Gamma_g}$. We say that the connection Γ is *variational* with respect to the metric g, if the mechanical system (M, g, F), defined by Γ is variational.

Using the Theorem 6.7.2 we obtain the following classification theorem for variational semispray connections on semi-finslerian manifolds (O. Krupková [8], see also [1,2]).

10.4.8. Theorem. *Let (M, g) be a semi-finslerian manifold, let Γ be a semispray connection on (M, g). The connection Γ is variational with respect to the metric g if and only if*

$$\Gamma = \Gamma_g + s \tag{10.37}$$

where Γ_g is the canonical connection of g and

$$\tilde{g}(s) = (\beta_\sigma + \alpha_{\sigma\nu})\, dq^\nu \wedge dt, \tag{10.38}$$

where β_σ, $\alpha_{\sigma\nu}$ are functions on $V \subset R \times M$, satisfying the conditions

$$\alpha_{\sigma\nu} = -\alpha_{\nu\sigma}, \qquad \frac{\partial \alpha_{\sigma\nu}}{\partial q^\rho} + \frac{\partial \alpha_{\rho\sigma}}{\partial q^\nu} + \frac{\partial \alpha_{\nu\rho}}{\partial q^\sigma} = 0, \qquad \frac{\partial \beta_\sigma}{\partial q^\nu} - \frac{\partial \beta_\nu}{\partial q^\sigma} = \frac{\partial \alpha_{\sigma\nu}}{\partial t}. \tag{10.39}$$

Notice that Γ is variational with respect to g if and only if the corresponding dynamical form $F = \tilde{g}(s)$ is locally variational, i.e., in correspondence with Chapter 1, iff the force F is *potential*.

Recall that if a force F is potential then any first-order Lagrangian for F is called *potential energy*.

From the above theorem we easily get the following corollaries:

10.4.9. Corollary. *A force F is potential if and only if it is of the form*

$$F = (\beta_\sigma + \alpha_{\sigma\nu})\, dq^\nu \wedge dt,$$

where

$$\alpha_{\sigma\nu} = -\alpha_{\nu\sigma}, \qquad \frac{\partial \alpha_{\sigma\nu}}{\partial q^\rho} + \frac{\partial \alpha_{\rho\sigma}}{\partial q^\nu} + \frac{\partial \alpha_{\nu\rho}}{\partial q^\sigma} = 0, \qquad \frac{\partial \beta_\sigma}{\partial q^\nu} - \frac{\partial \beta_\nu}{\partial q^\sigma} = \frac{\partial \alpha_{\sigma\nu}}{\partial t}.$$

We remind the reader that if, in particular $m = 3$, the above formulas define the familiar Lorentz-type force (see also Chapters 1, 6).

10.4.10. Corollary. *A linear torsion-free connection on M is variational if and only if it is metrizable.*

Proof. To see this it is sufficient to note that to every linear connection γ on M there is uniquely assigned a semispray connection Γ on $R \times TM$ by

$$\Gamma^i = -\gamma^i_{jk} \dot{x}^i \dot{x}^k,$$

where Γ^i (resp. γ^i_{jk}) are the components of Γ (resp. γ), and that metrizability of γ means that there exists a metric $g(x^i)$ on M, such that γ coincides with the Levi-Civita connection ∇ of g. Now, applying Theorem 10.4.8 we get $\Gamma = \Gamma_g$, where, by (10.31), Γ_g are the Christoffel symbols of the metric g. □

The results of Corollary 10.4.9 (resp. 10.9.10) are due to E. Engels and W. Sarlet [1], and J. Novotný [1] (resp. J. Klein [1], and D. Krupka and A. Sattarov [1]).

Some other results on metrizability of connections can be found in D. Krupka and A. Sattarov [1] (Finsler connections), O. Krupková [8], E.Martínez, J. F. Cariñena and W. Sarlet [1] (semispray connections), and on variationality of sprays in J. Klein [1,2], I. Anderson and G. Thompson [1], and E.Martínez, J. F. Cariñena and W. Sarlet [1].

Bibliography

R. Abraham and J. E. Marsden

1. *Foundations of Mechanics*, 2nd Ed., The Benjamin/Cummings Publ. Comp., Reading, 1978.

P. L. Antonelli and R. Miron

1. *Lagrange and Finsler Geometry*, P. L. Antonelli and R. Miron, eds., Kluwer, Dordrecht, 1996.

V. Aldaya and J. de Azcárraga

1. *Vector bundles, r-th order Noether invariants and canonical symmetries in Lagrangian field theory*, J. Math. Phys. **19** (1978), 1876–1880.
2. *Variational principles on r-th order jets of fibre bundles in field theory*, J. Math. Phys. **19** (1978), 1869–1875.
3. *Geometric formulation of classical mechanics and field theory*, Rev. Nuovo Cimento **3** (1980), 1–66.
4. *Higher order Hamiltonian formalism in field theory*, J. Phys. A: Math. Gen. **13** (1982), 22545–2551.

I. Anderson

1. *Aspects of the inverse problem to the calculus of variations*, Arch. Math. (Brno) **24** (1988), 181–202.
2. *The Variational Bicomplex*, Preprint, Utah State University, 1989.

I. Anderson and T. Duchamp

1. *On the existence of global variational principles*, Am. J. Math. **102** (1980) 781–867.

I. Anderson and G. Thompson

1. *The Inverse Problem of the Calculus of Variations for Ordinary Differential Equations*, Memoirs of the AMS **98**, No. 473 (1992) 110 pp.

L. C. de Andres, M. de León and P. R. Rodrigues

1. *Connections on tangent bundles of higher order*, Demonstratio Mathematica **22** (1989), 607–632.

J. Barcelos-Neto and N. R. F. Braga

1. *Symplectic analysis of a Dirac constrained theory*, J. Math. Phys. **35** (1994), 3497–3503.

C. Batlle, J. Gomis, J. M. Pons and N. Román-Roy

1. *Equivalence between the Lagrangian and Hamiltonian formalism for constrained systems*, J. Math. Phys. **27** (1986), 2953–2962.
2. *On the Legendre transformation for singular Lagrangians and related topics*, J. Phys. A: Math. Gen. **20** (1987), 5113–5123.
3. *Lagrangian and Hamiltonian constraints for second-order singular Lagrangians*, J. Phys. A: Math. Gen. **21** (1988), 2693–2703.

S. Benenti

1. *Symplectic relations in analytical mechanics*, in: *Modern Developments in Analytical Mechanics I: Geometrical Dynamics*, Proc. IUTAM-ISIMM Symposium, Torino, Italy 1982, S. Benenti, M. Francaviglia and A Lichnerowicz, eds. (Accad. delle Scienze di Torino, Torino, 1983), 39–91.
2. *L'intégration de l'equation d'Hamilton–Jacobi par séparation des variables: histoire et résultats récents*, in: *La Mécanique Analytique de Lagrange et son héritage*, Suppl. al. No 124 (1990) Atti Accad. Sci. Torino, Torino 1990, 119-144.

D. E. Betounes

1. *Extension of the classical Cartan form*, Phys. Rev. D **29** (1984), 599–606.

R. L. Bryant, S. S. Chern, R. B. Gardner, H. L. Goldschmidt and P. A. Griffiths

1. *Exterior Differential Systems*, Springer-Verlag, New York, 1991.

F. Cantrijn

1. *On the geometry of degenerate Lagrangians*, in: *Differential Geometry and Its Applications* (*Communications*), Proc. Conf., Brno, Czechoslovakia, 1986, D. Krupka and A. Švec, eds. (J. E. Purkyně University, Brno, 1986), 43–56.

F. Cantrijn, J. F. Cariñena, M. Crampin and L. A. Ibort

1. *Reduction of degenerate Lagrangian systems*, J. Geom. Phys. **3** (1986), 353–400.

F. Cantrijn and W. Sarlet

1. *Symmetries and conservation laws for generalized Hamiltonian systems*, Int. J. Theor. Phys. **20** (1981), 645–670.

C. Carathéodory

1. *Über die Variationsrechnung bei mehrfachen Integralen*, Acta Szeged **4** (1929), 193–216.

J. F. Cariñena

1. *Theory of singular Lagrangians*, Fortschr. Phys. **38** (1990), 641–679.

J. F. Cariñena, J. Fernández-Núñez

1. *Geometric theory of time-dependent singular Lagrangians*, Fortschr. Phys. **41** (1993), 517–552.

J. F. Cariñena and L. A. Ibort

1. *Geometric theory of equivalence of Lagrangians for constrained systems*, J. Phys. A: Math. Gen. **18** (1985), 3335–3341.

J. F. Cariñena and C. López

1. *The time-evolution operator for higher-order singular Lagrangians*, Int. J. Mod. Phys. A **7** (1992), 2447–2468.

J. F. Cariñena, C. López and E. Martínez

1. *Sections along a map applied to higher-order Lagrangian mechanics. Noether's Theorem*, Acta Appl. Math. **25** (1991), 127–151.

J. F. Cariñena, C. López and M. F. Rañada

1. *Geometric Lagrangian approach to first-order systems and applications*, J. Math. Phys. **29** (1988), 1134–1142.

J. F. Cariñena, C. López and N. Román-Roy

1. *Geometric study of the connection between the Lagrangian and Hamiltonian constraints*, J. Geom. Phys. **4** (1987), 315–334.
2. *Origin of the Lagrangian constraints and their relation with the Hamiltonian formulation*, J. Math. Phys. **29** (1988), 1143–1149.

J. F. Cariñena, E. Martínez and J. Fernández-Núñez

1. *Noether's theorem in time-dependent Lagrangian mechanics*, Rep. Math. Phys. **31** (1992), 189–203.

J. F. Cariñena and M. F. Rañada

1. *Blow-up regularization of singular Lagrangians*, J. Math. Phys. **25** (1984), 2430–2435.
2. *Noether's theorem for singular Lagrangians*, Lett. Math. Phys. **15** (1988), 305–311.
3. *Non-linear generalization of the gauge ambiguities of the Lagrangian formalism*, Int. J. Mod. Phys. A **6** (1991), 737–748.
4. *Lagrangian systems with constraints: a geometric approach to the method of Lagrange multipliers*, J. Phys. A: Math. Gen. **26** (1993), 1335–1351.

É. Cartan

1. *Leçons sur les Invariants Intégraux*, Hermann, Paris, 1922.

R. Cawley

1. *Determination of the Hamiltonian in the presence of constraints*, Phys. Rev. Lett. **42** (1979), 413–416.

M. Chaichian and D. L. Martinez

1. *On the Noether identities for a class of systems with singular Lagrangians*, J. Math. Phys. **35** (1994), 6536–6545.

D. Chinea, M. de León and J. C. Marrero

1. *The constraint algorithm for time-dependent Lagrangians*, J. Math. Phys. **35** (1994), 3410–3447.

J. Chrastina

1. *Formal Calculus of Variations on Fibered Manifolds*, J. E. Purkyně University, Brno, 1989.

M. Crampin, W. Sarlet and F. Cantrijn

1. *Higher order differential equations and higher order Lagrangian mechanics*, Math. Proc. Camb. Phil. Soc. **99** (1986), 565–587.

M. Crampin, W. Sarlet, E. Martínez, G. B. Byrnes and G. E. Prince

1. *Towards a geometrical undestanding of Douglas's solution of the inverse problem of the calculus of variations*, Inverse Problems **10** (1994), 245–260.

G. Darboux

1. *Leçons sur la théorie générale des surfaces et les applications géometriques du calcul infinitésimal*, Paris, 1899.

P. Dedecker

1. *Sur les intégrales multiples du Calcul des Variations*, Compt. rend. du IIIe congres Nat. des Sci., Bruxelles, 1950, 29–35.
2. *Calcul des variations, formes différentielles et champs géodésiques*, in: Coll. internat. du C.N.R.S., Strasbourg, 1953 (C.N.R.S. Paris, 1954), 17–34.

3. *Calcul des Variations et Topologie Algebrique*, Thesis, Univ. de Liège, Faculté des Sciences, 1957.
4. *On the generalization of symplectic geometry to multiple integrals in the calculus of variations*, in: Lecture Notes in Math. **570** (Springer, Berlin, 1977), 395–456.
5. *Intégrales complétes de l'équation aux dérivées partielles de Hamilton-Jacobi d'une intégral multiple*, C. R. Acad. Sci. Paris **285**, Sér. A (1977), 123–126.
6. *Problèmes variationelles dégénérés*, C. R. Acad. Sci. Paris **286**, Sér. A (1978), 547–550.
7. *Le théorème de Helmholtz-Cartan pour une intégrale simple d'ordre supérieur*, C. R. Acad. Sci. Paris **288**, Sér. A (1979), 827–830.
8. *Existe-t-il, en calcul des variations, un formalisme de Hamilton-Jacobi-E. Cartan pour les intégrales multiples d'ordre supérieur?*, C. R. Acad. Sci. Paris **298**, sér. 1 (1984), 397–400.

P. Dedecker and W. M. Tulczyjew

1. *Spectral sequences and the inverse problem of the calculus of variations*, Internat. Coll. on Diff. Geom. Methods in Math. Phys., Salamanca 1979, in: *Lecture Notes in Math.* **836** (Springer, Berlin, 1980), 498–503.

P. A. M. Dirac

1. *Generalized Hamiltonian dynamics*, Canad. J. Math. II (1950), 129–148.

V. V. Dodonov, V. I. Manko and V. D. Skarzhinsky

1. Nuovo Cim. B (1982) **69**, 185–205.

D. Dominici and J. Gomis

1. *Poincaré-Cartan integral invariant and canonical transformations for singular Lagrangians*, J. Math. Phys. **21** (1980), 2124–2130.
2. D. Dominici and J. Gomis, *Poincaré-Cartan integral invariant and canonical transformations for singular Lagrangians: An addendum*, J. Math. Phys. **23** (1982), 256–257.

D. Dominici, G. Longhi, J. Gomis and J. M. Pons

1. *Hamilton-Jacobi theory for constrained systems*, J. Math. Phys. **25** (1984), 2439–2452.

Th. De Donder

1. *Théorie Invariantive du Calcul des Variations*, Gauthier–Villars, Paris, 1930.

J. Douglas

1. *Solution of the inverse problem of the calculus of variations*, Trans. Amer. Math. Soc. **50** (1941), 71–128.

S. V. Duzhin and V. V. Lychagin

1. *Symmetries of distributions and quadrature of ordinary differential equations*, Preprint.

Ch. Ehresmann

1. *Sur la théorie des espaces fibrés*, in: *Topologie Algébrique* (Éditions du CNRS, Paris, 1947), 3–35.
2. *Les connections infinitésimales dans un espace fibré différentiable*, in: Coll. Topologie, Bruxelles 1950 (Liége, 1951), 29–55.
3. *Les prolongements d'une variété différentiable*: 1. *Calcul des Jets, prolongement principal*, C. R. Acad. Sci. Paris **233** (1951), 598–600.
4. *Extension du calcul des Jets aux Jets non-holonomes*, C. R. Acad. Sci. Paris **239** (1954), 1762–1764.
5. *Applications de la notion de Jet non-holonome*, C. R. Acad. Sci. Paris **240** (1955), 397–399.

6. *Les prolongements d'une space fibré différentiable*, C. R. Acad. Sci. Paris **240** (1955), 1755–1757.

E. Engels and W. Sarlet

1. *General solution and invariants for a class of Lagrangian equations governed by a velocity-dependent potential energy*, J. Phys. A: Math. Gen. **6** (1973), 818–825.

C. Ferrario and A. Passerini

1. *Symmetries and constants of motion for constrained Lagrangian systems: a presympectic version of the Noether theorem*, J. Phys. A: Math. Gen. **23** (1990), 5061–5081.
2. *Dynamical symmetries in constrained systems: a Lagrangian analysis*, J. Geom. Phys. **9** (1992), 121–148.

M. Ferraris

1. *Fibered connections and global Poincaré–Cartan forms in higher order calculus of variations*, in: *Geometrical Methods in Physics*, Proc. Conf. on Diff. Geom. and Appl. Vol. 2, Nové Město na Moravě, Sept. 1983, D. Krupka, ed. (J. E. Purkyně Univ. Brno, Czechoslovakia, 1984), 61–91.

M. Ferraris and M. Francaviglia

1. *On the global structure of the Lagrangian and Hamiltonian formalisms in higher order calculus of variations*, in: *Geometry and Physics*, Proc. Int. Meeting, Florence, Italy 1982, M. Modugno ed. (Pitagora Ed., Bologna, 1983), 43–70.
2. *On the globalization of the Lagrangian and Hamiltonian formalisms in higher order mechanics*, in: *Modern Developments in Analytical Mechanics I: Geometrical Dynamics*, Proc. IUTAM-ISIMM Symposium, Torino, Italy 1982, S. Benenti, M. Francaviglia and A Lichnerowicz eds. (Accad. delle Scienze di Torino, Torino, 1983), 109–125.

C. A. P. Galvão and N. A. Lemos

1. *On the quantization of constrained generalized dynamics*, J. Math. Phys. **29** (1988), 1588–1592.

P. L. Garcia

1. *The Poincaré-Cartan invariant in the calculus of variations*, Symposia Math. **14** (1974), 219–246.

P. L. Garcia and J. Muñoz

1. *On the geometrical structure of higher order variational calculus*, in: *Modern Developments in Analytical Mechanics I: Geometrical Dynamics*, Proc. IUTAM-ISIMM Symposium, Torino, Italy 1982, S. Benenti, M. Francaviglia and A Lichnerowicz eds. (Accad. delle Scienze di Torino, Torino, 1983), 127–147.

P. L. Garcia and A. Pérez-Rendón

1. *Symplectic approach to the theory of quantized fields, II*, Arch. Rational Mech. Anal. **43** (1971), 101–124.

I. M. Gelfand and L. A. Dikii

1. *The asymptotics of the resolvent of the Sturm-Liouville equations and the algebra of the Korteweg-de-Vries equation*, Uspekhi Mat. Nauk **30** (1975), 67–100.

G. Giachetta

1. *Jet methods in nonholonomic mechanics*, J. Math. Phys. **33** (1992), 1652–1665.

M. Giaquinta and S. Hildebrandt

1. *Calculus of Variations, I, II*, Springer, Berlin, 1996.

C. Godbillon

1. *Géométrie Différentielle et Mécanique Analytique*, Hermann, Paris, 1969.

H. Goldschmidt and S. Sternberg

1. *The Hamilton–Cartan formalism in the calculus of variations*, Ann. Inst. Fourier, Grenoble **23** (1973), 203–267.

J. Gomis, J. Llosa and N. Román

1. *Lee Hwa Chung theorem for presymplectic manifolds. Canonical transformations for constrained systems*, J. Math. Phys. **25** (1984), 1348–1355.

M. J. Gotay

1. *On the validity of Dirac's conjecture regarding first-class secondary constraints*, J. Phys. A: Math. Gen. **16** (1983), L141–L145.
2. *An exterior differential systems approach to the Cartan form*, in: Proc. Internat. Colloquium "Géometrie Symplectique et Physique Mathématique", P. Donato, C. Duval, J. Elhadad and G. M. Tuynman, eds. (Birkhäuser, Boston, 1991).

M. J. Gotay and J. M. Nester

1. *Presymplectic Lagrangian systems I: the constraint algorithm and the equivalence theorem, II: the second order equation problem*, Ann. Inst. H. Poincaré, Sect. A, **30** (1979), 129–142; **32** (1980), 1–13.

M. J. Gotay, J. M. Nester and G. Hinds

1. *Presymplectic manifolds and the Dirac–Bergmann theory of constraints*, J. Math. Phys. **19** (1978), 2388–2399.

X. Grácia, J. M. Pons and N. Román-Roy

1. *Higher order Lagrangian systems: Geometric structures, dynamics and constraints*, J. Math. Phys. **32** (1991), 2744–2763.

P. A. Griffiths

1. *Exterior Differential Systems and the Calculus of Variations*, Progress in Math. **25**, Birkhäuser, Boston, 1983.

J. Grifone

1. *Structure presque-tangente et connexions, I.*, Ann. Inst. Fourier, Grenoble **22** (1972), 287–334.

D. R. Grigore

1. *Generalized Lagrangian dynamics and Noetherian symmetries*, Int. J. Mod. Phys. A **7** (1992), 7153–7168.
2. *Higher order Lagrangian theories and Noetherian symmetries*, to be published in Roum. J. Phys.
3. *Variationally trivial Lagrangians and locally variational differential equations of arbitrary order*, Preprint, 1997.

D. R. Grigore and O. T. Popp

1. *On the Lagrange–Souriau form in classical field theory*, to be published in Math. Bohemica.

L. Haine

1. *La suite spectrale d'Euler-Lagrange*, Sem. Math. Pure, Inst. de Math. Pure et Appl., Université Catholique de Louvain (1980), XIII, 1–31.

P. Havas

1. *The range of application of the Lagrange formalism I*, Nuovo Cimento Suppl. **5** (1957), 363–388.
2. *The connection between conservation laws and invariance groups: folklore, fiction, and fact*, Acta Phys. Austr. **38** (1973), 145–167.

C. F. Hayes and J. M. Jankowski

1. *Quantization of generalized mechanics*, Nuovo Cimento B **58** (1968), 494–497.

H. Helmholtz

1. *Ueber die physikalische Bedeutung des Prinzips der kleinsten Wirkung*, J. für die reine u. angewandte Math. **100** (1887), 137–166.
2. *Das Prinzip des kleinsten Wirkung in der Elektrodynamik*, Wied. Ann., Bd. XLVII (1892), 1–25; Wiss. Abh., Bd. III (J. Barth, Leipzig, 1895), 476–504.
3. *Studien zur Statik monocyklischer Systeme*, Wiss. Abh., Bd. III (J. Barth, Leipzig, 1895), 119–141.
4. *Die physikalische Bedeutung des Prinzips der kleinsten Wirkung*, Wiss. Abh., Bd. III (J. Barth, Leipzig, 1895), 207.

M. Henneaux

1. *Equations of motion, commutation relations and ambiguities in the Lagrangian formalism*, Ann. Phys. **140** (1982), 45–64.
2. *On the inverse problem of the calculus of variations*, J. Phys. A **15** (1982), L93–L96.

M. Henneaux and L. C. Shepley

1. *Lagrangians for spherically symmetric potentials*, J. Math. Phys. **23** (1982), 2101–2107.

R. Hermann

1. *Differential Geometry and the Calculus of Variations*, Academic Press, New York, 1968.

D. Hilbert

1. *Die Grundlagen der Physik*, Götting. Nachrichten **1** (1915), 395.

R. Hojman, S. Hojman and J. Sheinbaum

1. *Shortcut for constructing any Lagrangian from its equations of motion*, Phys. Rev. D **28** (1983), 1333–1336.

S. A. Hojman, L. C. Shepley

1. *No Lagrangian? No quantization!*, J. Math. Phys. **32** (1991), 142–146.

S. Hojman and L. F. Urrutia

1. *On the inverse problem of the calculus of variations*, J. Math. Phys. **22** (1981), 1896–1903.

M. Horák and I. Kolář

1. *On the higher order Poincaré–Cartan forms*, Czechoslovak Math. J. **33 (108)** (1983), 467–475.

J. Hrivňák

1. *Symmetries and first integrals in higher-order mechanics*, Thesis, Silesian University, Opava, 1995 (in Czech).

R. Jackiw

1. *(Constrained) quantization without tears*, in: *Constraint Theory and Quntization Methods*, F. Colomo, L. Lusanna and G. Marmo, eds. (World Scientific, Singapore, 1994), 163–175.

Y. Kaminaga

1. *Quantum mechanics of higher derivative systems and total derivative terms*, J. Phys. A: Math. Gen. **29** (1996), 5049–5088.

H. A. Kastrup

1. *Canonical theories of Lagrangian dynamical systems in physics*, Phys. Rep. **101** (1983), 1–167.

J. Kijowski

1. *On a new variational principle in general relativity and the energy of the gravitational field*, Gen. Rel. Grav. **9** (1978), 857–877.

L. Klapka

1. *Integrals of motion and semiregular Lepagean forms in higher order mechanics*, J. Phys. A: Math. Gen. **16** (1983), 3783–3794.
2. *Euler–Lagrange expressions and closed two–forms in higher order mechanics*, in: *Geometrical Methods in Physics*, Proc. Conf. on Diff. Geom. and Appl. Vol. 2, Nové Město na Moravě, Sept. 1983, D. Krupka, ed. (J. E. Purkyně Univ. Brno, Czechoslovakia, 1984) 149–153.

J. Klein

1. *Geometry of sprays. Lagrangian case. Principle of least curvature*, in: *Modern Developments in Analytical Mechanics I: Geometrical Dynamics*, Proc. IUTAM-ISIMM Symposium, Torino, Italy 1982, S. Benenti, M. Francaviglia and A. Lichnerowicz, eds. (Accad. delle Scienze di Torino, Torino, 1983), 177–196.
2. *On variational second order differential equations*, in: Proc. Conf. on Differential Geometry and Its Applications, August 1992, Opava (Czechoslovakia), O. Kowalski and D. Krupka, eds. (Silesian University, Opava, Czech Republic, 1993), 449–459.

J. Koiller

1. *Reduction of some classical non-holonomic systems with symmetry*, Arch. Rational Mech. Anal. **118** (1992), 113–148.

I. Kolář

1. *Some geometric aspects of the higher order variational calculus*, in: *Geometrical Methods in Physics*, Proc. Conf. on Diff. Geom. and Appl. Vol. 2, Nové Město na Moravě, Sept. 1983, D. Krupka, ed. (J. E. Purkyně University, Brno, Czechoslovakia, 1984), 155–166.
2. *A geometric version of the higher order Hamilton formalism in fibered manifolds*, J. Geom. Phys. **1** (1984), 127–137.

I. Kolář, P. W. Michor and J. Slovák

1. *Natural Operators in Differential Geometry*, Springer, Berlin, 1993.

D. Krupka

1. *Some geometric aspects of variational problems in fibered manifolds*, Folia Fac. Sci. Nat. UJEP Brunensis **14** (1973), 1–65.
2. *A geometric theory of ordinary first order variational problems in fibered manifolds. I. Critical sections, II. Invariance*, J. Math. Anal. Appl. **49** (1975), 180–206; 469–476.
3. *A map associated to the Lepagean forms of the calculus of variations in fibered manifolds*, Czechoslovak Math. J. **27** (1977), 114–118.

4. *On the local structure of the Euler-Lagrange mapping of the calculus of variations*, in: Proc. Conf. on Diff. Geom. and Its Appl. 1980, O. Kowalski, ed. (Universita Karlova, Prague, 1981), 181–188.
5. *Lepagean forms in higher order variational theory*, in: *Modern Developments in Analytical Mechanics I: Geometrical Dynamics*, Proc. IUTAM-ISIMM Symposium, Torino, Italy 1982, S. Benenti, M. Francaviglia and A. Lichnerowicz, eds. (Accad. delle Scienze di Torino, Torino, 1983), 197–238.
6. *On the higher order Hamilton theory in fibered spaces*, in: *Geometrical Methods in Physics*, Proc. Conf. on Diff. Geom. and Appl., Vol. 2, Nové Město na Moravě, Sept. 1983, D. Krupka, ed. (J. E. Purkyně University, Brno, Czechoslovakia, 1984), 167–183.
7. *Regular lagrangians and Lepagean forms*, in: *Differential Geometry and Its Applications*, Proc. Conf., Brno, Czechoslovakia, 1986, D. Krupka and A. Švec, eds. (D. Reidel, Dordrecht, 1986), 11–148.
8. *Geometry of Lagrangean structures* 2, 3, Arch. Math. (Brno) **22** (1986), 211–228; Proc. 14th Winter School on Abstract Analysis, Jan. 1986, Srní (Czechoslovakia), Suppl. ai rend. del Circ. Mat. di Palermo **14** (1987), 178–224.
9. *Variational sequences on finite order jet spaces*, in: *Differential Geometry and Its Applications*, Proc. Conf., Brno, Czechoslovakia, 1989, J. Janyška and D. Krupka, eds. (World Scientific, Singapore, 1990), 236–254.
10. *Topics in the calculus of variations: Finite order variational sequences*, in: Proc. of the 5th International Conference on Differential Geometry and Its Applications, August 1992, Opava, Czechoslovakia, O. Kowalski and D. Krupka, eds. (Silesian University, Opava, Czech Republic, 1993), 473–495.
11. *The contact ideal*, Diff. Geom. Appl. **5** (1995), 257–276.
12. *Lectures on Variational Sequences*, Open Education and Sciences, Opava, 1995.
13. *Variational sequences in mechanics*, Calculus of Variations and PDE, to be published.

D. Krupka and J. Musilová

1. *Hamilton extremals in higher order mechanics*, Arch. Math. (Brno) **20** (1984), 21–30.
2. *Trivial Lagrangians in field theory*, Diff. Geom. Appl., to be published.

D. Krupka, A. E. Sattarov

1. *The inverse problem of the calculus of variations for Finsler structures*, Math. Slovaca **35** (1985), 217–222.

D. Krupka and O. Štěpánková

1. *On the Hamilton form in second order calculus of variations*, in: *Geometry and Physics*, Proc. Int. Meeting, Florence, Italy 1982, M. Modugno, ed. (Pitagora Ed., Bologna, 1983), 85–102.

O. Krupková

1. *A note on the Helmholtz conditions*, in: *Differential Geometry and Its Applications*, Proc. Conf., August 1986, Brno, Czechoslovakia (J. E. Purkyně University, Brno, Czechoslovakia, 1986), 181–188.
2. *Lepagean 2-forms in higher order Hamiltonian mechanics, I. Regularity, II. Inverse problem*, Arch. Math. (Brno) **22** (1986), 97–120; **23** (1987), 155–170.
3. *Hamilton-Jacobi distributions*, Preprint, Dept. of Math., Masaryk University, Brno, 1990.
4. *Variational analysis on fibered manifolds over one-dimensional bases*, Thesis, Silesian University, Opava, 1992, 67 pp.
5. *On the inverse problem of the calculus of variations for ordinary differential equations*, Math. Bohemica **118** (1993), 261–276.

6. *Liouville and Jacobi theorems for vector distributions*, in: Proc. of the 5th International Conference on Differential Geometry and Its Applications, August 1992, Opava, Czechoslovakia, O. Kowalski and D. Krupka, eds. (Silesian University, Opava, 1993), 75–87.
7. *A Geometric Theory of Variational Ordinary Differential Equations*, Preprint, Silesian University, Opava, 1993, 89 pp.; Thesis, Silesian University, Opava, 1995, 165 pp.
8. *Variational metrics on $R \times TM$ and the geometry of nonconservative mechanics*, Math. Slovaca, **44** (1994), 315–335.
9. *A geometric setting for higher order Dirac–Bergmann theory of constraints*, J. Math. Phys. **35** (1994), 6557–6576.
10. *Symmetries and first integrals of time-dependent higher-order constrained systems*, J. Geom. Phys. **18** (1996), 38–58.
11. *Higher-order constrained systems on fibered manifolds: An exterior differential systems approach*, in: *New Developments in Differential Geometry*, J. Tamássy and J. Szenthe, eds. (Kluwer, Dordrecht, 1996), 255–278.
12. *Noether theorem and first integrals of constrained Lagrangean systems*, Math. Bohemica, to be published.
13. *Mechanical systems with non-holonomic constraints*, J. Math. Phys. **38** (1997), to be published.

O. Krupková and A. Vondra

1. *On some integration methods for connections on fibered manifolds*, in: Proc. of the 5th International Conference on Differential Geometry and Its Applications, 1992, Opava, Czechoslovakia, O. Kowalski and D. Krupka, eds. (Silesian University, Opava, 1993), 89–101.

U. Kulshreshtha

1. *Hamiltonian formulation of a theory with constraints*, J. Math. Phys. **33** (1992), 633–638.

B. Kupershmidt

1. *Geometry of jet bundles and the structure of Lagrangian and Hamiltonian formalisms*, in: *Lecture notes in Math.* **775** (Springer-Verlag, Berlin, 1980), 162–217.

M. de León and D. M. de Diego

1. *Classification of symmetries for higher order Lagrangian systems II: the non-autonomous case*, Extracta Mathematicae **8** (1994), 111–114.
2. *Symmetries and constants of the motion for higher order Lagrangian systems*, J. Math. Phys. **36** (1995), 4138–4161.
3. *On the geometry of non-holonomic Lagrangian systems*, J. Math. Phys. **37** (1996) 3389–3414.

M. de León, D. M. de Diego and P. Pitanga

1. *A new look at degenerate Lagrangian dynamics from the viewpoint of almost product structures*, J. Phys. A: Math. Gen **28** (1995), 4951–4971.

M. de León, J. Marín-Solano and J. C. Marrero

1. *The constraint algorithm in the jet formalism*, Diff. Geom. Appl. **6** (1996), 275–300.

M. de León and J. C. Marrero

1. *Degenerate time-dependent Lagrangians of second order: the fourth order differential equation problem*, in: Proc. 5th Int. Conf. on Differential Geometry and Its Applications, 1992, Opava, Czechoslovakia, O. Kowalski and D. Krupka, eds. (Silesian Univ., Opava, 1993), 497–508.

M. de León, J. C. Marrero and D. M. de Diego

1. *Constrained time-dependent Lagrangian systems and Lagrangian submanifolds*, J. Math. Phys. **34** (1993), 622–644.

2. *Time-dependent mechanical systems with non-linear constraints*, in: Proc. Conf. on Diff. Geom., Budapest, 1996, to be published.
3. *Non-holonomic Lagrangian systems in jet manifolds*, J. Phys. A: Math. Gen. **30** (1997), 1167–1190.

M. de León and P. R. Rodrigues

1. *Generalized Classical Mechanics and Field Theory*, North-Holland, Amsterdam, 1985.
2. *Almost tangent geometry and higher order mechanical systems*, in: *Differential Geometry and Its Applications*, Proc. Conf., Brno, Czechoslovakia, 1986, D. Krupka and A. Švec, eds. (D. Reidel, Dordrecht, 1986), 179–195.
3. *Methods of Differential Geometry in Analytical Mechanics*, North-Holland, Amsterdam, 1989.

Th. Lepage

1. *Sur les champs géodésiques du Calcul des Variations*, Bull. Acad. Roy. Belg., Cl. des Sciences **22** (1936), 716–729.

P. Libermann

1. *Sur quelques proprietés des varietés symplectiques*, in: Proc. Conf. on Diff. Geom. and Its Appl. 1980, O. Kowalski, ed. (Universita Karlova, Prague, 1981), 137–159.
2. *Symplecticly regular foliations*, in: *Modern Developments in Analytical Mechanics I: Geometrical Dynamics*, Proc. IUTAM-ISIMM Symposium, Torino, Italy 1982, S. Benenti, M. Francaviglia and A. Lichnerowicz, eds. (Accad. delle Scienze di Torino, Torino, 1983), 239–246.
3. *Lie algebroids and mechanics*, Arch. Math. (Brno) **32** (1996) 147–162.

P. Libermann and Ch.-M. Marle

1. *Symplectic Geometry and Analytical Mechanics*, Mathematics and Its Applications, D. Reidel, Dordrecht, 1987.

L. Lusanna

1. *Classical observables of gauge theories from the multitemporal approach*, Contemporary Math. **132** (1992), 531–549.

V. Lychagin

1. *Lectures on Geometry of Differential Equations, Part I, Part II*, Consiglio Nazionale delle Ricerche (G.N.F.M.), Roma, 1992.

L. Mangiarotti and M. Modugno

1. *Fibered spaces, jet spaces and connections for field theories*, Proc. of the Meeting "Geometry and Physics", 1982, Florence (Pitagora, Bologna, 1982), 135–165.
2. *Some results on the calculus of variations on jet spaces*, Ann. Inst. H. Poincaré **39** (1983), 29–43.

Yu. I. Manin

1. *Algebraic aspects of non-linear differential euqations*, Itogi nauki i tekniki (VINITI, Ser. Sovremennye problemy matematiki) **11** (1979), 5–152 (in Russian).

C. M. Marle

1. *Contact manifolds, canonical manifolds and the Hamilton-Jacobi method in analytical mechanics*, in: *Modern Developments in Analytical Mechanics I: Geometrical Dynamics*, Proc. IUTAM-ISIMM Symposium, Torino, Italy 1982, S. Benenti, M. Francaviglia and A. Lichnerowicz eds. (Accad. delle Scienze di Torino, Torino, 1983), 255–272.

2. *Reduction of constrained mechanical systems and stability of relative equilibria*, Comm. Math. Phys. **174** (1995), 295–318.

G. Marmo, G. Mendella, W. M. Tulczyjew

1. *Symmetries and constants of the motion for dynamics in implicit form*, Ann. Inst. Henri Poincaré, Phys. Theor. **57** (1992), 147–166.

E. Martínez, J. F. Cariñena, W. Sarlet

1. *Derivations of differential forms along the tangent bundle projection I, II*, Diff. Geom. Appl. **2** (1992), 17–43; **3** (1993), 1–29.

M. Marvan

1. *On global Lepagean equivalents*, in: *Geometrical Methods in Physics*, Proc. Conf. on Diff. Geom. and Appl. Vol. 2, Nové Město na Moravě, Czechoslovakia, 1983, D. Krupka, ed. (J. E. Purkyně University, Brno, 1984), 185–190.

M. Matsumoto

1. *Foundations of Finsler Geometry and Special Finsler Spaces*, Kaiseisha Press, Shigaken, 1986.

A. Mayer

1. *Die Existenzbedingungen eines kinetischen Potentiales*, Ber. Ver. Ges. d. Wiss. Leipzig, Math.-Phys. Kl. **48** (1896), 519–529.

S. G. Mikhlin

1. *Variational Methods in Mathematical Physics*, Nauka, Moscow, 1970 (in Russian).

H. Minkowski

1. Nachr. d. K. Geselsch. d. Wissensch. zu Göttingen, Math.-Phys. Kl. (1908), 53-111.

R. Miron

1. *Spaces with higher order metric structures*, Tensor **53** (1993), 1–23.

R. Miron and M. Anastasiei

1. *The Geometry of Lagrange Spaces: Theory and Applications*, Kluwer, Dordrecht, 1994.

R. Miron and G. Atanasiu

1. *Compendium sur les espaces Lagrange d'ordre supérieur*, Preprint, Univ. Timisoara, 1994.
2. *Differential geometry of the k-osculator bundle*, Rev. Roum. Math. Pure et Appl. **XLI** (1996), 205–236.
3. *Prolongation of Riemannian, Finslerian and Lagrangian structures*, Rev. Roum. Math. Pure et Appl. **XLI** (1996), 237–250.
4. *Higher order Lagrange spaces*, Rev. Roum. Math. Pure et Appl. **XLI** (1996), 251–262.

M. Modugno

1. *Torsion and Ricci tensor for non-linear connections*, Diff. Geom. Appl. **1** (1991), 177–192.

P. Morando and S. Pasquero

1. *Singular Lagrangians and the Hamel-Appell system*, Acta Mechanica **96** (1993), 55–66.

J. Musilová

1. *Variational sequence in higher-order mechanics*, in: Proc. Conf. Diff. Geom. Appl., August 1995, Brno, Czech Republic, J. Janyška and I. Kolář, eds. (Masaryk Univ., Brno, 1996), 611–624.

V. V. Nesterenko

1. *Singular Lagrangians with higher derivatives*, J. Phys A: Math. Gen. **22** (1989), 1673–1687.

V. V. Nesterenko and A. M. Chervyakov

1. *Some properties of constraints in theories with degenerate lagrangians*, Theor. Math. Phys. **64** (1985), 82–91 (in Russian).

E. Noether

1. *Invariante Variationsprobleme*, Nachr. kgl. Ges. Wiss. Göttingen, Math. Phys. Kl. (1918), 235–257.

T. Nôno and F. Mimura

1. *Dynamical symmetries, I*, Bull. Fukuoka Univ. of Education **25** (1976), 9–26.

J. Novotný

1. *On the inverse variational problem in the classical mechanics*, in: Proc. Conf. on Diff. Geom. and Its Appl. 1980, O. Kowalski, ed. (Universita Karlova, Prague, 1981), 189–195.
2. *On the geometric foundations of the Lagrange formulation of general relativity*, in: *Differential Geometry*, Coll. Math. Soc. J. Bolyai 31, Budapest, 1979 (North-Holland, Amsterdam, 1982), 503–509.

P. J. Olver

1. *Equivalence and the Cartan form*, Acta Appl. Math. **31** (1993), 99–136.

Z. Oziewicz

1. *Classical mechanics: inverse problem and symmetries*, Rep. Math. Phys. **22** (1985), 91–111.

Z. Oziewicz and W. Gruhn

1. *On Jacobi's theorem*, Hadronic J. **6** (1983), 1579–1605.

H. Poincaré

1. *Sur la dynamique de l'électron*, Rend. Pal. **21** (1906), 129.

J.-F. Pommaret

1. *Spencer sequence and variational sequence*, Acta Appl. Math., to bc published.

M. F. Rañada

1. *Extended tangent bundle formalism for time dependent Lagrangian systems*, J. Math. Phys. **32** (1991), 500–505.
2. *Alternative Lagrangians for central potentials*, J. Math. Phys. **32** (1991), 2764–2769.
3. *Extended Legendre transformation approach to the time-dependent Hamiltonian formalism*, J. Phys. A: Math. Gen. **25** (1992), 4025–4035.
4. *Time-dependent Lagrangian systems: A geometric approach to the theory of systems with constraints*, J. Math. Phys. **35** (1994), 748–758.

R. Ray

1. *Lagrangians and systems they describe—how not to treat dissipation in quantum mechanics*, Am. J. Phys. **47** (1979), 626–629.

K. Rektorys

1. *Variational Methods in Engineering and Mathematical Physics*, SNTL, Prague, 1974 (in Czech).

N. Román-Roy

1. *Dynamics of constrained systems*, Int. J. Theor. Phys. **27** (1988), 427–431.

H. Rund

1. *A Cartan form for the field theory of Carathéodory in the calculus of variations of multiple integrals*, Lect. Notes in Pure and Appl. Math. **100** (1985), 455–469.

Y. Saito, R. Sugano, T. Ohta and T. Kimura

1. *A dynamical formalism of singular Lagrangian system with higher derivatives*, J. Math. Phys. **30** (1989), 1122.
2. *Addendum to a dynamical formalism of singular Lagrangian system with higher derivatives*, J. Math. Phys. **34** (1993), 3775–3779.

W. Sarlet

1. *On the transition between second-order and first-order systems within the context of the inverse problem of the Newtonian mechanics*, Hadronic J. **2** (1979), 407–432.
2. *The Helmholtz conditions revisited: A new approach to the inverse problem of Lagrangian dynamics*, J. Phys. A **15** (1982), 1503–1517.
3. *Noether's Theorem and the inverse problem of Lagrangian mechanics*, in: *Modern Developments in Analytical Mechanics*, Proc. IUTAM-ISIMM Symposium, Torino, Italy 1982, S. Benenti, M. Francaviglia and A. Lichnerowicz, eds. (Accad. delle Scienze di Torino, Torino, 1983), 737–751.
4. *Symmetries, first integrals and the inverse problem of Lagrangian mechanics I, II*, J. Phys. A: Math. Gen. **14** (1981), 2227–2238; **16** (1983), 1383–1396.
5. *Note on equivalent Lagrangians and symmetries*, J. Phys. A: Math. Gen. **16** (1983), L229–L233.
6. *Symmetries and alternative Lagrangians in higher-order mechanics*, Phys. Lett. **108A** (1985), 14–18.
7. *Geometrical structures related to second-order equations*, in: *Differential Geometry and Its Applications*, Proc. Conf., Brno, Czechoslovakia, 1986, D. Krupka and A. Švec, eds. (D. Reidel, Dordrecht, 1986), 279–288.
8. *A direct geometrical construction of the dynamics of non-holonomic Lagrangian systems*, Extracta Mathematicae **11** (1996), 202–212.
9. *The geometry of mixed first and second-order differential equations with applications to non-holonomic mechanics*, in: Differential Geometry and Applications, Proc. Conf., Brno, Czech Republic, 1995, J. Janyška and I. Kolář, eds., (Masaryk University, Brno, 1996), 641–650.

W. Sarlet and F. Cantrijn

1. *Generalizations of Noether's Theorem in classical mechanics*, SIAM Review **23** (1981), 467–494.
2. *Higher order Noether symetries and constants of the motion*, J. Phys. A: Math. Gen. **14** (1981), 479–492.

W. Sarlet, F. Cantrijn and M. Crampin

1. *A new look at second-order equations and Lagrangean mechanics*, J. Phys. A: Math. Gen. **17** (1984), 1999–2009.

W. Sarlet, F. Cantrijn and D. J. Saunders

1. *A geometrical framework for the study of non-holonomic Lagrangian systems*, J. Phys. A: Math. Gen. **28** (1995), 3253–3268.

W. Sarlet, A. Vandecasteele, F. Cantrijn and E. Martínez

1. *Derivations of forms along a map: the framework for time-dependent second-order equations*, Diff. Geom. Appl. **5** (1995), 171–203.

D. J. Saunders

1. *An alternative approach to the Cartan form in Lagrangian field theories*, J. Phys. A: Math. Gen. **20** (1987), 339–349.
2. *Jet fields, connections and second-order differential equations*, J. Phys. A: Math. Gen. **20** (1987), 3261–3270.
3. *The Geometry of Jet Bundles*, London Math. Soc. Lecture Notes Series 142, Cambridge Univ. Press, Cambridge, 1989.
4. *A note on Legendre transformations*, Diff. Geom. Appl. **1** (1991), 109–122.
5. *The regularity of variational problems*, in: *Mathematical Aspects of Classical Field Theory*, Proc. of the AMS-IMS-SIAM Joint Summer Research Conf. 1991, M. J. Gotay, J. E. Marsden and V. Moncrief, eds., Contemporary Math. **132** (AMS, Providence, 1992), 573–593.

D. J. Saunders, W. Sarlet and F. Cantrijn

1. *A geometrical framework for the study of non-holonomic Lagrangian systems: II*, J. Phys. A: Math. Gen. **29** (1996), 4265–4274.

R. S. Schechter

1. *The Variational Method in Engineering*, McGraw-Hill, New York, 1967.

W. F. Shadwick

1. *The Hamiltonian formulation of regular r-th order Lagrangian field theories*, Letters in Math. Phys. **6** (1982), 409–416.

V. D. Skarzhinsky

1. *Ambiguites of quantizatuin procedure and choice of the Hamiltonian*, in: *Differential Geometry and Its Applications*, Proc. Conf., Brno, Czechoslovakia, 1986, D. Krupka and A. Švec, eds. (D. Reidel, Dordrecht, 1986), 329–338.

J. Sniatycki

1. *On the geometric stucture of classical field theory in Lagrangian formulation*, Proc. Camb. Phil. Soc. **68** (1970), 475–484.

N. Ya. Sonin

1. Warsaw Univ. Izvestiya, No. 1–2 (1886), (in Russian).

J.-M. Souriau

1. *Structure des Systèmes Dynamiques*, Dunod, Paris, 1970.

J. Štefánek

1. *Odd variational sequences on finite-order jet spaces*, in: Proc. of the 5th International Conference on Differential Geometry and Its Applications, August 1992, Opava, Czechoslovakia, O. Kowalski and D. Krupka, eds. (Silesian University, Opava, 1993), 523–538.
2. *A Representation of the Variational Sequence by Forms*, Thesis, Silesian University, Opava, 1995, 36 pp.
3. *A representation of the variational sequence in higher-order mechanics*, in: Proc. Conf. on Differential Geometry and Its Applications, August 1995, Brno, J. Janyška and I. Kolář, eds. (Masaryk University, Brno, 1996), 469–478.

M. Štefanová and O. Štěpánková

1. *Variationality and invariance of systems of partial differential equations*, Scripta Fac. Sci. Nat. UJEP Brunensis **15** (Physica) (1985), 283–288.

O. Štěpánková

1. *The inverse problem of the calculus of variations in mechanics*, Thesis, Charles University, Prague, 1984, 63 pp. (in Czech).

S. Sternberg

1. *Lectures on Differential Geometry*, Prentice Hall, New York, 1964.
2. *Some preliminary remarks on the formal variational calculus of Gelfand-Dikii*, Lecture Notes in Math. **676** (Springer-Verlag, Berlin, 1978), 399-407.

E. C. G. Sudarshan and N. Mukunda

1. *Classical Dynamics: A Modern Perspective*, John Wiley & Sons, New York, 1974.

K. Sundermeyer

1. *Constrained Dynamics*, Springer-Verlag, Berlin, 1982.

H. Sussmann

1. *Orbits of families of vector fields and integrability of distributions*, Trans. Amer. Math. Soc. **180** (1973), 171–188.

W. Szczyrba

1. *On the geometric structure of the set of solutions of Einstein equations*, Dissertationes Math. **150**, Polska Akademia Nauk, Warszawa 1977.

B. Tabarrok and F. P. J. Rimrott

1. *Variational Methods and Complementary Formulations in Dynamics*, Kluwer, Dordrecht, 1994.

F. Takens

1. *A global version of the inverse problem of the calculus of variations*, J. Diff. Geom. **14** (1979), 543–562.

E. Tonti

1. *Variational formulation of nonlinear differential equations I, II*, Bull. Acad. Roy. Belg. Cl. Sci. **55** (1969), 137–165, 262–278.

A. Trautman

1. *Noether equations and conservation laws*, Comm. Math. Phys. **6** (1967), 248–261.
2. *Invariance of Lagrangian systems*, in: *General Relativity*, Papers in honour of J. L. Synge (Clarendon Press, Oxford, 1972), 85–99.

T. Tsujishita

1. *On variational bicomplex associated to differential equations*, Osaka J. Math. **19** (1982), 311–363.

W. M. Tulczyjew

1. *The Lagrange differential*, Bull. Acad. Polon. Sci. Ser. Sci. Math., Astron., Phys., **24** (1976), 1089–1096.
2. *The Lagrange complex*, Bull. Soc. Math. France **105** (1977), 419–431.
3. *The Euler-Lagrange resolution*, Internat. Coll. on Diff. Geom. Methods in Math. Phys., Aix-en-Provence 1979, in: *Lecture Notes in Math.* **836** (Springer, Berlin, 1980), 22–48.

4. *Geometric foundations of Lagrangian mechanics*, in: *Modern Developments in Analytical Mechanics I: Geometrical Dynamics*, Proc. IUTAM-ISIMM Symp., Torino, Italy 1982, S. Benenti, M. Francaviglia and A. Lichnerowicz, eds. (Accad. Sci. Torino, Torino, 1983), 419–442.

M. M. Vainberg

1. *Variational methods in the theory of nonlinear operators*, GITL, Moscow, 1959 (in Russian).

A. L. Vanderbauwhede

1. *Potential operators and variational principles*, Hadronic J. **2** (1979), 620–641.

L. Van Hove

1. *Sur la construction des champs de De Donder-Weyl par la méthode des caracteristiques*, Acad. Roy. Belg. Bull. Sci. **31** (1945), 278–285.

V. P. Viflyantsev

1. *Frobenius Theorem for distributions of non-constant rank*, Uspekhi Mat. Nauk **197** (1977), 177–178 (in Russian).
2. *On completely integrable distributions of non-constant rank*, Mat. Zametki **26** (1979), 921–929 (in Russian).

A. M. Vinogradov

1. *On the algebro-geometric foundations of Lagrangian field theory*, Soviet Math. Dokl. **18** (1977), 1200–1204.
2. *A spectral sequence associated with a non-linear differential equation, and algebro-geometric foundations of Lagrangian field theory with constraints*, Soviet Math. Dokl. **19** (1978), 144–148.
3. *The $\mathcal{C}$-spectral sequence, Lagrangian formalism, and conservation laws. I. The linear theory. II. The nonlinear theory*, J. Math. Anal. Appl. **100** (1984), 1–40, 41–129.

A. M. Vinogradov and B. A. Kupershmidt

1. *Structure of Hamiltonian mechanics*, Uspekhi Matemat. Nauk **32** (1975), 175–236 (in Russian).

R. Vitolo

1. *Some aspects of variational sequences in mechanics*, in: Proc. Conf. on Differential Geometry and Its Applications, August 1995, Brno, Czech Republic, J. Janyška and I. Kolář, eds. (Masaryk University, Brno, 1996), 487–494.

A. Vondra

1. *Connections in the Geometry of Non-Autonomous Regular Higher-Order Dynamics*, Thesis, Masaryk University, Brno, 1991, 114 pp.
2. *Semisprays, connections and regular equations in higher order mechanics*, in: *Differential Geometry and Its Applications*, Proc. Conf., Brno, Czechoslovakia, 1989, J. Janyška and D. Krupka eds. (World Scientific, Singapore, 1990), 276–287.
3. *Towards a Geometry of Higher-Order Partial Differential Equations Represented by Connections on Fibered Manifolds*, Thesis, Military Academy, Brno, 1995.

A. Weinstein

1. *Lectures on symplectic manifolds*, CBMS Regional Conference Series in Math. **29** (American Mathematical Society, Providence, 1977), 48 pp.

E. T. Whittaker

1. *A Treatise on Analytical Dynamics of Particles and Rigid Bodies*, The University Press, Cambridge, 1917.

Index